Studies in Rhythm Engineering

This is a multi-disciplinary book series, ranging from astrophysics to biology, chemistry, mathematics, geophysics and materials science. Its primary scope is the fundamental science and associated engineering wherever cyclic and rhythmic oscillations are observed.

Time neither being an entity nor a process is unmeasurable and undefined, although a clock only measures the passage of time. The clock drove recurring processes are observed in the biological rhythms, astrophysical and geophysical environments. Always, the clocks are nested, arranged in a geometric shape to govern a phenomenon in nature. These clocks are made of atoms, molecules, their circuits, and complex networks. From biology to astrophysics, the clocks have enriched the science and engineering and the series would act as a catalyst and capture the forthcoming revolution of time cycles expected to unfold in the 21st Century. From the cyclic universe to the time crystal, the book series makes a journey through time to explore the path that time follows.

The series publishes monographs and edited volumes.

More information about this series at https://link.springer.com/bookseries/16136

Anirban Bandyopadhyay · Kanad Ray
Editors

Rhythmic Advantages in Big Data and Machine Learning

Springer

Editors
Anirban Bandyopadhyay
National Institute for Materials Science
Tsukuba, Japan

Kanad Ray
Department of Physics
AMITY University Rajasthan
Jaipur, India

ISSN 2524-5546 ISSN 2524-5554 (electronic)
Studies in Rhythm Engineering
ISBN 978-981-16-5722-1 ISBN 978-981-16-5723-8 (eBook)
https://doi.org/10.1007/978-981-16-5723-8

This Springer imprint is published by the registered company Springer Nature Singapore Pte Ltd.
The registered company address is: 152 Beach Road, #21-01/04 Gateway East, Singapore 189721, Singapore

Preface

The current book in the series of Systems in Rhythm Engineering, SRE, compiles rhythms from Big data in pure computation, astrophysics to basic biological structures. In the earlier book edited by us, we concentrated primarily on the biophysics related studies, because most rhythm related researches are carried out in biophysics. Then Tanusree Dutta edited a book on economics, before getting into the field of rhythms in economic data, it was an effort to dwell on various aspects of rhythms where we could advance rhythm engineering. In future, we have a plan to discuss on share market rhythms and economic disaster predictions. Here we have made an effort to the field of astrophysics and big data. Our target is to create world's most authentic and rigorous database of rhythms in every single field possible. This book is the first comprehensive message that we wish to deliver to the world.

In Chap. 1, Rajdeep and Somesh have made an effort to discuss big data and how rhythm engineering could be used in understanding and fathoming the intricate dynamic patterns in a big data. An architecture of rhythm that we advocate as the fundamental information structure of nature could play pivotal role in the understanding of big data.

In Chap. 2, Joseph Singuinetti has described how brain rhythms could be perturbed to understand its true nature. Especially by concentrating on the transcranial ultrasound perturbation to neural oscillation, the author has made an effort to understand how brain rhythms are organized in an internal architecture that would regulate human cognition and behavior. Focused ultrasound is getting popular day by day and Joseph's documentation would help us understanding how rhythms organize in a human brain.

In Chap. 3, Mario Pinehiro has described vortex dynamics as a foundation of energy minimization process and entropy maximization process. It is amazing to think that loop of fields is engaging to create this universe. Mario has gone to the fundamental details of this mechanism to understand how rhythms could alone construct the foundation of this universe.

In Chap. 4, Shiva Kumar Singh and Marcos a Avila have taken thermal flow as a foundation to understand fractal superstructures contributing to the thermoelectric applications. Thermoelectric materials are very important because they resemble the biomaterials that harvest electrical energy from thermal noise using fractal structures. Thermal signals are electromagnetic and electrical signals are part of electromagnetic field. Such conversions are key governing features of energy harvesting and splitting the two parts of electromagnetic energy.

In Chap. 5, we return to brain modeling again with Taakaki Musha, he has been working on superluminal particles that travel faster than the velocity of light. Are they really produced in the brain? Do they take part in the brain cognition? Explore it here with Taakaki.

In Chap. 6, Pushpendra Singh and others have reported their work on building experimentally an artificial cortex with more than 10000 cortical columns. Real brain has roughly 200000 cortical columns constituting the cortex where in 47 Brodmann's regions, brain's major cognitive information processing takes place.

In Chap. 7, Max Rempel has taken an effort to look into the DNA resonance in a very different way. Normally, we look at dielectric property of a material directly, however, different symmetries have different contributions to the formation of a resonance band. Max has looked into the DNA as an assembly of a tiny semi-circular block and made an effort to understand the mystery of junk DNA. Do they really have codes that we really don't understand?

In Chap. 8, Max Rempel has returned with a topic on quantum consciousness, we have always made an effort to keep one topic on consciousness in our edited compilation of rhythm engineering. Quantum consciousness is interesting, exciting, and extremely debated in the scientific community.

In Chap. 9, Max Calliguri takes us one step further into the domain of microtubules and he advocates a hypothesis that brain carries out hypercomputation using water crystals inside the microtubules. This is where we think our book should end.

We invite all scholars to have a critical debate on the special selections we have made to explore the systems in rhythm engineering.

Jaipur, India Kanad Ray
Tsukuba, Japan Anirban Bandyopadhyay

Contents

Editors and Contributors

About the Editors

Anirban Bandyopadhyay is Senior Principal Scientist at the National Institute for Materials Science (NIMS), Tsukuba, Japan. He received Ph.D. in Supramolecular Electronics at the Indian Association for the Cultivation of Science (IACS), Kolkata, on 2005. From 2005 to 2008, he was ICYS Research Fellow at the ICYS, NIMS, Japan, and worked on the brain-like bio-processor. In 2008, he joined as a permanent scientist at NIMS, working on the time crystal model of human brain and design-synthesis of brain-like organic jelly, written a book "Nanobrain: The making of an artificial brain from a time crystal," on 2020. From 2013 to 2014, he was a visiting scientist at the Massachusetts Institute of Technology (MIT), USA. He has received Hitachi Science and Technology Award, 2010, Inamori Foundation Award, 2011–2012, Kurata Foundation Award, Inamori Foundation Fellow (2011), and Sewa Society International Member, Japan.

Kanad Ray (Senior Member, IEEE) received the M.Sc. degree in Physics from Calcutta University and the Ph.D. degree in Physics from Jadavpur University, West Bengal, India. He has been Professor of Physics and Electronics and Communication and is presently working as Head of the Department of Physics, Amity School of Applied Sciences, Amity University Rajasthan (AUR), Jaipur, India. His current research areas of interest include cognition, communication, electromagnetic field theory, antenna and wave propagation, microwave, computational biology, and applied physics. He has been serving as Editor for various Springer book series. He was Associate Editor of the Journal of Integrative Neuroscience (The Netherlands: IOS Press). He has visited several countries such as Netherlands, Turkey, China, Czechoslovakia, Russia, Portugal, Finland, Belgium, South Africa, Japan, Singapore, Thailand, and Malaysia for various academic missions.

Contributors

B. Aswathy Materials Science and Technology Division, CSIR-National Institute for Interdisciplinary Science and Technology (CSIR-NIIST), Thiruvanathapuram, Kerala, India

Marcos A. Avila CCNH, Universidade Federal do ABC (UFABC), Santo André, SP, Brazil

Anirban Bandyopadhyay International Center for Materials and Nanoarchitectronics (MANA), Research Center for Advanced Measurement and Characterization (RCAMC), NIMS, Tsukuba, Ibaraki, Japan

Rajdeep Banerjee Data Scientist, Strategic Growth Realisation, Bangalore, India

Somesh Kr. Bhattacharya National Institute for Materials Science, Tsukuba, Japan;
United Network of Professionals Pvt. Ltd, FD Block, Kolkata, India

Luigi Maxmilian Caligiuri Foundation of Physics Research Center (FoPRC), Cosenza, Italy

Daisuke Fujita International Center for Materials and Nanoarchitectronics (MANA), Research Center for Advanced Measurement and Characterization (RCAMC), NIMS, Tsukuba, Ibaraki, Japan

Subrata Ghosh Chemical Science and Technology Division, CSIR-North East Institute of Science and Technology, NEIST, Jorhat, Assam, India;
Academy of Scientific and Innovative Research (AcSIR), CSIR-NEIST Campus, Jorhat, Assam, India

Richard Alan Miller OAK, Inc., Grants Pass, OR, USA

Takaaki Musha Advanced Science-Technology Research Organization, Yokohama, Japan;
Foundation of Physics Research Center (FoPRC), Cosenza, Italy

Max Myakishev-Rempel DNA Resonance Research Foundation, San Diego, CA, USA;
Localized Therapeutics, San Diego, CA, USA

Mario J. Pinheiro Department of Physics, Instituto Superior Técnico-IST, Universidade de Lisboa, Lisboa Codex, Portugal

Kanad Ray Amity School of Applied Sciences, Amity University Rajasthan, Kant Kalwar, Jaipur, Rajasthan, India

Pathik Sahoo International Center for Materials and Nanoarchitectronics (MANA), Research Center for Advanced Measurement and Characterization (RCAMC), NIMS, Tsukuba, Ibaraki, Japan

Joseph L. Sanguinetti Center for Consciousness Studies, University of Arizona, Tucson, Arizona, USA

Ivan Viktorovich Savelev DNA Resonance Research Foundation, San Diego, CA, USA;
Localized Therapeutics, San Diego, CA, USA

Pushpendra Singh International Center for Materials and Nanoarchitectronics (MANA), Research Center for Advanced Measurement and Characterization (RCAMC), NIMS, Tsukuba, Ibaraki, Japan;
Amity School of Applied Sciences, Amity University Rajasthan, Kant Kalwar, Jaipur, Rajasthan, India

Shiva Kumar Singh CCNH, Universidade Federal do ABC (UFABC), Santo André, SP, Brazil

Chapter 1
Data: Periodicity and Ways to Unlock Its Full Potential

Rajdeep Banerjee and Somesh Kr. Bhattacharya

1 Introduction

A set of features or information collected through observation is known as "**Data**" [1]. Technically speaking, data is a set of numerical values, or images, or videos, or even text of about one or more persons or objects. In 2006, British mathematician Clive Humby coined the phrase "Data being the new oil" and we are increasingly seeing how true it is. Data is a vital resource that powers the information economy. Additionally, data isn't just abundant, it is cumulative. Finally, data is a non-rival good as the same can be used elsewhere for beneficial purposes. However, the data to be used should be prepared in a timely manner and must be reliable, accurate, complete, and relevant. However, raw data isn't much meaningful and requires processing. Data when processed and presented in a contextual format becomes information [2]. An important property of data is the presence of periodicity. Anything which doesn't have periodicity is "*noise*". In almost in every field of science, engineering, finance, and even music, periodicity is observed. The planetary motion, phases of the moon, solar bursts are a few examples of periodicity in astronomical data. Similarly, meteorological as well as seismological data have periodicity too. The highs and lows of the stock market have both seasonality and cyclical factors embedded in it giving rise to periodicity. Even in health care, one can see the periodic nature. The appearance of diseases related to seasonal changes is a very good example of that. Lastly, periodicity can be seen easily in music. The nodes repeat themselves at regular intervals giving rise to periodicity. Detecting periodicity in data is a field in itself

R. Banerjee
Data Scientist, Strategic Growth Realisation, Cerner India, Bangalore 560064, India

S. Kr. Bhattacharya (✉)
National Institute for Materials Science, 1-2-1 Sengen, Tsukuba 3050047, Japan

United Network of Professionals Pvt. Ltd, FD Block, Sector-III, Bidhan Nagar, Kolkata 700106, India

A. Bandyopadhyay and K. Ray (eds.), *Rhythmic Advantages in Big Data and Machine Learning* , Studies in Rhythm Engineering,
https://doi.org/10.1007/978-981-16-5723-8_1

and includes methods like Discrete Fourier Transform (DFT) for frequency domain analysis [3] or computing the periodogram of a signal and using autocorrelation to find the periodicity [4]. We will discuss the periodicity later in this chapter.

In today's world, anything and everything is data and we generate voluminous data almost every passing second which actually gave birth to "Big Data". The phrase "Big Data" which buzzes around us everywhere, is applied to a specific type of data that has certain traits. However, this phrase has been over used and often incorrectly, which is why itis difficult to gauge its true meaning. For commoners, it is very difficult to understand if big data is a tool or a technology or just a buzzword used by data scientists to scare us. Another concern is if big data really has the potential to usher in dramatic changes or will the hype fade away with time. In any case, over the past few years, big data has become an integral part of several industries and in many cases has shown the potential to be a game-changer.

2 What is Big Data?

To put it simply, big data describes the massive volume of both structured and unstructured data which is so enormous that processing this voluminous data requires special techniques. In short, big data is simply a whole lot of data. This concept is a relatively new one and it constitutes not only the increasing amount but also the diverse types of data that gets collected. As more and more information moves online and gets digitized, it paves way for the analysts to start using it as data. For instance, social media posts, online books, music, videos, along with their reviews, as well as the increased number of sensors have all added to the staggering increase in the amount of data that has become available for analysis. Everything we do now gets stored and tracked as data. Starting from online shopping to browsing movies generates data about our requirements, choices, and lifestyles. Our smartphone constantly uploads data about our location, movement as well as the apps we use.

So far, we have talked about big data from the perspective of its volume. However, big data isn't just the volume of data we generate, it's also about the different types of

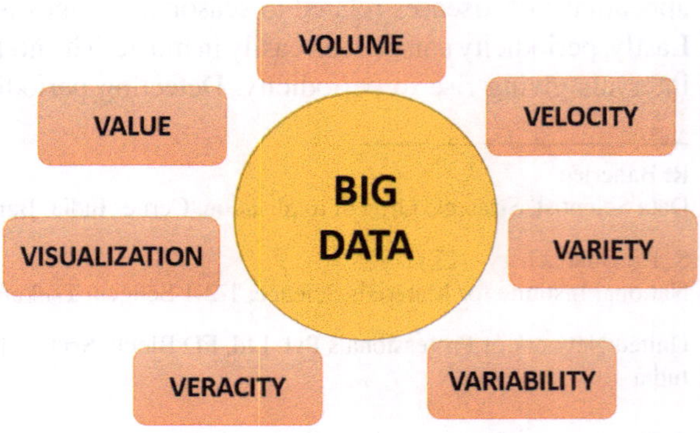

Fig. 1 The seven Versus of big data

data, viz. *text, video, search logs, sensor logs, customer transactions,* etc. Formally, big data is the data that satisfies the following ***seven Vs***: (Fig. 1)

- **Volume**: In every sense, big data is really big! With the dramatic growth of the Internet, mobile devices, social media, and Internet of Things (IoT) technology, the amount of data generated from all these sources has grown accordingly.
- **Velocity**: In addition to volume, the generation of data and the organizations' ability to process it is accelerating.
- **Variety**: In earlier times, most data types could be neatly captured in rows on a structured table. In the big data world, data often comes in unstructured formats like social media posts, server log data, lat-long geocoordinates, photos, audio, video, and free text.
- **Variability**: The meaning of words in the text (unstructured data) can change based on the context.
- **Veracity**: With many different data types and data sources, data quality issues invariably pop up in big data sets. Veracity deals with exploring a data set for data quality and systematically cleansing that data to be useful for analysis.
- **Visualization**: Data visualization is a great tool for data analysis. Exploratory data analysis (EDA) can provide quick and in-depth insight to the end-users.
- **Value**: Data must be combined with rigorous processing and analysis to be useful.

3 Big Data Terminologies

The uncanny feeling with big data comes from the variety of new terms that are associated with it. Here is a quick run-down of the most popular ones:

A. Algorithm: This is the backbone of big data analytics. The underlying mathematical intricacies help in gaining insights and forms the core of decision science.

B. Cloud computing: As big data is voluminous, processing it using the algorithms often specializes in structure. Cloud computing has emerged as a front runner in this and these remote servers have an edge over local machines.

C. Data scientist: A breed of professionals who analyze data and extract vital insights from it.

D. Hadoop: A collection of programs that allow for the storage, retrieval, and analysis of very large data sets.

E. Internet of Things (IoT): IoT refers to objects like sensors that collect, analyze and transmit their data (often without human interference).

F. Predictive analytics: The science of using data analytics to predict trends for future events.

G. Structured versus unstructured data: Structured data is anything that can be organized in a table so that it relates to other data in the same table. Unstructured data is everything that isn't structured like texts, images, and videos.

H. Web scraping: The tool of collection and structuring of data from websites usually through writing scripts.

4 Significance of Big Data

The importance of big data has grown proportionally with the volume of data being produced every day. Businesses have leveraged big data in a big way to understand a number of factors to drive their pursuit of success. Companies have used big data analytics to learn their customer's demand, choices, their best customers, and what they are missing out on conversions. Organizations gather useful insights into their sales and marketing questions so that they can optimize their campaigns and learn more about customer behavior.

Another area where big data analytics is handy for businesses is to make confident decisions. Big data can help companies make choices with confidence, based on an in-depth analysis of what they know about your marketplace, industry, and customers. One of the biggest benefits of big data is its accuracy. Big data analytics provide a complete overview of everything that has been learned so far as organizations grow and suggest positive changes for the company.

Knowledge is power. This concept is at the heart of big data analytics. Big data technologies like cloud computing and machine learning help companies to stay ahead of the curve by identifying inefficiencies and opportunities in company practices.

Finally, as a new generation of technology leaders enter the marketplace, big data delivers the agility and innovation top-tier talent needs from their employer. For instance, millennials are natural technology natives. The younger people in the team will expect access to technology that allows them to make useful decisions rapidly. By constantly collecting and analyzing information, one can create an agile culture that's ready to evolve to suit the latest trends.

5 Applications of Big Data

Big data technologies are leveraged to boost efficiency and design data-driven decisions. Big data finds applications in several industries where it is mandatory to analyze historical data to predict/forecast future performances. Though big data has penetrated almost every field, we mention few domains which leverage big data regularly:

Health Care

As discussed in the previous sections, big data involves data that is high in the four Vs, viz., volume, velocity, variety, and veracity. Since the digitization of data in the form of electronic medical records has begun, health care, because of its inherent need in society, has become one of the major domains of big data, both in terms of source and utilization. In healthcare, the source of the data, in general, determines its utilization. When we talk about utilization it generally means how this data is gobbled up using models and metrics to provide better insights and predictions. In general, the use-cases involve two types of analysis of the data—first, algorithmic––simple logic-based strategy development. These mainly involve a lot of domain knowledge and a limited number of parameters to consider. On the other hand, when the number of parameters at hand is large and in some cases their dependencies are not well defined, we turn to the second approach of machine learning and deep learning—artificial intelligence based predictions and strategies.

In the following few paragraphs, we would like to focus on different sources of big data in health care and see how the utilization differs based on the source. The primary sources of big data in health care can be broadly classified (non-exhaustive) into the following categories:

1. Electronic medical/health records (EMR/EHR),
2. Diagnostic tests data,
3. Social survey data,
4. Insurance claims data, and
5. Inventory management data.

We will briefly describe the kind of data each of these categories holds, and will then go into the possible use-cases in each category, that the data can provide insights and solutions to.

1. Electronic medical/health records (EMR/EHR):
   This is the type of data that the hospitals and the primary healthcare centers (PHCs) collect from patients, which include specifics about admission records, past illness history, patient demographics, present illness, diagnosis, doctor's prescription, etc. In general, these kinds of data are filled in through client (PHCs)-specific software, and follow some general schema, and therefore are structured, and easy to use for further processing.
   EMR/EHRs can help discover phenotype-genotype associations, improve clinical trial protocols, automate adverse drug event detection and prevention, and accelerate precision medicine research [5], using either an algorithmic or machine learning approach. One of the major use cases that has come up recently in this domain is to use machine learning or deep learning models to predict or reduce patient readmissions in health centers [6].
2. Diagnostic tests data:
   These types of data can range from tabular formats, such as lab tests, to image data such as X-ray, CT-scans, MRI, USG, ECG, EEG, etc. Generally, lab test results are inputted through specific software are tabular and structured. But image data sources need special care, especially images of the same use-case, from different sources differ a lot. It should also be noted that in cases where results are hand-written/filled in a form, digitization may include handwritten

text based natural language processing methods, before can be used for any kind of machine learning models.

Recently, deep learning based models are being used extensively for medical image processing and classification [7]. Some of these models are already showing potential to prove as an aide to the medical professionals in understanding and diagnosis of ailments. The use of chat-bots is also growing as primary detection and identification of diseases.

3. Social survey data:

   A large chunk of these types of data are mainly social determinants of health (SDOH) data, resulting from various government and non-government surveys, which are conducted based on geographic locations. SDOH data, in general, come into the following five categories: (i) economic stability; (ii) education access and quality; (iii) social and community context; (iv) neighborhood and built environment; and (v) healthcare access and quality.

   The SDOH has numerous use cases, out of which it has already shown significant impact in improving mortality, morbidity, life expectancy, healthcare expenditures, health status, and functional limitations [8]. Detection of disease prevalence based on age, living conditions, social or community status, etc., can also help in developing a more targeted approach to health care.

4. Insurance claims data:

   As the name suggests, these types of data contain mainly the financial records, such as hospital charges, bills paid, dues, amount claimed through insurance, type of insurance, and so on.

   The primary use cases include predicting claims rejection ratio and risk adjustment by insurers for efficient care management for high-risk individuals [9]. A better understanding of such data can improve insurance policies. Sometimes, where the is a mass insurance policy available, the claims data can be merged with SDOH data to improve the predictive power of the models.

5. Inventory management:

   Types of inventory management include the storage, utilization, and order of medicines, medical equipment, or any kind of resources related to patient care or hospitality.

   Inventory management use-cases can range from reducing medicine wastage, supply-chain management, to predicting high-demand scenarios beforehand, understanding the seasonality of diseases to stockpile specific medications and healthcare equipment, predicting medication usage based on historical data [10], etc.

As we can see, based on the sources, how the use-cases of the data differ. This leads to the variety of approaches that we can take advantage of. In the past few decades, the human race has made great strides in medicines and health care which is evident from the increase in average life expectancies over the years. But at the same time, we are also witnessing more and more disease outbreaks due to global warming, man-wild interactions and conflicts, etc. Therefore, we must make use of the huge power of the data that is ever-growing to tilt the balance toward us, humans, a little more.

Banking and Finance

Banking and Financial Service (BFS) is one of those domains which heavily employs and utilizes big data technology and its associated platforms regularly. Among various applications of big data in BFS, the two most important ones are: (a) Credit card fraud detection, and (b) Stock price prediction (forecasting). Credit card fraud detection provides a unique challenge in big data analytics. Previously, credit card companies use to call up customers to verify their transactions, a method that is both inefficient and expensive. Tracking customer spending patterns and identifying fraud-affected areas, big data technology can help to minimize these frauds.

Similarly, the stock price forecast is also critical for companies involved in the investment business. Taking into account the historical data and latest company news, big data can make reasonable predictions of the trend in the stock market. This allows investors to make strategic decisions about their investments.

Meteorology

Meteorology (weather forecast) is another critical area where big data is used extensively. Rainfall forecast is critical for countries that are heavily dependent on agriculture. Additionally, the weather forecast is also essential for several businesses for daily operations. It also helps in reducing losses caused by natural calamities like floods.

Seismology

The prime objective of seismic exploration is to develop an image of the subsurface geology. Seismographs can help us to understand the intricacies of subsurface geology. Additionally, using big data platforms like Hadoop, Hive, we can forecast seismological events using time series analysis [11].

Public sector.

Federal, as well as local governments, can use big data to plan their projects. It can be used in several public sectors like education, hospital, insurance, etc. Big data can be used to analyze the requirements in public sectors, and understand the need to implement projects.

Sports

Big data has been used to improve training and understanding competitors, using sport sensors. Big data has also been leveraged to predict winners in several sports like NBA, soccer, baseball, American soccer, etc. Additionally, the future performance of players can be predicted too. This helps in determining a players' value and salary by collecting data throughout the season [12].

In Formula One races, the use of technology goes a notch up. Racing cars are fitted with hundreds of sensors that generate terabytes of data. These sensors collect data from tire pressure to fuel burn efficiency [13]. https://en.wikipedia.org/wiki/Big_

data-cite_note-134 Based on these data, engineers and data analysts decide whether adjustments should be made in order to win a race. Besides, using big data, race teams try to predict the time they will finish the race beforehand, based on simulations using data collected over the season [14].

6 Dealing with Numeric Data, Images, and Videos

As discussed in Sect. 3, data can be in form of numbers, images or videos, and even text. Today in the realm of big data, these different types of data are analyzed to extract deep insights. There is a huge volume of work that deals with different algorithms used to analyze these data types. These algorithms collectively form the most talked about phrase of our time "Artificial Intelligence".

The field of AI began in the 1950s with Alan Turing's question: "Can machines think?" [15]. The quest to answer this led to the foundation of AI which since then has been progressing rapidly with applications to a wide variety of areas. The main aim of AI is to enable a machine to think and take decisions, perform repetitive tasks, and eliminate human errors by imitating human behavior and abilities. AI is broadly classified into three categories:

(a) **Artificial Narrow Intelligence (ANI)**: ANI is the most common and the weakest type of AI present today which is programmed to perform only a single specific task. Few examples of ANI are Siri, Autopilot in airplanes, chatbots, self-driving cars, and so on [16] (Fig. 2).
(b) **Artificial General Intelligence (AGI)**: AGI machines, conceptualized but not realized yet, will have the ability to learn, understand, and act in a way that is indistinguishable from a human in a given situation. AGI machines can perform multitasking under certain conditions [17].
(c) **Artificial Super Intelligence (ASI)**: ASI is a hypothetical AI where machines will exhibit intelligence superseding human capabilities. ASI will be a multi-faceted intelligent machine with far superior power to solve and decide compared to human beings.

One most common and widespread subset of AI is Machine Learning (ML), which uses statistical learning algorithms that have the ability to learn and improve from experiences without explicit programming. From recommender systems on Netflix or YouTube or Spotify, to search engines like Google or Yahoo to even voice assistants like Amazon Alexa, all employ ML to perform their respective tasks. The ML algorithms are trained with a lot of data, which allows them to process information and learn from it.

ML algorithms can be broadly classified into three types:

(a) Supervised Learning: In supervised learning, the data has two components—an input variable/feature and an output variable. The ML algorithm learns to map between the input and output variables. In other words, a supervised learning

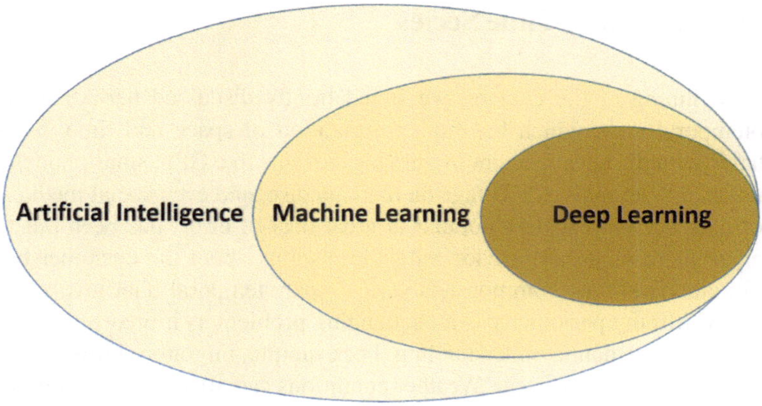

Fig. 2. Schematic representation of Artificial Intelligence and its subsets, viz., Machine Learning and Deep Learning

algorithm takes a known set of input datasets and its known responses to the data (output) to learn and perform the regression or classification model.

(b) Unsupervised Learning: Contrary to supervised learning, unsupervised learning does not have a labeled data. In this case, the algorithm tries to find similarities, differences and infer patterns in the data without any specific reference to outputs.

(c) Reinforcement Learning: The third type of ML which is completely distinct from the previous two is reinforcement learning. In reinforcement learning, the algorithm learns continuously by interacting with the environment. There is an agent which learns from the environment via trial and error by using feedback from its previous actions and experiences. Thus, using this reward-penalty method the algorithm learns to perform the correct actions.

Another subset of machine learning is "**deep learning**" which is inspired by the way our brain processes information. The theoretical base of deep learning is based on the concept of neural networks proposed independently by Alexander Bain [18] and William James [19]. In 1958, psychologist Frank Rosenblatt created the "perceptron" algorithm for pattern recognition based on a two-layer learning computer network using simple addition and subtraction [20]. The backpropagation algorithm created by Werbos allowed the neural networks to learn the relationship between complex, non-linear, and process parameters and update themselves from their failures [21]. Using the formulations of perceptron and backpropagation, Yann LeCun, YoshuaBengio, and Geoffrey Hinton formulated the "deep learning" architecture [22]. With the advancement in algorithms, deep learning has gained tremendous popularity over time. Deep learning has penetrated almost every domain like the field of computer vision, speech recognition, natural language processing, bioinformatics, drug design, etc. It is expected to match human intelligence or even surpass it although at present it is still limited to artificial narrow intelligence.

7 Periodicity and Time Series

At the beginning of the chapter, we stated briefly discussed periodicity in data. Spatiotemporal data, which has both information of space and time, has gained popularity with the advancement in tracking systems like GPS, smartphones, sensor tags attached to animals, RFID tags on merchandise, and even social media. As this type of data becomes widely popular, a great deal of effort has been put forward to analyze them as it can provide valuable insights about the customer behavior, wildlife, etc. The most common analysis of spatiotemporal data involves mining periodicity. Mining periodicity is a challenging problem as it involves bridging the raw data and its semantic understanding. For example, migratory birds never follow the exact same path every year. Weather conditions can affect their migration routes strongly. Similarly, it's quite common to have sparse or even incomplete observations as all the migratory birds are not attached to sensors. Among the existing periodicity mining methods, Discrete Fourier Transform (DFT) and autocorrelation are the most common types. Fourier Transform maps a function of time into a new function whose argument is frequency. In the case of a periodic function, DFT can be simplified to the calculation of a discrete set of complex amplitudes, called Fourier series coefficients. Given a sequence $x(n), n = 0, 1, ..., N - 1$, the normalized DFT is a sequence of complex numbers $X(f)$ is given by

$$X(f) = \frac{1}{\sqrt{N}} \Sigma_{n=0}^{N-1} x(n) e^{-\frac{i2\pi kn}{N}}$$

where the subscript k/N denotes the frequency that each coefficient captures. Vlachos et al. proposed a non-parametric approach to predict the periodicity by introducing a new periodic distance measure technique for time-series sequences [23]. Li et al. designed Fourier Transform based method to mine periodicity in spatiotemporal data of moving objects [24]. Yuan et al. proposed a Bayesian-based non-parametric approach to detect periodic patterns in spatiotemporal data of social media users [25]. For biological systems which are fast, ultradian, circadian, circalunar, and yearly time domains, determining periodicity can be troublesome. Amariei et al. proposed a Fourier-based approach that generates de-noised waveform from multiple frequencies [26]. Though we have discussed briefly periodicity and techniques, we haven't discussed time series yet (Fig. 3). Below we discuss time series data and its analysis.

In sectors like banking and finance, healthcare, meteorology, seismology, and even retails, we encounter a special type of data that is indexed in time order. This type of data is known as time series where the data points are sequenced taken at successively spaced points in time giving rise to a sequence of discrete-time data. Time series data have a natural temporal ordering and the analysis of these time series is significantly different from cross-sectional studies in which there is no natural ordering of the observation points.

A time series data can exhibit a repetitive pattern with elapsed time which is known as periodicity. For instance, data that has a sinusoidal character is periodic.

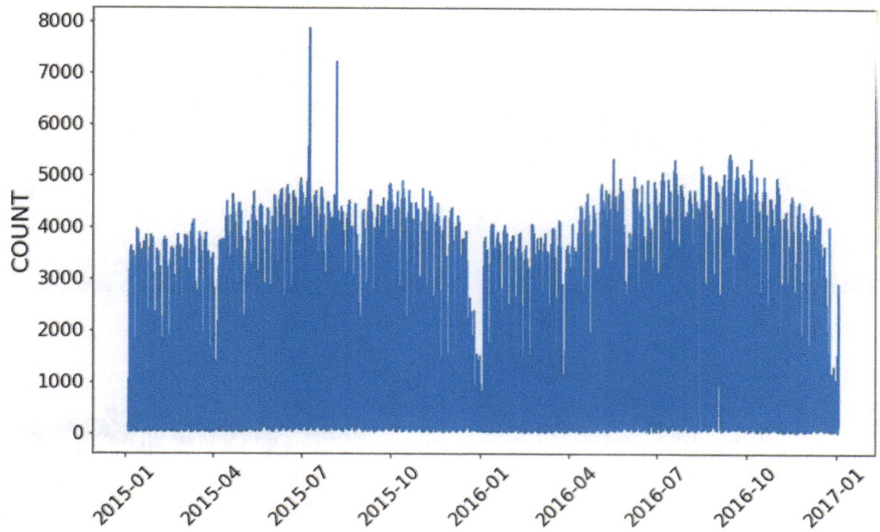

Fig. 3. An example of a time series plot

The period can vary from few minutes (as in the case of an electrocardiograph) to years (solar activity). However, the important point is that the pattern must repeat itself after the time period as shown in the Fig. 3.

For the sake of convenience, periodicity has been divided into two broad categories: (a) Seasonal and (b) Cyclical. Both are extremely important as vital insights can be obtained by analyzing the underlying periodicity. However, before diving into this, let us look at a much simpler and general concept of data called ***"trend"***. Almost all data exhibit a trend. A trend occurs if there is a tendency in the time series data to move upward over time or downward. Thus, we say that the data has a general tendency to either move up (increasing trend) or down (decreasing trend). However, the term "trend" is not well defined mathematically and may not give a lot of insight. On the other hand, periodicity is mathematically well defined. Seasonality occurs when there is a regular pattern in the data related to the calendar. A seasonal data can exhibit patterns which can either be daily, weekly, monthly, quarterly, or even annually. For example, sales of air conditioners or ice cream may peak every year during the summer months and this observation can be observed every year. Similarly, people who are prone to allergies may show allergic reactions with seasonal changes. Whenever the time series is influenced in a periodic manner by the calendar, we call such time series to have a seasonality. On the other hand, a cyclic pattern exists when the data exhibit highs and lows over a period of time that is not fixed. Figure 4 shows the plot corresponding to seasonality, upward and downward trend as well as a time series with none of these characteristics. The duration of these peaks and troughs may vary for years. For instance, business cycles may usually last several years, but in this case, the length of the current cycle is unknown beforehand.

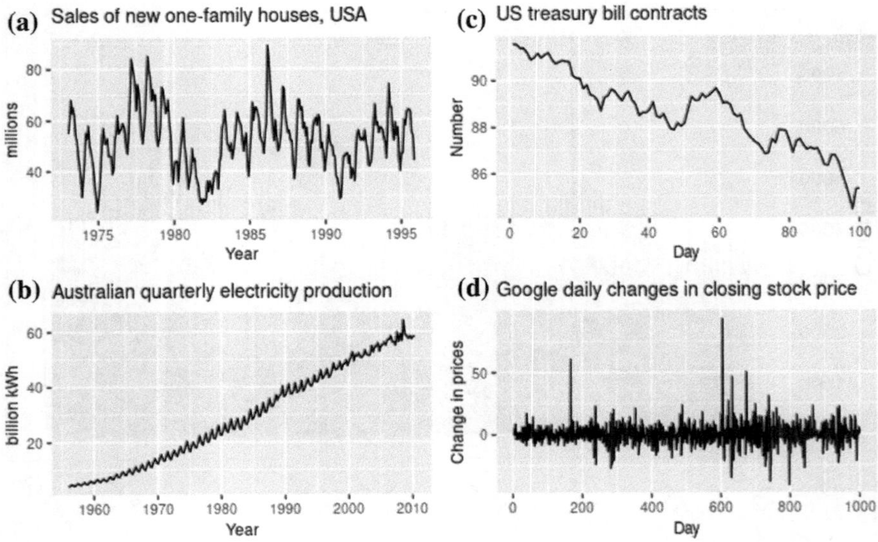

Fig. 4 Plots for **a** seasonality, **b** upward trend, **c** downward trend, and **d** time series with any trend or seasonality

Although the word seasonal and cyclic may be confusing at the first glance, there are important differences between them. While the seasonal pattern is for a constant length of time, a cyclic pattern occurs over a variable length of time. The average length of a cycle is longer than the length of a seasonal pattern. Moreover, the magnitude of a cyclic patternis more variable than the magnitude of a seasonal pattern. Finally, while the timing of peaks and troughs are predictable with seasonal data, it is much harder to predict a cyclic event.

In order to discover the periodicities in a time series data, it is possible to use the periodogram to estimate the spectral density of the signal. The periodogram P can be obtained by squaring the length of each Fourier coefficient as.

$$P\left(f_{\frac{k}{N}}\right) = \left|X_{\frac{k}{N}}\right|^2$$

for $k = 0, 1, 2 \ldots \frac{(N-1)}{2}$. At this point it is also imperative to briefly discuss the subtle difference between prediction and forecast. Prediction is everything about guessing a number (value) of something. This "*something*" can be anything ranging from the price of a house in a city as is done using the Boston Housing or Airbnb dataset to the label of a photo to label a photo as cat or dog etc. On the other hand, forecasting is a special type of prediction where the data has a temporal component. In forecasting, one looks at the past/historical data and makes a prediction for a future time. The weather for future days/weeks, sales of a product in upcoming days/months, the stock market in coming days can be clubbed under the category forecasting as the future is predicted using the historical data.

The statistical art of dealing with time series data is called time series analysis. There are several techniques that are bundled together in time series analysis. Next, we will briefly look into some of these classical techniques of time series analysis for forecasting followed by more modern approaches which are based on deep learning.

7.1 Classical Time Series Forecasting

The traditional techniques of time series forecasting include:

(i) Moving average(MA),
(ii) Autoregression (AR),
(iii) Autoregressive Moving Average (ARMA),
(iv) Autoregressive integrated moving average (ARIMA),
(v) Seasonal autoregressive integrated moving average (SARIMA),
(vi) Seasonal autoregressive integrated moving average with exogenous regressors (SARIMAX),
(vii) Vector autoregression (VAR),
(viii) Vector autoregression moving average (VARMA),
(ix) Vector autoregression moving average with exogenous regressors (VARMAX),
(x) Simple exponential smoothing (SES),
(xi) Holt Winter's exponential smoothing (HWES).

Depending on the type of data, domain, and requirement, these abovementioned techniques are used. However, it is important to note that each of these models has their own limitations depending on the nature of the series, viz., seasonality, stationarity, etc.

To keep things simple, we will briefly outline only a few of these techniques without diving deep into the mathematical formulations. A detailed description of each of these techniques is beyond the scope of the present work. For details, we refer to the book by Brockwell and Davis on Time Series [27, 28].

(i) **Moving Average (MA).**

A moving average or rolling average or running average is a way to analyze data points by creating a series of averages of different subsets of the full dataset. It is a type of finite impulse response filter and simple moving average, cumulative, weighted, and exponential moving average are different types used depending on the type of dataset.

(ii) **Autoregressive Mode**

As the name suggests, the autoregressive (AR) model is a regression model where the forecasting is performed using the historical data. The autoregressive model is an extremely simple yet effective model where the output or the forecasted value depends linearly on its own previous values. Mathematically speaking, the AR model of order p can be written as:

$$X_t = \beta_0 + \sum_{i=1}^{p} \alpha_i X_{t-1} + \varepsilon_t \tag{1}$$

where X_t and X_{t-1} are the observations at time *t* and *t-1*, respectively, *c* is a constant and ε_t is the white noise with mean zero and standard deviation …

A fundamental principle of autoregression is that the series has a correlation with a delayed (lagged) copy of itself. There exists two important correlations in this context : (a) Autocorrelation [29], and (b) Partial Autocorrelation [29]. Let us understand this by taking a simple example related to the sales of ice cream. Let us denote the volume of sales of ice cream as $X = \{X_1, X_2, \ldots\ldots, X_N\}$ for time stamps $\{t_1, t_2, \ldots., t_N\}$. It is possible that the volume of sales of any given month may have an effect on the volume of sales for the next month or the month later. In other words, we can think of having a relationship like $\{X_1 \rightarrow X_2 \rightarrow X_3 \rightarrow X_4 \rightarrow X_5 \rightarrow X_6 \ldots and soon\}$, where the sales of the any given time point t is directly influenced by the sales of the previous time point $t - 1$. Additionally, we can have a scenario where the present month's sales are influenced by the sales from two time points earlier. For example, the sales at time point t_3 (X_3) can be influenced by the sales at time point $t_1(X_1)$. In this case, there are two ways the sales X_3 can get affected by the sales in X_1. One way is the direct influence that X_1 has on X_3. while the other is the indirect influence v X_2 denoted by $X_1 \rightarrow X_2 \rightarrow X_3$. The first case is what is known as autocorrelation or serial correlation while the latter is known as partial autocorrelation. The autocorrelation is a bit intuitive as it is mathematically given by the Pearson correlation function. If $\{X\}$ is a series of observable at time $\{t_i\}$, then the lag p autocorrelation function (ACF) is defined as

$$ACF = \frac{\sum_{i=1}^{N-k}(X_i - \underline{X})(X_{i+p} - \underline{X})}{\sum_{i=1}^{N}(X_i - \underline{X})} \tag{2}$$

ACF is a correlation coefficient, but instead of being the correlation coefficient between two different variables, the correlation is calculated between two values of the same variable at times X_i and X_{i+k}. Hence, the name autocorrelation.

For the partial autocorrelation function ($PACF$), we can write the linear regression expression

$$X_t = \Phi_{21} X_{i-1} + \Phi_{22} X_{t-2} + \varepsilon_t \tag{3}$$

where X_t is the value of time t while X_{t-1}, and X_{t-2} are the lagged values at time $t-1$ and $t-2$, respectively, and ε_t is the error term. The coefficient "Φ_{22}" is the $PACF$ in this case and it gives the effect of X_{t-2} on X_t. Overall, the $PACF$ help us to build a good time series forecasting model. It not only helps us to decide if the AR model is appropriate for the data, but also in deciding the order of the AR model.

To apply the AR model, it is critical for the time series data to be stationary. The stationarity condition is satisfied if the mean, the variance, the autocorrelation does not change over time and there are no periodic fluctuations (seasonality). If the series is non-stationary, then there will be a non-declining effect of the previous values on the current value. It simply means that the correlation with increasing lags should ideally decrease.

Going back to our ice cream sales data, the series will have non-stationarity if the sales at time point $t_6(X_6)$ is more influenced by the sales at time point $t_1(X_1)$ or $t_2(X_2)$.

(iii) **Autoregressive Integrated Moving Average (ARIMA).**

Consider the hypothetical sales data shown in Fig. 5. It shows a clear upward trend in sales with time. In strict mathematical terminology, this data is not stationary. Thus, if we want to predict the sales for the year 2020, we cannot apply the AR model as in this case the mean is not constant. To apply autoregression, the first step is to make the data stationary. Stationarity can be achieved by calculating the difference between successive time points. If the sales data time series $\{X\}$ mentioned earlier has stationarity, we can perform a transformation and define a new series Z_i such that

$$Z_i = X_{i+1} - X_i \tag{4}$$

where $\{Z_i\}$ satisfies the condition of stationarity. This is because if the original time series has linearity, then the difference between two such timestamps will be linear barring the fluctuations around a constant mean. Once we have the stationarity in our data, we can freely use the AR model with more confidence. We can combine the AR and MA models and create a new model called Autoregressive Integrated Moving Average (ARIMA) for forecasting. ARIMA models are characterized by three parameters, (p, d, q),where p is the lag of the AR model and q is the order of the MA model. The parameter d is the order of the difference which we perform to make the data stationary. Thus, we can define the ARIMA model with parameters (p, q, d)

$$Z_i = \Phi_1 Z_{i-1} + \theta_1 \varepsilon_{i-1} + \varepsilon_i \tag{5}$$

In Eq. 5, term $\Phi_1 Z_{i-1}$ is the AR part while the term $\theta_1 \varepsilon_{i-1}$ is the MA part. The final term ε_i is the error term. The part corresponding to "integrated" in ARIMA comes from the fact that Z_i was created by taking the differences of successive terms in the series $\{X\}$. Finally, we would like to predict X_i and not the difference Z_i. If

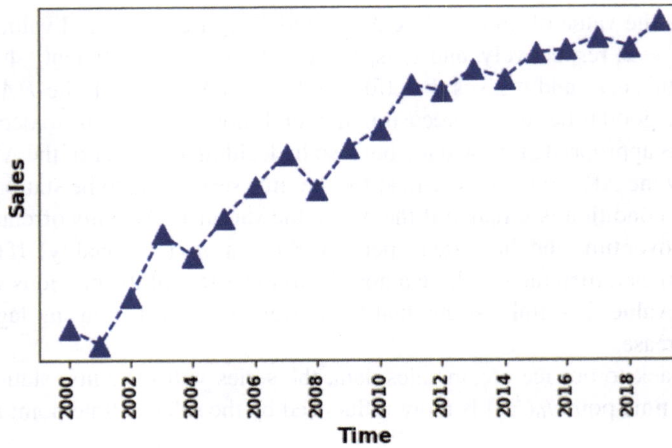

Fig. 5 Seasonality in sales data

X_l is the last term, in the series $\{X\}$,. then for a future time point τ, the value X_τ is given by

$$X_\tau = \sum_{i=1}^{k-l} Z_{k-i} + X_l \tag{6}$$

Finally, if we want to handle time series data with periodic characteristics, then we can use the SARIMA model which combines the seasonal differencing with the ARIMA model. SARIMA model is described by seven parameters $(p, d, q)(P, D, Q)m$.hile (p, d, m) are the same as the ARIMA model; the $(P, D, Q)m$ are the parameters that incorporate the seasonal elements.

7.2 Modern Techniques

Apart from the classical time series analysis, there are few modern recurrent neural network (RNN)-based techniques that have gained popularity in recent times. We will briefly discuss it here. RNN-based [30] methods are capable of identifying the nonlinear and complex structures and patterns of data in time series forecasting. One such RNN architecture is the Long Short-Term Memory (LSTM) [30] (Fig. 6).

LSTM networks have an internal memory (like RNN) making them efficient to remember past events/data. The vanishing gradient problem associated with RNN is resolved in LSTM. These features make LSTM well suited for classification and time series forecasting with time lags of unknown duration. Figure 6 shows the schematic diagram of LSTM.

Three gates constitute the LSTM network architecture:

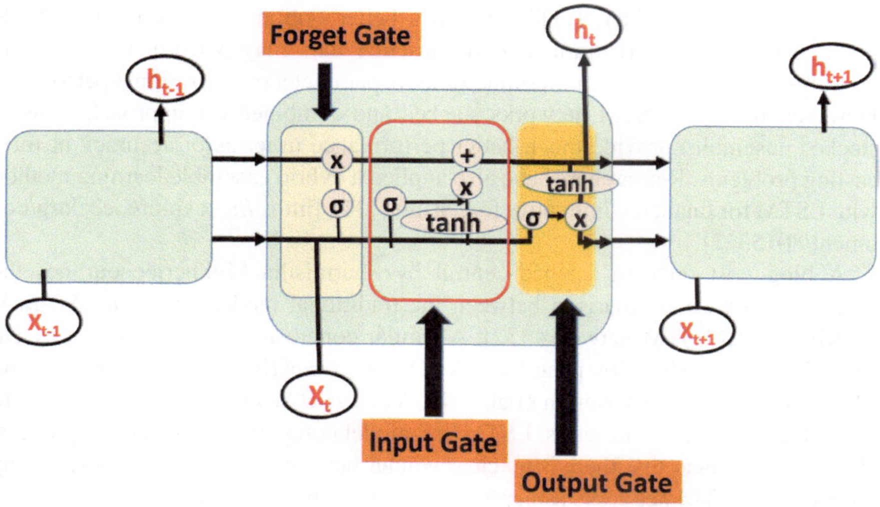

Fig. 6 Schematic representation of a LSTM network

(i) Forget gate—The forget gate is responsible for the removal of information from the cell state. The filter removes information that is less important for LSTM. As shown in the figure, this gate takes in two inputs h_{t-1} and X_t. While h_{t-1} is the hidden state from the previous cell, X_t is the input for the current time step. The input is passed through a sigmoid function which outputs a vector in the range of 0 to 1. This sigmoid function decides which values to keep and which one to discard.

(ii) Input gate—The input gate adds information to the cell state. Information addition happens in a three-step process:

- Regulating values that must be added to the cell state via a sigmoid function. This is very similar to the forget gate and acts as a filter for all the information from h_{t-1} and X_t.
- Creating a vector containing all possible values that can be added to the cell state using the tanh function. The output value ranges between –1 and +1.
- Multiply the value of the regulatory filter (sigmoid function) and the created vector (tanh function) and then adding this useful information to the cell state via an addition operation.

(iii) Output gate—The operations of the output gate can be summarized in three steps: -

- Apply tanh function to the cell state and create a vector.
- Employ sigmoid function to filter using the values of h_{t-1} and X_t. This step regulates the output from the vector created from the input gate.
- Multiply the value of this regulatory filter to the vector created in the first step of which becomes the hidden state for the next cell

Compared to ARIMA, LSTM provides better options for fine-tuning. While a single LSTM network trained on a particular data may perform poorly on an entirely different time series unless an extensive parameter optimization is performed. However, multiple LSTM networks stacked and combined (an approach called a stacked ensemble) provide much better performance in terms of accuracy in forecasting problems. Researchers have also applied a hybrid ensemble learning method with LSTM for financial time series forecasting [31]. [https://ieeexplore.ieee.org/document/8015322].

A blog post on data science central by Sharmistha Mukherjee outlines the performance-wise comparison between the traditional model, viz., AR, ARIMA, SARIMA, and LSTM networks [32]. A similar conclusion was also drawn by Dr. Jason Brownlee in his blog posts [33]. A detailed comparison between ARIMA and LSTM was provided by Namin et al. [34]. While AR/ARIMA models can perform better for univariate time series, LSTM can model complex multivariate sequences. However, it is clear that there is no clear winner between the traditional forecasting models and LSTM networks as per the studies performed to date.

Apart from the abovementioned methods, there are several new techniques that have been proposed for handling big data not restricted to time series data. Lea et al. proposed the Temporal Convolution Network (TCN) for video-based action-segmentation [35, 36].

8 Horses for Courses

From our discussion earlier, we see that there exist several methods which are either based on traditional statistics or deep learning. To solve a particular problem, one can often apply several techniques and choose the one which provides the best result. To forecast sales using univariate time series, we can try several methods as listed in Sect. 7.1 and select the best performing model. For multivariate time series, we can try ARIMA model, SARIMA model, or even LSTM. Finally, we can select the best-performing model. In general, this is normally a strategy applied not only for time series forecasting but also for any regression or classification. One applies a bunch of algorithms to a given problem and selects the best-performing model. This is both time consuming and inefficient. Firstly, one has to spend a lot of time fine-tuning the model parameters. Secondly, a lot of time is spent on training the model. Additionally, if the dataset gets updated (by addition of new data), the already trained model needs retraining which has been already trained earlier and is in (probably) in production adding burden to the cost of the project.

Over the past few years, several algorithms have been designed which could perform several types of complicated tasks. However, no single algorithm can perform all the tasks with equal efficiency. Consequently, the current approach employed globally is to try a few algorithms on the same problem and select the best one. However, this is also an inefficient strategy.

This raises an important question: Are these algorithms reading the data (information) improperly? Is it possible that the algorithms cannot grasp all the possible profiles/aspects of the data? For the present algorithms, it is not clear how much (training) data is adequate for training the model, which in turn will have a high accuracy for the test data. In other words, there is ample ambiguity to ascertain how much data is sufficient to build a reasonable model.

Another important aspect of the data is symmetry. Statistically speaking, data can follow different types of distribution, viz., normal, binomial, Poisson, etc. Additionally, time-series data can have periodicity. On top of these, there may exist both spatial and temporal invariance. The open and natural question at this juncture is whether the algorithms currently being employed can utilize the underlying symmetry, periodicity, and invariance of the data. Several algorithms have been developed to address these aspects but each of them has their own limitations.

9 Difference Between the Current Neural Nets and the Human Brain

Although the concept of neural network based deep learning has been inspired by the human brain, at present deep learning is still limited to artificial narrow intelligence. Despite recent successes, deep learning needs significant radical improvement in terms of the algorithm to achieve efficiency at the level of the human brain. The prime differences between the human brain and present deep learning are:

i. Form and function: While the human brain's form and function go together, deep learning is a software code in some coding language and runs on embedded silicon chips. The human brain has specialized compartments to perform specific tasks, while deep learning can perform only a specific task. Moreover, the human brain can perform parallel processing of information but the deep learning networks perform only serial processing.
ii. Learning Technique: The human brain can make generalized learning while deep learning is not good at generalization. The addition of small noise to the data can make the deep learning network yield inaccurate and incorrect results. While the human brain requires far less data for training, a huge volume of data is necessary for the training of deep learning networks.
iii. Size: The human nervous system is extremely complex in structure and consists of around eighty-six billion neurons while a typical deep learning network can have about a few hundred neurons.

10 A Holistic Grand Approach

An alternative and unique approach was proposed by Bandopadhyay. They proposed an approach, sans algorithm, to efficiently handle big data. The details are provided in the book "*Nanobrain*: The making of an artificial brain from time crystal"[37]. The approach by Bandopadhyay et al. follows the single-shot supervised learning on the training data and accurately predicts the test data. If we train a child by showing an image or few images of a cat with the appropriate training label "*cat*", the child can immediately recognize the new image of a cat as "*cat*" irrespective of its orientation or size, and even if the cat is in mid of other animals or objects. This is because the human brain interprets the invariance of the data. Similarly, the methodology of Bandopadhyay et al. recognizes and interprets the invariance in the data thereby unlocking the mystery of how information is processed by the human brain. This approach in a way mimics the human learning and hence is "*holistic*" and "*grand*" approach.

Summary

In summary, we have presented an overview of data, big data, and sectors leveraging big data. We also discussed techniques to mine periodicity in data, including time series data. Moreover, we briefly discussed artificial intelligence, machine, and deep learning. Data including time series data can be analyzed by the traditional statistical methods as well as modern machine and deep learning algorithms including some of the latest development. We also briefly point out the limitations of the problems with statistical and machine/deep learning algorithms. Finally, we briefly discuss an emerging method that is able to outperform current advanced deep learning algorithms. We call this emerging technique a holistic grand approach which not only exhibits a single-shot supervised learning behavior but also understands the invariance in the data mimicking the human learning behavior. We strongly believe that this holistic grand approach if fine-tuned properly can replace all the existing deep learning algorithms and pave way for artificial general intelligence. We also believe that this method can unlock the mysteries related to how humans learn, how information is stored in the universe, and many more.

Acknowledgements SKB is thankful to Dr. Pathik Sahoo, National Institute for Materials Science, Tsukuba, Japan, for privately sharing the results of the emerging holistic grand approach which helped us in understanding the human-like learning behavior.

References

1. Statistical language -what are data?". Australian Bureau of Statistics. 2013-07-13
2. Information-a very short introduction. Oxford University Press. ISBN 978-0-19-160954-1
3. Strang G (1994) Wavelets. Am Scientist 82(3):250–255. JSTOR 29775194
4. Chatfield C (2003) The analysis of time series: an introduction, 6th edn. Chapman & Hall/CRC Texts in Statistical Science, CRC Press
5. Shinozaki A (2020) Electronic medical records and machine learning in approaches to drug development. https://doi.org/10.5772/intechopen.92613
6. Mahmoudi E, Kamdar N, Kim N, Gonzales G, Singh K, Waljee AK et al (2020) Use of electronic medical records in development and validation of risk prediction models of hospital readmission: systematic review. BMJ 369:m958. https://doi.org/10.1136/bmj.m958

7. Suzuki K (2017) Overview of deep learning in medical imaging. Radiol Phys Technol 10:257–273. https://doi.org/10.1007/s12194-017-0406-5
8. Bompelli A, Wang Y, Wan R, Singh E, Zhou Y, Xu L, Oniani D, Kshatriya BSA, Joyce (Joy), Balls-Berry E, Zhang R (2021) Social determinants of health in the era of artificial intelligence with electronic health records: A systematic review. arxiv: 2102.04216
9. Irvin JA, Kondrich AA, Ko M et al (2020) Incorporating machine learning and social determinants of health indicators into prospective risk adjustment for health plan payments. BMC Public Health 20:608. https://doi.org/10.1186/s12889-020-08735-0
10. Galli L, Levato T, Schoen F, Tigli L (2020) Prescriptive analytics for inventory management in health care. J Oper Res Soc. https://doi.org/10.1080/01605682.2020.1776167
11. INDIACom (2016) Proceedings of the 3rd International Conference on 10th INDIACom "Computing for Sustainable Global Development, 16–18 Mar 2016. Earthquake Data Analysis and Visualization using Big Data Tool. Gaurav Gupta & Inder Singh Gupta
12. Regina Pazvakavambwa. Predictive analytics, big data transform sports. itweb.co.za
13. Dave Ryan. Sports: where big data finally makes sense. huffingtonpost.com
14. Bi F (2014) How formula one teams are using big data to get the inside edge. Forbes
15. Turing AM (1950) Computing Machinery and Intelligence. Mind LIX(236):433–460. https://doi.org/10.1093/mind/LIX.236.433
16. Frankish K, Ramsey WM (1960) The Cambridge handbook of artificial intelligence, 12 June 2014. Cambridge, UK, p. 342
17. Goertzel B (2014) Artificial general intelligence: concept, state of the art, and future prospects. J Artif Intell 5(1):1–46
18. Bain A (1873) Mind and body: the theories of their relation. D. Appleton and Company, New York
19. James W (1890) The principles of psychology. H. Holt and Company, New York
20. Rosenblatt F (1958) The perceptron: a probalistic model for information storage and organization in the brain. Psychol Rev 65(6):386408. https://doi.org/10.1037/h0042519. PMID 13602029
21. Werbos PJ (1974) Beyond regression: new tools for prediction and analysis in the behavioral sciences. Ph. D. Thesis, Harvard University, Cambridge
22. LeCun Y, Bengio Y, Hinton G (2015) Deep learning. Nature 521:436–444. https://doi.org/10.1038/nature14539
23. Vlachos M, Yu PS, Castelli V (2005) On periodicity detection and structural periodic similarity. In: Proceedings of the 2005 SIAM international conference on data mining. https://doi.org/10.1137/1.9781611972757.40
24. Li Z, Ding B, Han J, Kays R, Nye P (2010) Mining periodic behaviors for moving objects. In: Proceedings of the 16th ACM SIGKDD international conference on knowledge discovery and data
25. Yuan Q, Zhang W, Zhang C, Geng X, Cong G, Han J (2017) PRED: periodic region detection for mobility modeling of social media users in 10th international conference on web search and data mining. https://doi.org/10.1145/3018661.3018680
26. Amariei C, Tomita M, Murray DB (2019) Quantifying periodicity in omics data. Front Cell Dev Bio 2(40):1. https://doi.org/10.3389/fcell.2014.00040
27. Brockwell PJ, Davis RA (1991) Time series: theory and methods, 2nd ed. Springer-Verlag, New York
28. Cromwell JB, Hannan MJ, Labys WC, Terraza M (1994) Multivariate tests for time series models. Sage Publications, Thousand Oaks, CA
29. Box GEP, Jenkins G (1976) Time series analysis: forecasting and control. Holden-Day
30. Deep Learning with Python (2018) Francois Chollet. Manning Shelter Island
31. Zhu S, Zhao Y, Yang H, Tang Z, Wang, Hao J (2017) A real-time ensemble classification algorithm for time series data. In: 2017 IEEE international conference on agents (ICA), Beijing, China, 2017, pp 145–150. https://doi.org/10.1109/AGENTS.2017.8015322
32. ARIMA/SARIMA vs LSTM with Ensemble learning Insights for Time Series Data by Sharmistha Chatterjee in Data Science Central (2020). https://www.datasciencecentral.com/profiles/blogs/arima-sarima-vs-lstm-with-ensemble-learning-insights-for-time-ser

33. Comparing classical and machine learning algorithms for time series forecasting by jason brownlee in deep learning for time series (2018).https://machinelearningmastery.com/findings-comparing-classical-and-machine-learning-methods-for-time-series-forecasting/
34. Temür AS, Akgün M, Temür G (2019) Predicting housing sales in turkey using ARIMA, LSTM and hybrid models. J Bus Econ Manag 20:920-938. https://doi.org/10.3846/jbem.2019.10190
35. Lea C et al (2016) Temporal convolutional networks: a unified approach to action segmentation. European Conference on Computer Vision. Springer, Cham
36. Lea C et al (2017) Temporal convolutional networks for action segmentation and detection. In: Proceedings of the IEEE conference on computer vision and pattern recognition
37. Bandyopadhyay A (2019) Nanobrain: The making of an artificial brain from time crystal. CRC Press

Chapter 2
Modulating Neural Oscillations with Transcranial Focused Ultrasound

Joseph L. Sanguinetti

1 Introduction

Hans Berger [3] first discovered neural oscillations by recording electrical activity from subjects' heads, known as the electroencephalogram (EEG). Berger noticed that when participants closed their eyes, the EEG increased at 10 Hz, which was later called the "Berger wave," or alpha rhythm. Although Berger's discovery was initially controversial, the alpha wave is now accepted alongside the many frequency-specific oscillations that emerge from brain activity. The current understanding is that as groups of neurons fire in synchrony, the sum of their postsynaptic potentials generates the rhythmic patterns observed in the EEG [33]. Almost 100 years of research shows that modulations in the amplitudes of specific brain oscillations correlate to cognitive, perceptual, and motor processes and tasks [2]. Alpha oscillations, for example, co-occur in conditions when top-down attention is applied to a visual stimulus, leading researchers to claim that alpha serves as an attentional suppression mechanism to shape perception [16, 34]. How could rhythmic variations in neural activity help the brain carry out complex tasks like top-down attention?

All biological systems exhibit rhythmic behavior, from cytoskeletal oscillations that promote cellular function [28] to circadian rhythms that modulate sleep in mammals [43]. Oscillations play a critical role in orchestrating biological processes over space and time [8]. Biological systems use oscillations as timekeepers, or clocks, to organize activity and function. Given their prevalence in living systems, it comes as no surprise that the brain exhibits oscillatory behavior as well. Many current theories of brain oscillations claim that the cyclic activity of neurons creates a temporal reference frame—like a clock—that allows the brain to organize the information

J. L. Sanguinetti (✉)
Center for Consciousness Studies, University of Arizona, Tucson, Arizona, USA
e-mail: sanguine@email.arizona.edu

A. Bandyopadhyay and K. Ray (eds.), *Rhythmic Advantages in Big Data and Machine Learning* , Studies in Rhythm Engineering,
https://doi.org/10.1007/978-981-16-5723-8_2

processing between groups of neurons and their interaction with the environment [62].

Neurons are connected in vast hierarchical networks, spanning 86 billion neurons and 100 trillion connections [14]. There must be a mechanism to orchestrate this massive amount of activity and information flow. The emerging view is that neural oscillations perform this function by synchronizing the activity of neural elements and organizing them into functional groups [7]. There is now growing evidence that frequency-specific neural rhythms relate to brain functions like perception, attention, and memory [2, 57]. Take, for example, the problem that visual neurons face when processing a vase. Many neurons must work together to process the vase while other neural elements segment the vase's background or other objects. How do the "vase neurons" organize their activity to ensure they are all working on the same parts (or features) of the figure? One clue is that brain oscillations in the gamma frequency band (35–100 Hz) are prominent when participants view visual stimuli. This finding led to the idea that gamma may help neural elements synchronize their firing across cortical areas that are processing the parts of the figure ("feature binding;" [58]). Visual information for figures spans multiple levels, for example, across V1 and V2 in the visual cortex, and activity must be organized across these scales as well. Oscillations may play a role by helping to shape the brain's functional architecture, as they reflect nested networks that help neural systems integrate their outputs across multiple scales [57]. Thus, brain oscillations provide the temporal structure for neural firing, helping the brain coordinate neural activity across networks [72].

Although oscillations have been linked to almost every form of cognitive, perceptual, and motor activity, much of the current evidence is correlational. Investigators have historically explored the role of brain oscillations in behavior by recording EEG or magnetoencephalography (MEG) and asking participants to perform tasks; changes in behavior are correlated to changes in oscillatory dynamics, and inferences are made about their relationship. It is unclear whether these oscillations truly support cognitive processes like attention or are rather epiphenomena, i.e., electrical shadows of the underlying neural mechanisms. Another way to state the problem: Does rhythmic brain activity play a causal role in brain function and behavior? How can we determine whether brain oscillations provide a mechanistic account of neural function, or whether they are simply reflections of some other more fundamental computational process? Researchers have turned to interventions that allow for causal manipulation to address these questions. In particular, noninvasive brain stimulation has emerged as a promising approach to modulate oscillatory brain activity within experimental paradigms. The thinking goes like this: If there is a causal relationship between brain oscillations and behavior, then modulating oscillations should cause changes in performance. For example, an intervention that directly alters alpha activity in the visual cortex should alter visual experience or behaviors that rely on visual attention. Such an investigation would reveal a causal link between visual alpha oscillations and experience or behavior. In the following sections, I unpack how noninvasive brain stimulation, or neuromodulation, can help elucidate causal links between time-keeping brain oscillations and behavior. In the final sections, I introduce ultrasonic neuromodulation as an ideal tool to address these issues.

2 Moving from Correlation to Causation with Neuromodulation

Noninvasive neuromodulation has shifted the paradigm in studying brain oscillations [71]. Two methods in particular, transcranial magnetic stimulation (TMS) and transcranial alternating current stimulation (tACS), reliably modulate the excitably of neurons in the neocortex and neural oscillations. For example, Griškova and colleagues [19] pulsed TMS at 10 Hz–known as *repetitive* TMS (rTMS)—to the left dorsolateral prefrontal cortex and found an overall increase in delta power (4 Hz), suggesting these stimulation parameters altered brain oscillations but did not necessary entrain them. Later experiments found that the pulse frequency of rTMS could directly entrain brain oscillations with different experimental paradigms (e.g., Thut et al. [64]). In addition to driving brain rhythms, TMS elicits widespread changes in functional connectivity [26], suggesting that network dynamics are also altered. Another form of neuromodulation, tACS, induces a weak alternating current via sponge electrodes attached to the scalp. tACS generates a periodic electrical signal that readily entrains brain oscillations in the neocortex in a frequency-specific manner[23], making it an ideal candidate to modulate brain rhythms.

TMS and tACS have emerged as reliable methods to modulate brain oscillations (including direct entrainment) in specific frequency bands and impact network activity that relies on brain oscillations. Does manipulating rhythmic brain activity alter behavior in a reliable way to reveal a causal relationship between brain activity and behavior?

Let's consider the functional nature of the alpha wave and its role in behavior as an example. Researchers have proposed that alpha waves in the visual cortex shape perception [16, 34, 45]. Stimuli are less likely to be perceived when the alpha amplitude is higher in the visual cortex [30], leading to proposals that top-down attention increases alpha power in task-irrelevant brain regions [59]. Although many empirical studies support this view, there was no causal evidence that pre-stimulus alpha shapes perception, leading to alternative perspectives that instead alpha indexes information processing in general [18]. One way to ask whether alpha oscillations play a causal role in perception is to enhance pre-stimulus alpha over the visual cortex during a perceptual task. If the behavior that relies on perception is altered, one could conclude that alpha plays a causal role in perception (see [25, 61, 71], for reviews). Indeed, Romei et al. [53] demonstrated that pre-stimulus alpha shapes perception by employing rTMS in the ranges of theta (5 Hz), alpha (10 Hz), and beta (20 Hz) just before visual stimuli were presented. Inducing alpha, but not theta or beta, over the visual cortex inhibited visual detection of a target in the visual field opposite of stimulation but enhanced detection on the same side. This finding draws a causal link between pre-stimulus alpha amplitude and perception by using rTMS and showing that the amplitude of the alpha wave shapes what the participant sees.

Likewise, tACS has been helpful to test the causal role of brain activity and behavior. One example is Strüber and colleagues [60], who established a causal role for gamma in the process by which neurons synchronize their activity to the same features of a stimulus (i.e., "feature binding"). When participants view stimuli that have multiple interpretations (bistable stimuli), researchers observe EEG gamma-band modulations, which suggests that changes in gamma power may relate to the moment when neural units switch their coordination from one set of features to another, leading to the change in perception of the figure. Strüber and colleagues [60] induced in-phase or anti-phase gamma oscillations between the right and left visual hemispheres during a bistable motion task. These authors found that motion perception was disrupted following anti-phase gamma tACS but not in-phase gamma or 6 Hz (theta) stimulation. These results indicate that interhemispheric gamma is needed to bind visual features necessary for stable motion perception of objects, revealing a causal link between gamma and feature binding.

Although noninvasive brain stimulation is proving vital for testing the links between brain activity and behavior, the established electromagnetic methods like TMS and tACS are limited in their ability to modulate the brain. There are three issues with concern to answering questions of causality using brain stimulation: (1) Spatial resolution (are we affecting the brain areas we think we are?); (2) Depth (are we able to reach deep targets in the brain?); and (3) electrophysiological artifacts (does the stimulation modality induce impeding artifacts in the EEG?). While TMS is relatively focal, with 0.5 to 1 cm spatial resolution [6], this method only reaches about 2 cm into the brain [54]. The development of new TMS coils (like the H-coil) may allow TMS to reach deeper into the brain. Still, the depth of penetration comes at the expense of higher and wider electric fields in superficial areas, reducing applicability by reducing spatial resolution. In contrast, the spatial resolution of transcranial electrical stimulation such as tACS is low, with the current spreading over many centimeters and between stimulating electrodes [29] and making it difficult to know which brain regions are affected. Also, like TMS, tACS technologies are unable to reach deep targets in the brain. While temporally interfering electric fields may allow tACS to reach deeper into the brain [20], the method must be further developed and tested for safety before widespread adoption. So, although TMS has a relatively high spatial resolution, deeper brain targets like the thalamus, amygdala, or hippocampus are not practical to reach with either method at the present time.

The third issue of applicability is electrical artifact during stimulation. Investigators need to record EEG concurrently with neuromodulation and behavioral tasks to draw a causal link between brain oscillations and behavior. This poses a challenge for electromagnetic-based stimulation modalities like TMS and tACS because active stimulation overwhelms the EEG signal with high-voltage artifacts, making it virtually impossible to record EEG at the exact moment of interest (during stimulation). Several groups have developed real-time and post-hoc artifact removal methods to address this problem (e.g., [69], but each has limitations [4], which are outside the scope of this chapter. Upcoming technologies that rely on induced electric currents also have drawbacks in this respect. For example, although the temporally interfering method allows tACS to reach deeper into the brain with more spatial specificity [20], this method will not overcome the issue of causing artifacts in the EEG.

While the present electromagnetic brain stimulation methods are valuable tools for establishing causal links between brain and behavior, they have unique limitations, leaving researchers searching for other noninvasive neuromodulation technologies to fill the gaps. A novel method involving ultrasound (acoustic waves above the threshold for human hearing), called transcranial focused ultrasound (tFUS), has emerged as a possible candidate to modulate brain oscillations alongside the established methods. Low-intensity focused ultrasound passes through the human skull with high spatial resolution, reaches deep brain targets, and produces small or manageable artifacts in the EEG [15]. Below I will introduce ultrasonic neuromodulation and discuss the possibility for tFUS to control brain oscillations, much like TMS/tACS.

3 Ultrasonic Neuromodulation

Over 80 years of evidence indicates that low-energy ultrasound can selectively and reversibly excite or suppress neuronal activity (for review, see [15]). Suppression of cortical evoked potentials in felines with unfocused ultrasound was shown 50 years ago [17]. Advances in ultrasound technology and the relatively recent demonstration that ultrasound induces action potentials [67] led to a resurgence in interest in ultrasonic neuromodulation. Safe and effective cortical stimulation using ultrasound has now been demonstrated in rodents [32, 47, 65, 74], pigs [11], sheep [75], and monkeys [13]. The excellent safety profile of ultrasonic neuromodulation in animals led to the first human trials in 2013 [21] and many groups have now demonstrated that ultrasound is an effective method to modulate human brain activity [38, 39, 41, 42, 51, 55], with an excellent safety record [40].

Ultrasound neuromodulation offers many advantages for human neurostimulation relative to TMS and tACS. Transducers can be fashioned that create a focused acoustic beam. Focused ultrasound passes through the human skull with a high spatial resolution [15], known as transcranial focused ultrasound (tFUS). TFUS can modulate neural targets at any depth in the brain, with millimeter resolution [41], including deep brain targets out of reach of established methods like TMS [66]. One example of a deep target is the thalamus, and several groups have used ultrasound to modulate thalamic activity in humans[1, 41, 48]. Another significant advantage of tFUS is that researchers can use it concurrently with neuroimaging like fMRI [74] and EEG [42] with much less signal distortion to the imaging modality relative to TMS and tACS. Although the mechanisms of tFUS are still unknown[5], given the ability for high spatial resolution and low signal distortion with concurrent EEG and mounting evidence that tFUS safely modulates neural activity, focused ultrasound may be an appropriate tool to probe the oscillatory dynamics of the brain.

TFUS addresses many of the problems faced by TMS and tACS, like spatial specificity and depth of penetration, but does it modulate the electrophysiological signals and the oscillatory dynamics of neural activity? In terms of animal studies, several lines of evidence indicate that ultrasound can modulate brain oscillations. Electrophysiological effects produced by ultrasonic stimulation were first documented in 1929 by Edmund Harvey, who showed that muscle and nerve fibers could be stimulated with unfocused ultrasound in frogs [22]. Later, William Fry et al. [17] found that sensory-evoked potentials, recorded from EEG electrodes placed on the scalp over the occipital cortex in felines, could be reversibly suppressed by targeting the thalamus with ultrasound. Several animal studies demonstrate tFUS-induced changes in brain activity as recorded by local field potentials [10, 65, 72] or on evoked potentials as recorded by EEG [32, 74]. Notably, over three decades ago, Velling and Shklyaruk [70] showed that ultrasound modulated rhythmic activity of the electrocorticography (ECoG)in lower frequency bands in felines, demonstrating that ultrasound could indeed alter the oscillatory dynamics of brain activity. More recently, hippocampal oscillations in theta, alpha, and gamma were modulated in rodents [77], showing that tFUS has the potential to modulate the oscillatory dynamics of targeted brain regions.

Human electrophysiological evidence that tFUS could modulate evoked potentials has begun to emerge over the past decade. Legon and colleagues [42] found that tFUS targeting the somatosensory cortex (S1) modulated EEG somatosensory evoked potentials elicited from median nerve stimulation and enhanced sensory discrimination. Importantly, these researchers performed a spectral decomposition of the EEG data: tFUS decreased power in the alpha- (7–12 Hz), beta- (13–30 Hz), and gamma-bands (30–55 Hz). The impact of tFUS on EEG frequency bands was one of the first demonstrations that tFUS could modulate sensory-evoked brain oscillations across multiple frequency bands in humans. Later, the same group targeted the thalamus during a median nerve stimulation paradigm in humans and found reductions in ongoing alpha and beta, as well as reductions in gamma that was time locked to the stimulus [41]. Mueller and colleagues [49] found that tFUS modulated the phase distribution of frequency-specific oscillations (beta, but not gamma), suggesting that tFUS can preferentially modulate the phase dynamics of beta rhythms.

These experiments with tFUS show a small but growing body of evidence that focused ultrasound induces changes in evoked and ongoing oscillatory dynamics in the EEG in animals and humans. Furthermore, most of these studies recorded EEG concurrently with tFUS with minimal artifact introduced by ultrasound, which is not the case with TMS and tACS (see above). Taken together, the evidence suggests that tFUS is a potential candidate for the manipulation of brain oscillations.

Demonstrations that tFUS modulates evoked and endogenous brain oscillations are prima facie evidence that ultrasound is a valuable tool for interacting with brain oscillations. However, the field must address open questions before tFUS is comparable to established methods like TMS or tACS. In particular, what are the neural mechanisms of ultrasonic neuromodulation, and do different parameters modulate the effects on oscillations? Can tFUS enhance and decrease the amplitude of neural oscillations independent of entrainment? Is it possible to selectively modulate brain oscillations (like alpha but not theta/beta/gamma) with tFUS? And finally, can tFUS directly entrain brain oscillations? Researchers can make progress on these questions by conducting well-controlled studies to begin to address which ultrasound parameters lead to differential effects on neural oscillations, as well as an investigation into the neural mechanisms themselves. Below, I will discuss each of these questions sequentially and the impediments to using tFUS to study the causal relationship between neural rhythms and behavior.

3.1 Mechanisms and Parameters

What are the neural mechanisms of ultrasonic neuromodulation, and do different parameters modulate the effects on oscillations? The mechanisms of how ultrasound affects neural activity are not yet determined, and therefore it is not clear how ultrasound mediates cellular discharge. One issue in answering this question is the waveforms emitted by focused ultrasound transducers for neuromodulation are defined by several parameters: fundamental frequency, intensity, pulse duration, pulse repetition frequency (PRF), and duty cycle. Altering any one of the parameters may change how ultrasound mechanistically interacts with neural activity and influence excitability.

Different mechanisms have been proposed, whereby ultrasound alters membrane conductance by acting on the viscoelastic properties of neurons and their fluid environments [67]. The mechanical energy carried by ultrasound waves may alter membrane capacitance by acting on mechanosensitive ion channels [37, 68], expansion/contraction effects of the lipid bilayers [35], intra-neuronal microtubules [21], or by causing membrane waves [5]. Ultrasound may act on neurons through different or interacting mechanisms depending on the selected pulse parameters. Indeed, different stimulation parameters produce unique effects on neural excitability [5]. Whether ultrasound stimulation parameters can be selected to target, particular frequency bands must be studied systematically. Understanding the mechanisms and how various stimulation parameters act on those mechanisms will help researchers develop stimulation protocols to target neural oscillations.

3.2 Modulating Excitability

Can tFUS enhance or decrease the amplitude of neural oscillations by modulating neural excitability? Although the mechanisms of tFUS are not yet understood, their effect on neural excitability is well-documented [36]. Depending on how tFUS alters the electrical capacitance of the membrane, it may make spontaneous or task-related oscillations more or less likely to occur across groups of neurons by altering neural excitability. Such a process might occur independently of whether tFUS entrains brain rhythms (see Sect. 3.4). A similar phenomenon has been demonstrated with transcranial direct current stimulation (tDCS). TDCS increases neuronal excitability under the (anodal) stimulating electrode, and one study found that multiple frequency-specific oscillations are enhanced after anodal tDCS application [46]. Although the mechanisms of tDCS are not well understood either, this method does not induce a periodic signal; therefore, McDermott and colleagues' [46] finding demonstrate that simply altering baseline neural excitability can modulate endogenous brain oscillations. Therefore, tFUS could also be a useful tool for modulating neural oscillations and relating alterations in brain rhythms to behavioral changes, by controlling baseline neural excitability.

There are multiple reports of excitation and inhibition of neural activity with tFUS, which, taken together, indicate three factors influencing amplitude: duty cycle, duration of stimulation, and intensity. Providing evidence that duty cycle is an important parameter Yoo and colleagues [74] showed fMRI and electrophysiological signals are suppressed or enhanced in rabbits depending on the intensity and duty cycle, with short duty cycles leading to inhibition and longer duty cycles producing excitation. The magnitude of visual-evoked potentials (VEPs) in animals is suppressed with lower duty-cycle and lower-intensity pulses. In contrast, VEPs are enhanced with higher intensity, longer duty cycles [31]. Thus, the duty cycle appears to be an essential factor for altering excitability with tFUS [72].

The total duration of the ultrasound stimulation may play a role in whether neurons are excited or inhibited. Interestingly, researchers demonstrated that low-intensity ultrasound had an excitatory effect on the ECoG when the total duration was less than 10 s but desynchronized the signal with a duration of over 60 s [70], suggesting that total duration of stimulation is important for our discussion. More research needs to be conducted to understand how total duration may impact neural excitability. However, researchers should be cautious. One report found evidence of negative bioeffects of tFUS with highly repetitive ultrasound stimulations (more than 500 stimulations within one second for several minutes); however, the damage was not detected at longer inter-stimulus intervals [39].

Finally, the overall intensity of the ultrasound is relevant for neuromodulatory effects. In animals, as the intensity of ultrasound increases, local field potentials increase at different frequencies [76]. Yuan and colleagues [77] subsequently showed that phase-amplitude coupling (i.e., the synchrony between two neural oscillations) also follows this power relationship: Increasing ultrasound power leads to an increase in the phase-amplitude coupling between theta (4–8 Hz) and gamma (30–80 Hz), and between alpha (9–13 Hz) and ripple (81–200 Hz), in intact rat hippocampus. The elicitation of motor movements with tFUS in animals, which has been documented by several groups [32, 74], is an all or nothing phenomenon (i.e., requiring sufficient intensity to cause motor activation). From this evidence, we can conclude that the intensity of the stimulation is also an important factor for excitatory effects.

Based on the findings reviewed above, we can make some tentative recommendations about which pulse parameters may preferentially modulate brain rhythms. Making neurons more excitable with tFUS, and thereby enhancing oscillations, may require a new combination of pulse parameters, with higher intensity (but still within the FDA limits), higher duty cycles, or shorter duration leading to excitatory effects on neural activity. The extant data on tFUS inhibition suggests that inhibiting oscillations may occur with lower duty cycles (less than 50%), lower intensity, or greater total stimulation durations. Still, researchers should be cautious about highly repetitive ultrasound stimulations over longer periods of time, ensuring sufficient inter-stimulus intervals between pulses. The future of ultrasound research needs to create experimental paradigms to systematically measure how each parameter, or combination of parameters, influences the amplitude of neural oscillations to determine which pulse parameters best influence oscillatory dynamics.

3.3 Targeting Distinct Oscillations

Is it possible to selectively modulate frequency-specific brain oscillations (like alpha rhythms) with tFUS? Most of the published studies find that tFUS decreases the power of multiple frequency bands (e.g., [42], but it is unclear whether tFUS is a useful tool to investigate specific frequency bands, or if the mechanisms of action allow only for the non-specific modulation of rhythms. Indeed, just a few human studies to date have reported more frequency-band specific effects that were time-locked to stimulus presentation, like Mueller and colleagues [49] , but they did not investigate the entire EEG frequency spectrum. In a similar experimental paradigm but with a different brain target (thalamus), Legon and colleagues [41] found strong stimulus-locked gamma-band attenuations in humans and alpha/beta changes that were not time-locked to the stimulus. Taken together with previous findings that tFUS modulates the amplitude of sensory-evoked potentials, these findings suggest that more specific modulations of brain oscillations may be possible by investigating task-related oscillations or by time-locking the delivery of the tFUS stimulation to trials in which participants are engaged with a particular task.

Another possibility is to localize brain oscillators relevant to individual cognitive operations or behaviors and target them with focused ultrasound. TMS researchers have used this approach with some success, setting a precedent for research using ultrasound. For instance, Thut and colleagues [64] first demonstrated that rTMS could entrain the alpha rhythm by identifying a parietal alpha oscillator (using MEG) regulated by visual attention. They then showed that rTMS, pulsed at the generator's preferred frequency (10 Hz), could modulate alpha and visual perception. A similar approach could be taken with focused ultrasound. Researchers might identify brain oscillators correlated with task performance, record their preferred oscillations, and target those regions with focused ultrasound. A demonstration that reducing the brain oscillator's preferred frequency disrupts task performance or perception would suggest a causal relationship, even if other frequencies were also reduced. This type of paradigm would also raise the question of whether the PRF (see below) of the ultrasound stimulation could directly entrain regions like the parietal-alpha oscillator identified by Thut and colleagues.

Finally, a novel approach for modulating specific neural oscillations is closed-loop stimulation, in which the tFUS system triggers based on the phase of selected endogenous brain rhythms. Yang and colleagues [73] recently developed a closed-loop tFUS stimulation system for animals to pulse the ultrasound based on the phase of the ongoing theta rhythm as recorded by electrodes implanted in the hippocampus of rats. Triggering the tFUS system relative to the phase of endogenous theta modulated the absolute and relative power of the theta rhythm, demonstrating that closed-loop tFUS can manipulate frequency-specific neural oscillations. Real-time, closed-loop tFUS stimulation offers an exciting approach for targeting brain rhythms, and this may be the most promising way for tFUS to manipulate individual oscillations.

3.4 Driving Brain Rhythms

Can ultrasound directly entrain oscillations, much like TMS [52] and tACS [24]? And which ultrasound pulse parameters might lead to brain entrainment? The idea that tFUS might directly drive brain oscillations would give researchers a powerful tool alongside TMS and tACS to study whether the rhythmic dynamics of neural elements play a causative role in behavior. Controlled entrainment of brain oscillations requires periodicity in the input stream and synchronization (phase alignment) between the entraining stimulus and the targeted neural oscillators [63]. Several tFUS parameters are periodic, including the fundamental frequency (which is much too high to drive neural oscillations) and the PRF (typically between 100 and 1000 Hz in tFUS studies). This raises the question of whether the parameters of the ultrasound stimulus—like PRF—can entrain neural oscillations.

The picture of whether the PRF can drive oscillations is complicated. It has been demonstrated that various PRFs (namely, in frequencies higher than typical EEG signals) modulate the inhibitory effects of tFUS [44]. The PRF seems like an important parameter to affect neural excitability, but this finding does not address whether the PRF may entrain brain rhythms. Several studies outlined above suggest that tFUS may directly induce brain oscillations, but they used PRFs higher than the natural oscillations in the brain. For example, a study in mice showed tFUS targeting the hippocampus induced gamma (40 to 100 Hz) and sharp-wave ripples (160 to 200 Hz), but the PRF was 1,500 Hz [65]. Thus, it appears PRF did not directly entrain brain oscillations in that study and, instead, that some other mechanism was at play.

However, a recent experiment in rats complicates the story about whether the PRF can drive brain oscillations. Darvas et al. [12] induced EEG signals at the frequency of the PRF (1050 Hz), which is at the extreme end of the possible firing rate of neurons, along with increased gamma and decreased beta bands after demodulation of the measured electrophysiological signal of 1050 Hz. Although this finding suggests that the application of tFUS at 1050 Hz drove ultra-high-frequency oscillations, the mechanism of how this effect altered gamma rhythm remains a mystery. Furthermore, the reduction of beta suggests some mechanism other than direct brain entrainment. Researchers have not systematically investigated how varying pulse frequencies in the EEG range (below 250 Hz) modulate brain oscillations or what the mechanism of action might be for Darvas and colleagues' findings. More systematic experiments are needed before the questions of whether the PRF or other ultrasound parameters can directly entrain brain oscillations.

3.5 *Confounds*

Many factors modulate brain oscillations, including internal processes like the participant's internal state or other indirect processes. When attempting to modulate brain oscillations with ultrasound, researchers should keep these indirect factors in mind. For example, human brain oscillations can be modulated indirectly by auditory or attentional confounds. Ultrasound has a unique advantage over rTMS, which causes tactile sensations and clicks, or tACS, which causes only tactile sensations. On the other hand, tFUS creates no tactile sensations. Likewise, the frequency of ultrasound is much higher than the range of human hearing. However, some participants may have auditory sensations due to bone conduction linked to the PRF, and tFUS is known to cause widespread auditory activity via indirect mechanisms [56]. This may be especially true if researchers decide to lower their PRF into frequency ranges consistent with EEG frequencies (i.e., less than 100 Hz). Therefore, researchers need to pay attention to both auditory and attentional confounds (i.e., orienting to the sound), and these issues should be controlled in careful experimental designs.

4 Summary and Conclusion

The development of tFUS as a method to manipulate the brain's rhythmic dynamics would lead to significant advances in the brain sciences, as tFUS offers many advantages over established electromagnetic methodologies. The possibility to modulate deep brain circuits with millimeter spatial resolution is a clear advantage of tFUS over TMS and tACS. The advantages would allow researchers to ask questions about deep cortical rhythms, like the role of thalamocortical alpha rhythms in consciousness. For example, in a first-in-person report, targeting the thalamus with tFUS enhanced a coma patient's level of consciousness [48]. Modulating the thalamocortical alpha rhythm in patients could increase the efficacy of the intervention, targeting a core mechanism altered in the coma state. These interventions would also address fundamental questions about brain oscillations and consciousness [9] that are hard to study with established noninvasive neuromodulation methods.

Ultrasound can be used concurrently with EEG with a minimal artifact, and real-time stimulation relative to the ongoing EEG is possible. This opens the door to creating closed-loop systems that modulate the brain relative to the ongoing oscillations. Closed-loop systems will significantly accelerate research investigating the relationship between brain rhythms and behavior and open the door for promising therapeutic applications. For example, individualizing rTMS treatments for schizophrenia to the patient's peak alpha frequency increases clinical efficacy [27]. These types of effects could be replicated in real-time, with closed-loop tFUS. Finally, tFUS technology is developing rapidly and may become a cost-effective solution for many labs and clinics. The current price tag for tFUS systems is comparable to TMS systems or less expensive (see BrainSonix, Inc. for an example), already putting the technology in reach for many investigators. The development of multi-element arrays that could guide ultrasound stimulation with ultrasound imaging is possible. This reduces the need for MRI guidance, which is currently the standard for tFUS human research, allowing many more researchers to target the brain areas with precision [50].

In conclusion, tFUS is a rapidly emerging noninvasive neuromodulation methodology that influences neural function with high spatial precision and penetration depth, offering advantages over established electromagnetic tools. Intrinsic and evoked brain rhythms are modulated by tFUS, but the mechanisms are unclear, and researchers need to conduct systematic investigations into how different stimulation parameters influence oscillatory dynamics. Once researchers make advances in understanding how the mechanisms of ultrasound neuromodulation, and which parameters affect neural oscillations, tFUS has the potential to become a powerful intervention to manipulate and drive the brain's rhythms, especially if closed-loop systems are developed. TFUS will help probe the oscillatory dynamics of deep brain targets that are out of reach of established noninvasive neuromodulation technologies. Such technology will significantly advance our understanding of neural dynamics and help researchers draw causal links between brain oscillations and behavior.

Acknowledgements I owe my deep gratitude to Dr. Natalie Bryant for her careful reading and many helpful comments of the manuscript.

References

1. Badran BW, Caulfield KA, Stomberg-Firestein S, Summers PM, Dowdle LT, Savoca M, Li X, Austelle CW, Short EB, Borckardt JJ (2020) Sonication of the anterior thalamus with MRI-Guided transcranial focused ultrasound (tFUS) alters pain thresholds in healthy adults: A double-blind, sham-controlled study. Brain Stimul 13:1805–1812
2. Başar E, Başar-Eroglu C, Karakaş S, Schürmann M (2000) Gamma, alpha, delta, and theta oscillations govern cognitive processes. Int J Psychophysiol 39:241–248. https://doi.org/10.1016/S0167-8760(00)00145-8
3. Berger H (1929) Uber das elektrenkephalogramm des menschen. Arch Psychiatr Nervenkr 87:527–570
4. Bertazzoli G, Esposito R, Mutanen TP, Ferrari C, Ilmoniemi RJ, Miniussi C, Bortoletto M (2021) The impact of artifact removal approaches on TMS–EEG signal. bioRxiv
5. Blackmore J, Shrivastava S, Sallet J, Butler CR, Cleveland RO (2019) Ultrasound neuromodulation: a review of results, mechanisms and safety. Ultrasound Med Biol. https://doi.org/10.1016/j.ultrasmedbio.2018.12.015
6. Brasil-Neto JP (2012) Learning, memory, and transcranial direct current stimulation. Front Psychiatry 3. https://doi.org/10.3389/fpsyt.2012.00080
7. Buzsáki G, Watson BO (2012) Brain rhythms and neural syntax: Implications for efficient coding of cognitive content and neuropsychiatric disease. Dialogues Clin Neurosci. https://doi.org/10.31887/dcns.2012.14.4/gbuzsaki
8. Cao Y, Lopatkin A, You L (2016) Elements of biological oscillations in time and space. Nat Struct Mol Biol. https://doi.org/10.1038/nsmb.3320

9. Ching S, Cimenser A, Purdon PL, Brown EN, Kopell NJ (2010) Thalamocortical model for a propofol-induced α-rhythm associated with loss of consciousness. Proc Natl Acad Sci USA. https://doi.org/10.1073/pnas.1017069108
10. Choi JB, Lim SH, Cho KW, Kim DH, Jang DP, Kim IY (2013) The effect of focused ultrasonic stimulation on the activity of hippocampal neurons in multi-channel electrode. In: International IEEE/EMBS conference on neural engineering, NER. https://doi.org/10.1109/NER.2013.6696038
11. Dallapiazza RF, Timbie KF, Holmberg S, Gatesman J, Lopes MB, Price RJ, Miller GW, Elias WJ (2017) Noninvasive neuromodulation and thalamic mapping with low-intensity focused ultrasound. J Neurosurg. https://doi.org/10.3171/2016.11.JNS16976
12. Darvas F, Mehić E, Caler CJ, Ojemann JG, Mourad PD (2016) Toward deep brain monitoring with superficial EEG sensors plus neuromodulatory focused ultrasound. Ultrasound Med Biol 42(8):1834–1847. https://doi.org/10.1016/j.ultrasmedbio.2016.02.020
13. Deffieux T, Younan Y, Wattiez N, Tanter M, Pouget P, Aubry JF (2013) Low-intensity focused ultrasound modulates monkey visuomotor behavior. Curr Biol. https://doi.org/10.1016/j.cub.2013.10.029
14. DeWeerdt S (2019) How to map the brain. Nature 571:S6–S6
15. Fini M, Tyler WJ (2017) Transcranial focused ultrasound: a new tool for non-invasive neuromodulation. Int Rev Psychiatry. https://doi.org/10.1080/09540261.2017.1302924
16. Foxe JJ, Snyder AC (2011) The role of alpha-band brain oscillations as a sensory suppression mechanism during selective attention. Front Psychol 2:154. https://doi.org/10.3389/fpsyg.2011.00154
17. Fry FJ, Ades HW, Fry WJ (1958) Production of reversible changes in the central nervous system by ultrasound. Science 80. https://doi.org/10.1126/science.127.3289.83
18. Griffiths BJ, Mayhew SD, Mullinger KJ, Jorge J, Charest I, Wimber M, Hanslmayr S (2019) Alpha/beta power decreases track the fidelity of stimulus-specific information. Elife 8:e49562
19. Griškova I, Rukšenas O, Dapšys K, Herpertz S, Höppner J (2007) The effects of 10 Hz repetitive transcranial magnetic stimulation on resting EEG power spectrum in healthy subjects. Neurosci Lett. https://doi.org/10.1016/j.neulet.2007.04.030
20. Grossman N, Bono D, Dedic N, Kodandaramaiah SB, Rudenko A, Suk HJ, Cassara AM, Neufeld E, Kuster N, Tsai LH, Pascual-Leone A, Boyden ES (2017) Noninvasive deep brain stimulation via temporally interfering electric fields. Cell. https://doi.org/10.1016/j.cell.2017.05.024
21. Hameroff S, Trakas M, Duffield C, Annabi E, Gerace MB, Boyle P, Lucas A, Amos Q, Buadu A, Badal JJ (2013) Transcranial ultrasound (TUS) effects on mental states: A pilot study. Brain Stimul 6:409–415. https://doi.org/10.1016/j.brs.2012.05.002
22. Harvey E (1929) The effect of high frequency sound waves on heart muscle and other irritable tissues. Am J Physiol 91:284–290
23. Helfrich RF, Schneider TR, Rach S, Trautmann-Lengsfeld SA, Engel AK, Herrmann CS (2014) Entrainment of brain oscillations by transcranial alternating current stimulation. Curr Biol. https://doi.org/10.1016/j.cub.2013.12.041
24. Herrmann CS, Rach S, Neuling T, Strüber D (2013) Transcranial alternating current stimulation: a review of the underlying mechanisms and modulation of cognitive processes. Front Hum Neurosci 7:1–13. https://doi.org/10.3389/fnhum.2013.00279
25. Herrmann CS, Strüber D, Helfrich RF, Engel AK (2016) EEG oscillations: From correlation to causality. Int J Psychophysiol. https://doi.org/10.1016/j.ijpsycho.2015.02.003
26. Huerta PT, Volpe BT (2009) Transcranial magnetic stimulation, synaptic plasticity and network oscillations. J Neuroeng Rehabil. https://doi.org/10.1186/1743-0003-6-7
27. Jin Y, Potkin SG, Kemp AS, Huerta ST, Alva G, Thai TM, Carreon D, Bunney WE (2006) Therapeutic effects of individualized alpha frequency transcranial magnetic stimulation (αTMS) on the negative symptoms of schizophrenia. Schizophr Bull 32:556–561. https://doi.org/10.1093/schbul/sbj020
28. Jülicher F (2001) Mechanical oscillations at the cellular scale. Comptes Rendus l'Academie des Sci Ser. IV Phys Astrophys. https://doi.org/10.1016/S1296-2147(01)01228-8

29. Karabanov AN, Saturnino GB, Thielscher A, Siebner HR (2019) Can transcranial electrical stimulation localize brain function? Front Psychol. https://doi.org/10.3389/fpsyg.2019.00213
30. Kelly SP, Gomez-Ramirez M, Foxe JJ (2008) Spatial attention modulates initial afferent activity in human primary visual cortex. Cereb Cortex 18:2629–2636. https://doi.org/10.1093/cercor/bhn022
31. Kim H, Park MY, Lee SD, Lee W, Chiu A, Yoo SS (2015) Suppression of EEG visual-evoked potentials in rats through neuromodulatory focused ultrasound. NeuroReport. https://doi.org/10.1097/WNR.0000000000000330
32. King RL, Brown JR, Newsome WT, Pauly KB (2013) Effective parameters for ultrasound-induced in vivo neurostimulation. Ultrasound Med Biol 39:312–331. https://doi.org/10.1016/j.ultrasmedbio.2012.09.009
33. Kirschstein T, Köhling R (2009) What is the source of the EEG? Clin EEG Neurosci. https://doi.org/10.1177/155005940904000305
34. Klimesch W, Sauseng P, Hanslmayr S (2007) EEG alpha oscillations: the inhibition-timing hypothesis. Brain Res Rev 53:63–88. https://doi.org/10.1016/j.brainresrev.2006.06.003
35. Krasovitski B, Frenkel V, Shoham S, Kimmel E (2011) Intramembrane cavitation as a unifying mechanism for ultrasound-induced bioeffects. Proc Natl Acad Sci 108:3258–3263. https://doi.org/10.1073/pnas.1015771108
36. Kubanek J (2018) Neuromodulation with transcranial focused ultrasound. Neurosurg Focus 44:E14. https://doi.org/10.3171/2017.11.FOCUS17621
37. Kubanek J, Shi J, Marsh J, Chen D, Deng C, Cui J (2016) Ultrasound modulates ion channel currents. Sci Rep 6:1–14. https://doi.org/10.1038/srep24170
38. Lee W, Chung YA, Jung Y, Song IU, Yoo SS (2016a) Simultaneous acoustic stimulation of human primary and secondary somatosensory cortices using transcranial focused ultrasound. BMC Neurosci. https://doi.org/10.1186/s12868-016-0303-6
39. Lee W, Lee SD, Park MY, Foley L, Purcell-Estabrook E, Kim H, Fischer K, Maeng LS, Yoo SS (2016b) Image-guided focused ultrasound-mediated regional brain stimulation in sheep. Ultrasound Med Biol. https://doi.org/10.1016/j.ultrasmedbio.2015.10.001
40. Legon W, Adams S, Bansal P, Patel PD, Hobbs L, Ai L, Mueller JK, Meekins G, Gillick BT (2020) A retrospective qualitative report of symptoms and safety from transcranial focused ultrasound for neuromodulation in humans. Sci Rep. https://doi.org/10.1038/s41598-020-62265-8
41. Legon W, Ai L, Bansal P, Mueller JK (2018) Neuromodulation with single-element transcranial focused ultrasound in human thalamus. Hum Brain Mapp. https://doi.org/10.1002/hbm.23981
42. Legon W, Sato TF, Opitz A, Mueller J, Barbour A, Williams A, Tyler WJ (2014) Transcranial focused ultrasound modulates the activity of primary somatosensory cortex in humans. Nat Neurosci 17:322–329. https://doi.org/10.1038/nn.3620
43. Logan RW, McClung CA (2019) Rhythms of life: circadian disruption and brain disorders across the lifespan. Nat Rev Neurosci. https://doi.org/10.1038/s41583-018-0088-y
44. Manuel TJ, Kusunose J, Zhan X, Lv X, Kang E, Yang A, Xiang Z, Caskey CF (2020) Ultrasound neuromodulation depends on pulse repetition frequency and can modulate inhibitory effects of TTX. Sci Rep https://doi.org/10.1038/s41598-020-72189-y
45. Mathewson KE, Lleras A, Beck DM, Fabiani M, Ro T, Gratton G (2011) Pulsed out of awareness: EEG alpha oscillations represent a pulsed-inhibition of ongoing cortical processing. Front Psychol 2:99. https://doi.org/10.3389/fpsyg.2011.00099
46. McDermott TJ, Wiesman AI, Mills MS, Spooner RK, Coolidge NM, Proskovec AL, Heinrichs-Graham E, Wilson TW (2019) tDCS modulates behavioral performance and the neural oscillatory dynamics serving visual selective attention. Hum Brain Mapp. https://doi.org/10.1002/hbm.24405
47. Mehić E, Xu JM, Caler CJ, Coulson NK, Moritz CT, Mourad PD (2014) Increased anatomical specificity of neuromodulation via modulated focused ultrasound. PLoS ONE. https://doi.org/10.1371/journal.pone.0086939
48. Monti MM, Schnakers C, Korb AS, Bystritsky A, Vespa PM (2016) Non-invasive ultrasonic thalamic stimulation in disorders of consciousness after severe brain injury: a first-in-man report. Brain Stimul 9:940–941. https://doi.org/10.1016/j.brs.2016.07.008

49. Mueller J, Legon W, Opitz A, Sato TF, Tyler WJ (2014) Transcranial focused ultrasound modulates intrinsic and evoked EEG dynamics. Brain Stimul 7:900–908. https://doi.org/10.1016/j.brs.2014.08.008
50. Pashaei V, Dehghanzadeh P, Enwia G, Bayat M, Majerus SJA, Mandal S (2020) Flexible body-conformal ultrasound patches for image-guided neuromodulation. IEEE Trans Biomed Circuits Syst. https://doi.org/10.1109/TBCAS.2019.2959439
51. Reznik SJ, Sanguinetti JL, Allen JJB (n.d.) Transcranial ultrasound (TUS) reduces worry in a five-day double-blind pilot study. Change 6:10
52. Romei V, Bauer M, Brooks JL, Economides M, Penny W, Thut G, Driver J, Bestmann S (2016) Causal evidence that intrinsic beta-frequency is relevant for enhanced signal propagation in the motor system as shown through rhythmic TMS. Neuroimage. https://doi.org/10.1016/j.neuroimage.2015.11.020
53. Romei V, Gross J, Thut G (2010) On the role of prestimulus alpha rhythms over occipito-parietal areas in visual input regulation: Correlation or causation? J Neurosci. https://doi.org/10.1523/JNEUROSCI.0160-10.2010
54. Roth, B.J., Saypol, J.M., Hallett, M., Cohen, L.G., 1991. A theoretical calculation of the electric field induced in the cortex during magnetic stimulation. Electroencephalogr. Clin. Neurophysiol. Evoked Potentials. https://doi.org/10.1016/0168-5597(91)90103-5
55. Sanguinetti JL, Hameroff S, Smith EE, Sato T, Daft CMW, Tyler WJ, Allen JJB (2020) Transcranial focused ultrasound to the right prefrontal cortex improves mood and alters functional connectivity in humans. Front Hum Neurosci. https://doi.org/10.3389/fnhum.2020.00052
56. Sato T, Shapiro MG, Tsao DY (2018) Ultrasonic neuromodulation causes widespread cortical activation via an indirect auditory mechanism. Neuron 98:1031–1041.e5. https://doi.org/10.1016/j.neuron.2018.05.009
57. Siegel M, Donner TH, Engel AK (2012) Spectral fingerprints of large-scale neuronal interactions. Nat Rev Neurosci. https://doi.org/10.1038/nrn3137
58. Singer W (1999) Neuronal synchrony: a versatile code for the definition of relations? Neuron. https://doi.org/10.1016/S0896-6273(00)80821-1
59. Snyder AC, Foxe JJ (2010) Anticipatory attentional suppression of visual features indexed by oscillatory alpha-band power increases: a high-density electrical mapping study. J Neurosci 30:4024–4032. https://doi.org/10.1523/JNEUROSCI.5684-09.2010
60. Strüber D, Rach S, Trautmann-Lengsfeld SA, Engel AK, Herrmann CS (2014) Antiphasic 40 Hz oscillatory current stimulation affects bistable motion perception. Brain Topogr. https://doi.org/10.1007/s10548-013-0294-x
61. Thut G (2014) Modulating brain oscillations to drive brain function. PLoS Biol. https://doi.org/10.1371/journal.pbio.1002032
62. Thut G, Miniussi C, Gross J (2012) The functional importance of rhythmic activity in the brain. Curr Biol. https://doi.org/10.1016/j.cub.2012.06.061
63. Thut G, Schyns PG, Gross J (2011a) Entrainment of perceptually relevant brain oscillations by non-invasive rhythmic stimulation of the human brain. Front Psychol. https://doi.org/10.3389/fpsyg.2011.00170
64. Thut G, Veniero D, Romei V, Miniussi C, Schyns P, Gross J (2011b) Rhythmic TMS causes local entrainment of natural oscillatory signatures. Curr Biol. https://doi.org/10.1016/j.cub.2011.05.049
65. Tufail Y, Matyushov A, Baldwin N, Tauchmann ML, Georges J, Yoshihiro A, Tillery SIH, Tyler WJ (2010) Transcranial pulsed ultrasound stimulates intact brain circuits. Neuron 66:681–694. https://doi.org/10.1016/j.neuron.2010.05.008
66. Tyler WJ (2011a) Noninvasive neuromodulation with ultrasound? A continuum mechanics hypothesis. Neuroscientist 17:25-29
67. Tyler WJ (2011b) Noninvasive neuromodulation with ultrasound? A continuum mechanics hypothesis. Neuroscientist 17:25–36. https://doi.org/10.1177/1073858409348066
68. Tyler WJ, Tufail Y, Finsterwald M, Tauchmann ML, Olson EJ, Majestic C (2008) Remote excitation of neuronal circuits using low-intensity, low-frequency ultrasound. PLoS ONE 3. https://doi.org/10.1371/journal.pone.0003511

69. Varone G, Hussain Z, Sheikh Z, Howard A, Boulila W, Mahmud M, Howard N, Morabito FC, Hussain A (2021) Real-time artifacts reduction during TMS-EEG co-registration: A comprehensive review on technologies and procedures. Sensors (Switzerland). https://doi.org/10.3390/s21020637
70. Velling VA, Shklyaruk SP (1988) Modulation of the functional state of the brain with the aid of focused ultrasonic action. Neurosci Behav Physiol. https://doi.org/10.1007/BF01193880
71. Vosskuhl J, Strüber D, Herrmann CS (2018) Non-invasive Brain Stimulation: A Paradigm Shift in Understanding Brain Oscillations. Front Hum Neurosci. https://doi.org/10.3389/fnhum.2018.00211
72. Wang XJ (2010) Neurophysiological and computational principles of cortical rhythms in cognition. Physiol Rev. https://doi.org/10.1152/physrev.00035.2008
73. Yang H, Yuan Y, Wang X, Li X (2020) Closed-Loop Transcranial Ultrasound Stimulation for Real-Time Non-invasive Neuromodulation in vivo. Front Neurosci 14
74. Yoo SS, Bystritsky A, Lee JH, Zhang Y, Fischer K, Min BK, McDannold NJ, Pascual-Leone A, Jolesz FA (2011a) Focused ultrasound modulates region-specific brain activity. Neuroimage 56:1267–1275. https://doi.org/10.1016/j.neuroimage.2011.02.058
75. Yoon K, Lee W, Lee JE, Xu L, Croce P, Foley L, Yoo SS (2019) Effects of sonication parameters on transcranial focused ultrasound brain stimulation in an ovine model. PLoS ONE. https://doi.org/10.1371/journal.pone.0224311
76. Yuan Y, Lu CB, Li XL (2015) Effect of focused ultrasound stimulation at different ultrasonic power levels on the local field potential power spectrum. Chinese Phys B. https://doi.org/10.1088/1674-1056/24/8/088704
77. Yuan Y, Yan J, Ma Z, Li X (2016) Noninvasive focused ultrasound stimulation can modulate phase-amplitude coupling between neuronal oscillations in the rat hippocampus. Front Neurosci. https://doi.org/10.3389/fnins.2016.00348

Chapter 3
Ergontropic Dynamics: Contribution for an Extended Particle Dynamics

Mario J. Pinheiro

PACS 05.45.-a · 02.50.-r · 02.50.Tt · 24.60.-k · 45.00.00 · 47.32.-y · 52.27.-h · 89.20.Dd

1 Introduction

The classical mechanistic description of Sir Isaac Newton and Pierre-Simon Laplace presupposes the reversibility of time, i.e., time follows it course (irreversibly) but it didn't affect the possibility of the reverse motion. Historically, the mechanistic theory was called into question when the first anomalies appeared in heat experiments, namely, the fact that an warm body transmit heat to a cold body until their temperature equalize, but never the reverse happened, indicating the first striking example of irreversible phenomena. However, in Newtonian, or Galilean mechanics, an observer O on an inertial reference frame R, recognizes an equilibrium state A of a mechanical system M (that we may visualize as an ideal fluid or a "rigid body"), assigning to this state A an energy E and a number N of particles constituents (by virtue its mass). And, from these assumptions, it follows the thermodynamic implications of the isotropy and homogeneity of Newtonian space-time.

Here, we present an approach differing in fundamental aspects from the standard treatments, particularly not deriving irreversible thermodynamics from the reversible microscopic dynamics. By the contrary, we relate dynamics and thermodynamics based on the concept of energy, work, and entropy. At the core of the present theory is the second law of thermodynamics which has a form of causality as a universal

M. J. Pinheiro (✉)
Department of Physics, Instituto Superior Técnico-IST, Universidade de Lisboa, Av. Rovisco Pais, 1049-001 Lisboa Codex, Portugal
e-mail: mpinheiro@tecnico.ulisboa.pt

A. Bandyopadhyay and K. Ray (eds.), *Rhythmic Advantages in Big Data and Machine Learning* , Studies in Rhythm Engineering,
https://doi.org/10.1007/978-981-16-5723-8_3

property. Working out the general method here proposed, at the end, a system can be described in terms of dynamical quantities and free energy.

In a previous publication [1] we have shown that rotational motion and free energy gradients imply dissipation of energy, are sources of irreversible phenomena. We may place the gradient of rotational energy in the class of transport of angular momentum problems.

In this proposed program, we develop a standard technique for treating a physical system on the basis of an information-theoretic framework previously developed [1–5]. The proposed technique starts with the total entropy of the system composed of N particles (or bodies), and then the method of the Lagrange multipliers for the differential of entropy (0-form) $d\overline{S} = \nabla_k \overline{S} dx^k$ $(k = x, y, z)$ is applied; finally, the total entropy is inserted on the fundamental equation of entropy written under differential form, $\nabla_k U = T \nabla_k \overline{S} - F_k$, summing over the all ensemble of N particles. This method leads to a set of two first-order differential equations revealing the same formal symplectic structure shared by classical mechanics and thermodynamics [6, 7]. When the maximization of entropy is seek, it results on the well-known equations of (electro)dynamics (if electromagnetic entities enter into the system). This work leads to a reformulation of mechanics and electrodynamics, in particular, giving a new definition of the equation of motion that explicitly displays irreversibility.

Consider here a simple dynamical system constituted by a set of N discrete interacting mass-points (although macroscopic) $m^{(\alpha)}$ $(\alpha = 1, 2, \ldots, N)$ with x_i^α and v_i^α) $(i = 1, 2, 3; \alpha = 1, \ldots, N)$ denoting the coordinates and velocities of the mass point in a given inertial frame of reference. The inferior Latin index refers to the Cartesian components and the superior Greek index distinguishes the different mass-points.

Following the line of thought proposed in Ref. [5] the total energy of a, let's say, mechano-electric system constituted by N interacting particles, which includes here, for the purpose of illustration, the gravitational potential $\phi^{(\alpha)}$ for the species α, and the gravitational interaction $\phi^{(\alpha,\beta)}$ between specie α and β (all the other terms are well known in the technical literature) is given by:

$$
\begin{aligned}
E = & \sum_{\alpha=1}^{N} [U^{(\alpha)} + \frac{p^{(\alpha)2}}{2m^{(\alpha)}} + \frac{J^{(\alpha)2}}{2I^{(\alpha)}} + q^{(\alpha)} V^{(\alpha)} \\
& - q^{(\alpha)} (\mathbf{A}^{(\alpha)} \cdot \mathbf{v}^{(\alpha)}) + m^{(\alpha)} \phi^{(\alpha)}(\mathbf{r}) + m^{(\alpha)} \sum_{\beta=1}^{N} \phi^{(\alpha,\beta)}]. \qquad (1)
\end{aligned}
$$

The next fundamental task is the construction of the invariant entropy function $\overline{S}$, as a function of the, so-far, *known* fundamental invariants in a closed system. Hence, $\overline{S} = \overline{S}(E, \boldsymbol{p}, \boldsymbol{L})$, and we must construct the entropy as a function of the linearly independent components via the Lagrange multiplier form of entropy (LMFE) given by

$$\overline{S} = \sum_{\alpha=1}^{N} \left\{ S^{(\alpha)}(E^{(\alpha)} - \frac{(\mathbf{p}^{(\alpha)})^2}{2m^{(\alpha)}} - \frac{(\mathbf{J}^{(\alpha)})^2}{2I^{(\alpha)}} - q^{(\alpha)}V^{(\alpha)} + q^{(\alpha)}(\mathbf{A}^{(\alpha)} \cdot \mathbf{v}^{(\alpha)}) \right.$$
$$\left. -U_{mec}^{(\alpha)}) + (\mathbf{a} \cdot \mathbf{p}^{(\alpha)} + \mathbf{b} \cdot ([\mathbf{r}^{(\alpha)} \times \mathbf{p}^{(\alpha)}] + \mathbf{J}^{(\alpha)}) \right\} \tag{2}$$
$$= \sum_{\alpha=1}^{N} \mathfrak{S}^{(\alpha)}. \tag{3}$$

Here, **a** and **b** are Lagrange multipliers vectors, related to the velocity of translation $\mathbf{v}_e = T\mathbf{a}$, and the velocity of rotation $\boldsymbol{\omega} = T\mathbf{b}$. The method of Lagrange multipliers is applied to find extremal values of a function of several variables subject to one, or more, constraints. The absolute temperature is denoted by T, more realistically interpreted as the kinetic temperature, the equipartition expression in momenta. Thus, for an isolated extended particle in thermodynamic equilibrium, the translational and rotational motion is only possible as a rigid body [2].

The entropy, assumed an invariant to the observer O in a given reference frame (possibly under rotations), is an invariant function, with 8 linearly independent components: mass M, energy E, total linear momentum $\boldsymbol{p}$, total angular momentum $\boldsymbol{L}$.

The introduction of the electrostatic and magnetic energy term were also used in a previous publication [5], where some consequences of this formulation were explored in electrodynamics. The additional terms included in the LMFE are the total linear momentum and the total angular momentum about the origin, both with respect to a definite initial reference frame:

$$\sum_{\alpha} \mathbf{p}_{\alpha} = \mathbf{P} = const, \tag{4a}$$

$$\sum_{\alpha} ([\mathbf{r}_{\alpha} \times \mathbf{v}_{\alpha}] + \mathbf{J}_{\alpha}) = \mathbf{J} = const, \tag{4b}$$

with **J** denoting the total sum of the orbital and spin angular momentum, both equations due to the principles of energy and angular momentum conservation in an isolated body, constituted by all $\alpha = 1, 2, \ldots, N$ particles, or bodies, all at the same temperature to avoid, at this level of description, mathematical complexity. Equations 4a and 4b are the constraints equations referred above. The aim of the method of the Lagrange multipliers is to allow the determination of a particular set of points where a given function (here, the function $\overline{S}$) have minimal or maximal values.

Extremum conditions imposed on entropy or internal energy not only provide criteria for the evolution of the system but determine as well the stability of thermodynamic systems at equilibrium. It has been shown [8] that a state of mechanical equilibrium (or in the absence of convective motion) is onset when the entropy increase with distance from equilibrium, such as

$$\frac{\partial \overline{S}}{\partial \mathbf{r}} > 0. \tag{5}$$

Applying this concept to Eq. 2, the entropic flux in space is obtained, a kind of generalized force X_α [9], with its value given by

$$\frac{T}{2}\frac{\partial \overline{S}}{\partial \mathbf{r}_i^{(\alpha)}} = \nabla_{r_i} U^{(\alpha)} + m_i \frac{dv_i^{(\alpha)}}{dt} + \frac{1}{2}\nabla_{r_i}(\boldsymbol{\omega} \cdot \mathbf{J}_i), \tag{6}$$

where

$$U_{eq}^{(\alpha)} = m^{(\alpha)}\phi^{(\alpha)}(\mathbf{r}) + m^{(\alpha)} \sum_{\beta=1}^{N} \phi^{(\alpha,\beta)}, \tag{7}$$

represents the internal energy of a given mass-point α at equilibrium. When seeking for an extremal value, imposing $\nabla \overline{S} = 0$, it gives the fundamental hydrodynamic equation:

$$m^{(\alpha)} \frac{dv_i^{(\alpha)}}{dt} = -\nabla_r (m^{(\alpha)}\phi^{(\alpha)}) - m^{(\alpha)} \nabla \sum_{\beta=1}^{N} \phi^{(\alpha,\beta)} - \frac{1}{2} \boldsymbol{\nabla}_{r^{(\alpha)}} (\boldsymbol{\omega} \cdot \mathbf{J}^{(\alpha)}). \tag{8}$$

In conformity with the definition of temperature, we look at the gradient of the total entropy in momentum space multiplied by factor T, which gives

$$T \frac{\partial \overline{S}}{\partial \mathbf{p}_i} = \{-\frac{\mathbf{p}_i}{m_i} + \frac{q_i}{m_i}\mathbf{A} + \mathbf{v}_{rot} + [\boldsymbol{\omega} \times \mathbf{r}_i]\}. \tag{9}$$

The maximization of Eq. 9 leads to the total (canonical) momentum, given by

$$\mathbf{p}_i = m_i \mathbf{v}_{rot} + m_i [\boldsymbol{\omega} \times \mathbf{r}_i] + q_i \mathbf{A}. \tag{10}$$

However, in non-equilibrium conditions, the exact canonical momentum for each particle is

$$\mathbf{p} = m\mathbf{v}_{rot} + m[\boldsymbol{\omega} \times \mathbf{r}] + q\mathbf{A} - mT\boldsymbol{\nabla}_{\mathbf{p}}\overline{S}. \tag{11}$$

This variational approach, based on the entropy-energy principle, points to the dominance and interplay of the fundamental dualism:

Entropy: (i) do not explicitly depend on energy; (ii) is dependent of the particles configuration (or collectively dependent); (iii) it has a characteristic speed of "entropic propagation" (equal at maximum to the speed of light);

Energy: (i) decay over time and space; (ii) propagates with characteristic speed, at maximum the speed of light.

In the course of advancing knowledge the two have exchanged places on the understanding of natural processes. Entropy dictates how the all process evolves, while energy merely does the bookkeeping. Figure 1 shows the link between propensities and variables.

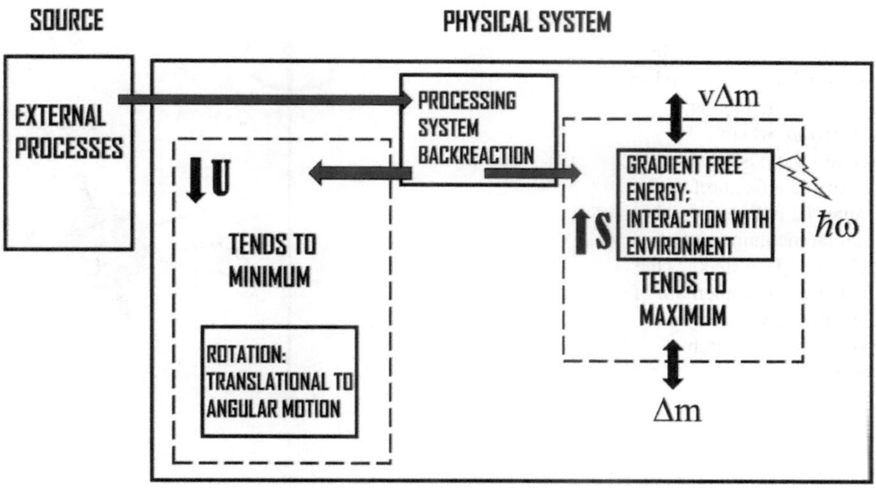

Fig. 1 System process chain—a physical system has, at least, two ways to process the external action (it is not enslaved to it), by means of two processes chain: energy tends to a minimum by means of conversion of linear or transversal motion to rotational motion; entropy tends to a maximum by means of decreasing free energy (e.g., ejection of mass, acoustic or electromagnetic radiation)

It was shown that in the case of an extended rotating charged sphere the canonical momentum is given by

$$\mathbf{p} = m\mathbf{v} + q\mathbf{A} + \frac{1}{2}\gamma I \nabla^2 \mathbf{A}. \tag{12}$$

Here, γ is the gyromagnetic ration and I is the moment of inertia [10]. We may be associated to the term $\nabla^2\mathbf{A}$ a spread field in space such as $-\frac{\mathbf{A}}{\lambda_e}$, with λ_e denoting the Compton wavelength, in which case it would appear a correction to the electron observed charge e_R. This correction results from the collective action of all the other charged electrons on the test particle and the correction, in a crude estimation, would be of the order, $e_r = e(1 - \alpha^2/16\pi^2)$. In view of this, we may observe that it is not correct to adopt the usual definition of a particle as a system without extension, as a system of negligible size. Continuous mechanics is a far-more general theory than the usual mass-point mechanics [11].

Now, let us assume that the total energy E is constant. We then obtain (letting down the index α) by taking the gradient of the Eq. 1:

$$\nabla_r U = m\mathbf{v} \cdot \nabla\mathbf{v} - \nabla\left(\frac{\mathbf{J}}{2I} - \boldsymbol{\omega} \cdot \mathbf{J}\right) - q\nabla V + q\nabla(\mathbf{A} \cdot \mathbf{v}). \tag{13}$$

You'll notice that the non-translational energy of the system U (or the internal energy) replaces the total energy, determines the actual existence of mechanical states.

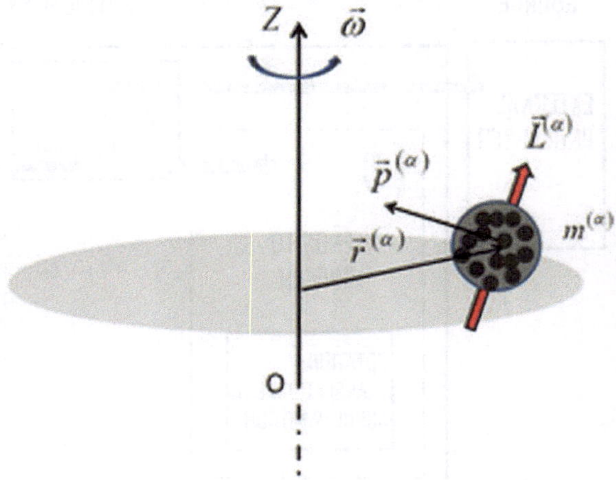

Fig. 2 An assortment of particles of mass $m^{(\alpha)}$, in rotational motion around an axis OZ with angular velocity $\boldsymbol{\omega}$, where $\mathbf{r}^{(\alpha)}$ denotes the position vector relative to a fixed reference frame, $\mathcal{R}$, $\mathbf{p}^{(\alpha)}$ denotes the total momentum (particle + field), and $\mathbf{L}^{(\alpha)}$ denotes the total angular momentum of the particle. Courtesy from Nature research Journal, Ref. 1

Applying the present formalism in a previous work [5], we obtained the ponderomotive forces acting on a charged particle. For a neutral particle (or rigid body) we can obtain from Eq. 41, when there is no flux of entropy, a kind of extended fundamental equation of dynamics for a given species (α):

$$m^{(\alpha)}\frac{d\mathbf{v}^{(\alpha)}}{dt} = \mathbf{f}^{(\alpha)} - \frac{1}{2}\nabla_{r^{(\alpha)}}(\boldsymbol{\omega}\cdot\mathbf{J}^{(\alpha)}). \quad (14)$$

Equation 41 has a new term because the body possess an intrinsic angular momentum, and this expression can be integrated according to the theorem work-energy in order to obtain the mechanical energy of the system, $E_{mec} = K + U + I\omega^2/2$. There is an interesting aspect related to Eq. 41: classical problems of mechanics, as, for example, rigid body rolling down an inclined plane (see, e.g., p. 97 of Ref. [12]) may be solved in an integrated manner. At this stage, it is worthwhile to refer that our procedure includes the effect of angular momentum (through Eq. 2), as it should be in a consistent theory, according to Curtiss [13] (Fig. 2).

1.1 Forces in an Accelerated Frame

Now remark that the term $\nabla_r(\boldsymbol{\omega}\cdot\mathbf{J})$ may be developed in the following manner (from now on we don't refer to the set of particles and discard the supra-index (α)):

$$\frac{1}{2}\boldsymbol{\nabla_r}(\boldsymbol{\omega}\cdot\mathbf{J}) = \frac{1}{2}[\boldsymbol{\omega}\times \text{curl }\mathbf{J}] + \frac{1}{2}(\boldsymbol{\omega}\cdot\nabla)\mathbf{J} + \frac{1}{2}[\mathbf{J}\times \text{curl }\boldsymbol{\omega}] + \frac{1}{2}(\mathbf{J}\cdot\nabla)\boldsymbol{\omega}. \quad (15)$$

Within this new formulation, the transformation from an inertial coordinate system to a rotating system shows the so-called "fictitious forces", although certainly not related to interacting fields, they are the outcome of the transport of angular momentum, they are real forces in the rotating medium. We should have in mind that there is active transformations (implying its motion, strongly or weakly coupled to other systems) and passive transformations (method of describing transitions to different reference systems, e.g., Galileo or Lorentz transformations). Any mathematical procedure can be reliable unless it is completely understood about the purpose and how it should be applied, otherwise becomes a source of confusion.

In order to justify what we have state above, one therefore has to work out the resultant term $\nabla_r(\boldsymbol{\omega} \cdot \mathbf{J})$ in Eq. 14, developing it in the following manner:

$$\frac{1}{2}\nabla_r(\boldsymbol{\omega} \cdot \mathbf{J}) = \frac{1}{2}[\boldsymbol{\omega} \times \nabla \times J] + \frac{1}{2}(\boldsymbol{\omega} \cdot \nabla)\mathbf{J} + \frac{1}{2}[\mathbf{J} \times \nabla \times \omega] + \frac{1}{2}(\mathbf{J} \cdot \nabla)\boldsymbol{\omega}. \tag{16}$$

We may test the above Eq. 16 with the well-known problem of rotating coordinate frame, without translation, considering a rotating system (x', y', z') whose origin coincides with the origin of an inertial system (x, y, z). We suppose in addition, that the z- and z'-axes always coincide and, therefore, the angular velocity of the rotating system, $\boldsymbol{\omega}$, lies along the z-axis. We take $\mathbf{J} = I\boldsymbol{\Omega}$ and $\boldsymbol{\Omega} = \boldsymbol{\omega}$, since we consider the "body" in rotational motion around the z-axis. An easy calculation (considering $I_z = mr^2$) shows that

$$\nabla \times \mathbf{J} = 2m\mathbf{v}_{rot}, \tag{17}$$

where $\mathbf{v}_{rot}$ denotes the particle velocity in the rotating frame. Also, following the analogy

$$\nabla \times \mathbf{v} = \ = 2\boldsymbol{\omega}, \tag{18}$$

and using a mathematical identity, we obtain

$$\nabla \times \left(\frac{\mathbf{J}}{r^2}\right) = \frac{1}{r^2}\nabla \times J + \left[\nabla\left(\frac{1}{r^2}\right) \times \mathbf{J}\right], \tag{19}$$

which gives

$$\nabla \times \left(\frac{\mathbf{J}}{r^2}\right) = \frac{2m}{r^2}(\mathbf{v}_{rot}) + [\boldsymbol{\omega} \times \mathbf{r}]. \tag{20}$$

Then

$$\frac{1}{2}\left[\mathbf{J} \times \nabla \times \left(\frac{\mathbf{J}}{r^2}\right)\right] = m\boldsymbol{\omega} \times [\mathbf{v}_{rot} + [\boldsymbol{\omega} \times \mathbf{r}]], \tag{21}$$

and, therefore, adding together all the terms, the well-known formula of Newtonian dynamics is finally obtained:

$$\left(m\frac{d\mathbf{v}}{dt}\right)_{rot} = \mathbf{F}^{ext} - 2m[\boldsymbol{\omega} \times \mathbf{v}_{rot}] - m(\boldsymbol{\omega} \times [\boldsymbol{\omega} \times \mathbf{r}]). \tag{22}$$

We stress that in Eq. 22 the force term is calculated in the non-inertial frame inserting the acceleration force inside the symbol $(\ldots)_{rot}$.

Euler's force term $(d\boldsymbol{\omega}/dt \times \mathbf{r})$ does not appear in Eq. 22 because the assumption behind Eqs. 41–46 is of extremum of entropy and $d\omega/dt = 0$. It is worth to note that Eq. 14 gives the force measured on the moving reference frame for a given part of mass $m^{(\alpha)}$ pertaining to an extended mass. In addition, it shows that the force exerted on an extended body maybe quite different from the one that it is usually expected, based on inferences that follow from the point particle dynamics. In particular, the changing of inertial moment of the extended particle along its motion may lead to new effects.

This means that Eq 14 is more general than the well-known Eq. 22, valid for an ideal particle, provided that the constriction Eq. 17 is obeyed for an extended body.

2 The Ergontropic Equation of Dynamics

Taking the gradient of $\overline{S}$ in Eq. 2, we obtain

$$\nabla \overline{S}_{r_i} = \frac{\partial S_i}{\partial U_i}\frac{\partial U_i}{\partial \mathbf{r}_i} + m\mathbf{a}_i \cdot \frac{\partial \mathbf{v}_i}{\partial \mathbf{r}_i} + \mathbf{b} \cdot \frac{\partial}{\partial \mathbf{r}_i}([\mathbf{r}_i \times \mathbf{p}_i] + \mathbf{J}_i). \tag{23}$$

Here, by definition of temperature, $\frac{1}{T} = \frac{\partial S_i}{\partial U_i}$, and $\mathbf{v}_e = \mathbf{a}T$, $\boldsymbol{\omega} = \mathbf{b}T$.
Then (now dropping the i-index to simplify), the following is obtained

$$T\nabla \overline{S}_r = \frac{\partial U}{\partial \mathbf{r}} + m\mathbf{v}_e \cdot \frac{\partial \mathbf{v}}{\partial \mathbf{r}} + \frac{\partial}{\partial \mathbf{r}}(\boldsymbol{\omega} \cdot [\mathbf{r} \times \mathbf{p}]) + \boldsymbol{\omega} \cdot \frac{\partial \mathbf{J}}{\partial \mathbf{r}}. \tag{24}$$

The internal energy is the summing up of the kinetic energy with the potential energy (e.g., gravitational, electromagnetic) and, therefore,

$$\frac{\partial U}{\partial \mathbf{r}} = -m\mathbf{v} \cdot \frac{\partial \mathbf{v}}{\partial \mathbf{r}} - m\frac{\partial \Phi}{\partial \mathbf{r}}. \tag{25}$$

Hence,

$$T\nabla \overline{S}_r = -m\mathbf{v} \cdot \frac{\partial \mathbf{v}}{\partial \mathbf{r}} - m\frac{\partial \Phi}{\partial \mathbf{r}} + m\frac{\partial}{\partial \mathbf{r}}(\mathbf{v} \cdot (\mathbf{v}_e + [\boldsymbol{\omega} \times \mathbf{r}])) + \frac{\partial}{\partial \mathbf{r}}(\boldsymbol{\omega} \cdot \mathbf{J}). \tag{26}$$

or,

$$T\nabla \overline{S}_r = -m\mathbf{v} \cdot \frac{\partial \mathbf{v}}{\partial \mathbf{r}} - m\frac{\partial \Phi}{\partial \mathbf{r}} + m\frac{\partial}{\partial \mathbf{r}}(\mathbf{v} \cdot \mathbf{v}) + \frac{\partial}{\partial \mathbf{r}}(\boldsymbol{\omega} \cdot \mathbf{J}). \tag{27}$$

Finally, getting them together, we get

$$T\nabla\overline{S}_r = m\mathbf{v}\cdot\frac{\partial\mathbf{v}}{\partial\mathbf{r}} - m\frac{\partial\Phi}{\partial\mathbf{r}} - \frac{\partial}{\partial\mathbf{r}}\left(\frac{J^2}{2I} - \boldsymbol{\omega}\cdot\mathbf{J}\right). \tag{28}$$

The first term on the RHS of Eq. 28 is the convective term and we know that

$$\frac{d\mathbf{v}}{dt} = \frac{\partial\mathbf{v}}{\partial t} + \mathbf{v}\cdot\frac{\partial\mathbf{r}}{\partial\mathbf{v}}. \tag{29}$$

However, the net force acting on the volume element of the body causes the acceleration $\frac{\partial\mathbf{v}}{\partial t}$, in which case we should replace

$$-m\mathbf{v}\cdot\frac{\partial\mathbf{v}}{\partial\mathbf{r}} \to m\frac{\partial\mathbf{v}}{\partial t}.$$

This amount to the recognition of the principle of equivalence, and for an observer sit on the extended particle center of mass, he is at rest. Then, it results

$$T\nabla\overline{S}_r = -m\frac{\partial\mathbf{v}}{\partial t} - m\frac{\partial\Phi}{\partial\mathbf{r}} - \frac{\partial}{\partial\mathbf{r}}\left(\frac{J^2}{2I} - \boldsymbol{\omega}\cdot\mathbf{J}\right). \tag{30}$$

We can rewrite Eq. 30 under the more suitable form

$$m\frac{\partial\mathbf{v}}{\partial t} = -m\frac{\partial\Phi}{\partial\mathbf{r}} - \frac{\partial}{\partial\mathbf{r}}\left(\frac{J^2}{2I} - \boldsymbol{\omega}\cdot\mathbf{J} + TS\right), \tag{31}$$

assuming, to simplify, that the process has no thermal gradients. Finally, and for more practical uses, we introduce instead the free energy term, in which case the final form of the Newtonian dynamics become more consistently represented in the complete form:

$$\boxed{m\frac{\partial\mathbf{v}}{\partial t} = -m\frac{\partial\Phi}{\partial\mathbf{r}} - \frac{\partial}{\partial\mathbf{r}}\left(\frac{J^2}{2I} - \boldsymbol{\omega}\cdot\mathbf{J} - F\right).} \tag{32}$$

This equation describes the combined motion of rotation and translation, with a rotation about an *instantaneous center* with angular velocity $\boldsymbol{\omega}(t)$. The free energy term inside Eq. 59 ensures the tendency of the free energy to the minimum, tending to the smallest possible value, connected to the term $-TS$ that actuates to give the maximum possible value, that is, ensuring that all possible states can be realized equally.

Such a formulation suggests that the external forces do not dictate the fate of a physical system per se, as in fact, due to entropy (and information entropy, via Shannon's theorem)-related phenomena, the physical system can organize itself, can tend towards novel behavioral patterns. Here I also wish to emphasize that the action-to-reaction law needs a reinterpretation, different from the deprecated form as appearing in standard text books. The reaction follows the action in a time lag that depends of the entropy (or information) gradients; the release from constraints follow the tendency towards new arrangement of the parts of the system.

2.1 Rolling Body on an Inclined Plan

We may check Eq. 41 in a standard example of classical mechanics: a rigid body of mass M rolling down an inclined plane making an angle θ with the horizontal (see, e.g., p. 97 of Ref. [12]). Equation 42 can be applied to solve the problem, with $\omega = 0$ (there is no rotation of the frame of reference) and considering that only the gravitational force acts on the rolling body, with inertial moment relative to its own center of mass given by $I_c = \beta M R^2$. Hence, we obtain

$$M\ddot{x} = Mg\sin\theta - \partial_x w. \tag{33}$$

Here, $w \equiv \frac{(J_c)^2}{2I_c}$. Assuming that the x-axis is directed along the inclined plane, and considering that the angular momentum relative to the rigid body center of mass is given by $J_c = I_c\omega'$, with $\omega' = d\theta/dt$, and noticing that $dx = v_x dt$ (holonomic constraint), it is readily obtained

$$M\ddot{x} = Ma_x = Mg\sin\theta - I_c\omega'\frac{d\omega'}{v_x dt} = Mg\sin\theta - \beta M R^2\frac{\omega'}{\omega' R}\alpha. \tag{34}$$

Since $\alpha = a/R$, then it results in the well-known equation (Fig. 3)

$$a_x = \frac{g\sin\theta}{(1+\beta)}. \tag{35}$$

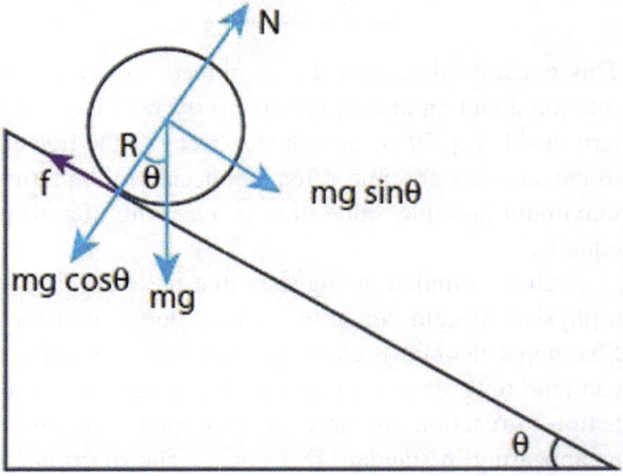

Fig. 3 Rolling body on an inclined plane

2.2 The Fluid Equation

The *internal mechanical energy* term, $U^{(\alpha)}_{mec}$, appearing in Eq. 2 may be defined by

$$U^{(\alpha)}_{mec} = m^{(\alpha)}\phi^{(\alpha)}(\mathbf{r}) + m^{(\alpha)}\sum_{\substack{\beta=1\\ \beta\neq\alpha}}^{N}\phi^{(\alpha,\beta)}. \tag{36}$$

Considering Eq. 36 and the definition of thermodynamic temperature, $\partial S^{(\alpha)}/\partial U^{(\alpha)} \equiv 1/T^{(\alpha)}$, it follows that

$$\begin{aligned}\nabla_{r^{(\alpha)}}U^{(\alpha)} = -m^{(\alpha)}\nabla\phi^{(\alpha)} - \frac{\mathbf{p}^{(\alpha)}}{m^{(\alpha)}}\cdot\nabla\mathbf{p}^{(\alpha)} - \nabla_{r^{(\alpha)}}\left(\frac{J^{(\alpha)}2}{2I^{(\alpha)}}\right)\\ -q^{(\alpha)}\nabla V^{(\alpha)} + q^{(\alpha)}\nabla(\mathbf{A}^{\alpha}\cdot\mathbf{v}^{(\alpha)}).\end{aligned} \tag{37}$$

Equation 37 contains the particle's self-energy and the particle interaction energy for the gravitational and electromagnetic fields, but it may also include other terms, such as terms included in Eq. 36, representing different occurring phenomena (exemplifying energy as a bookkeeping concept). We may recall that the entropic flux in space is a type of generalized force X_α [9, 14]; therefore, it can be shown that the following equation holds:

$$\begin{aligned}T\nabla_{r^{(\alpha)}}\mathfrak{S}^{(\alpha)} = -q^{(\alpha)}\nabla_{r^{(\alpha)}}V^{(\alpha)} + q^{(\alpha)}\nabla_{r^{(\alpha)}}(\mathbf{A}^{(\alpha)}\cdot\mathbf{v}^{(\alpha)})\\ + m^{(\alpha)}\mathbf{v}^{(\alpha)}\cdot\nabla\mathbf{v}^{(\alpha)} - \nabla_{r^{(\alpha)}}\left(\frac{(\boldsymbol{J}^{(\alpha)})^2}{2I^{(\alpha)}} - \boldsymbol{\omega}\cdot\boldsymbol{J}^{(\alpha)}\right),\end{aligned} \tag{38}$$

We can now write the fundamental equation of thermodynamics using the form of a space-time differential equation:

$$T\nabla\overline{S} + \sum_{\alpha} m^{(\alpha)}\frac{\partial\mathbf{v}^{(\alpha)}}{\partial t} = \sum_{\alpha}\nabla\mathbf{U}^{(\alpha)}. \tag{39}$$

Taking into account the convective derivative, $d\mathbf{v}^{(\alpha)}/dt \equiv \partial\mathbf{v}^{(\alpha)}/\partial t + \mathbf{v}^{(\alpha)}\cdot\nabla\mathbf{v}^{(\alpha)}$, we obtain

$$\begin{aligned}m^{(\alpha)}\frac{d\mathbf{v}^{(\alpha)}}{dt} = -T\nabla_{r^{(\alpha)}}\mathfrak{S}^{\alpha} - m^{(\alpha)}\nabla\phi^{(\alpha)} - q^{(\alpha)}\nabla V^{(\alpha)}\\ + q^{(\alpha)}\nabla(\mathbf{A}^{(\alpha)}\cdot\mathbf{v}^{(\alpha)}) - \nabla_{r^{(\alpha)}}\left(\frac{(\mathbf{J}^{(\alpha)})^2}{2I^{(\alpha)}} - \boldsymbol{\omega}\cdot\mathbf{J}^{(\alpha)}\right) + \mathbf{F}^{(\alpha)}_{ext}.\end{aligned} \tag{40}$$

For conciseness, the term $U^{(\alpha)}$ now includes all forms of energy inserted into the above Eq. 2. On the right-hand side (R.H.S.), the first term must be present whenever the mechanical and thermodynamical equilibrium conditions are not fulfilled, the second term is the gravitational force, the third and fourth terms constitute the Lorentz force, the fifth term is a new term that represents the transport of angular momentum, and the last term represents other external forces that are not explicitly included but still act on the particle with index (α).

The present formalism was applied in a previous article [5], and therein we obtained the ponderomotive forces acting on a charged particle. For a neutral particle or body in a gravitational field, Eq. 40 points to a type of extended fundamental equation of dynamics for a given species (α) at equilibrium and at a given point of space-time (Eulerian description):

$$m^{(\alpha)}\frac{\partial \mathbf{v}^{(\alpha)}}{\partial t} = -m^{(\alpha)}\nabla\phi^{(\alpha)} - \nabla_{r^{(\alpha)}}\left(\frac{\mathbf{J}^{(\alpha)2}}{2I^{(\alpha)}} - \boldsymbol{\omega}\cdot\mathbf{J}^{(\alpha)}\right). \quad (41)$$

Equation 41 gains a new term because the body possesses an intrinsic angular momentum. In a non-rotating frame of reference, we set $\omega = 0$, wherein we use the work-energy theorem to obtain the total mechanical energy of the system: $\mathcal{E}_{mec} = K + U + J_c^2/2I_c$. This is a common approach in classical mechanics. We are interested in the effect of a given force at a given space-time coordinate, not in its effect along the particle trajectory. It is worth noting that Eq. 41 was obtained through a variational procedure in contrast to the usual conservation theorem used, for example, in Refs. [15, 16].

Included in the internal energy term is the interpressure term (see Eq. 36; note that here we consider a homogeneous and isotropic fluid). The above-described framework (see also Ref. [5] for additional information) leads us to the well-known hydrodynamic equation for a given species (α):

$$m^{(\alpha)}\frac{d\mathbf{v}^{(\alpha)}}{dt} = -\nabla_{r^{(\alpha)}}(m^{(\alpha)}\phi^{(\alpha)}) - m^{(\alpha)}\nabla_{r^{(\alpha)}}\sum_{\substack{\beta=1\\ \beta\neq\alpha}}^{N}\phi^{(\alpha,\beta)}$$
$$-\nabla_{r^{(\alpha)}}\left(\frac{J^{(\alpha)2}}{2I^{(\alpha)}} - \boldsymbol{\omega}\cdot\mathbf{J}^{(\alpha)} + T\mathfrak{S}^{(\alpha)}\right). \quad (42)$$

In the RHS of Eq. 42, we explicitly introduce external force terms, eventually present in open systems.

Using the following correspondence from particle to fluid descriptions

$$\sum_{\alpha} m^{(\alpha)} \to \int_V d^3x' \rho_v(\mathbf{x}'), \quad (43)$$

and an analogous relation for the electric charge

$$\sum_{\alpha} q^{(\alpha)} \to \int_V d^3x' \rho(\mathbf{x}'), \tag{44}$$

we can rewrite Eq. 42 using the form of the Euler (governing) equation:

$$\rho_v \frac{d\mathbf{v}}{dt} = -\rho_v \nabla_r \phi - \nabla_r p - \nabla_r \Phi_J - \nabla_r f. \tag{45}$$

Here, as usual, the total interparticle pressure term (e.g., Ref. [1]) is given by

$$p(\mathbf{r}) = \sum_{\alpha} m^{(\alpha)} \sum_{\substack{\beta=1 \\ \beta \neq \alpha}}^{N} \phi^{(\alpha,\beta)}(\mathbf{r}). \tag{46}$$

To simplify, we introduce a functional integral of the form of an intrinsic angular momentum energy density (comprising the "interaction energy term", $\boldsymbol{\omega} \cdot \mathbf{J}$), Φ_J:

$$\sum_{\alpha} \left[\frac{J^{(\alpha)2}}{2I^{(\alpha)}} - \boldsymbol{\omega} \cdot \mathbf{J}^{(\alpha)} - (\Delta F)^{(\alpha)} \right] \to \int [\Phi_J(\mathbf{x}') + f(\mathbf{x}')] d^3x', \tag{47}$$

considering that the intrinsic angular momentum density refers to a given blob of fluid (with inertial momentum I, a measure of the local rotation, (i.e., spin, of the fluid element), and its associated free energy (per unit volume), $f = f_0 - Ts$. Equation 45 also suggests that the function $S^{(\alpha)}$ (the *field integral of* $\mathbf{r}^{(\alpha)}$) is constant along the integral curves of the space field $\mathbf{r}^{(\alpha)}$. The gradient of the free energy, f, of the out-of-equilibrium state is the source of the spontaneous change from an unstable state to a more stable state while performing work. For example, a common source of free energy in a collisionless plasma is an electric current [17]; in a magnetically confined plasma, several classes of free energy sources are available to drive instabilities, e.g., the relaxation of a non-Maxwellian, non-isotropic velocity distribution [18]. At this stage, it is worth noting that our procedure includes the treatment of the effect of angular momentum (through Eq. 2), a necessary inclusion in a consistent theory, according to Curtiss [13].

2.3 Application to Electrodynamics

The development of the theoretical framework in the previous sections can be easily extended to the electrodynamics, showing to be a useful pedagogical device. In addition, the method is fruitful since a new fourth term of force appears, the topological torsion term of force, of the type $q[\boldsymbol{\omega} \times \boldsymbol{A}]$ (to be discussed later).

First, we may note that the mathematical identity holds:

$$\begin{aligned}\nabla(\mathbf{A}^{(\alpha)}\cdot\mathbf{v}^{(\alpha)}) = & \ (\mathbf{A}^{(\alpha)}\cdot\nabla)\mathbf{v}^{(\alpha)} + (\mathbf{v}^{(\alpha)}\cdot\nabla)\mathbf{A}^{(\alpha)} + [\boldsymbol{A}^{(\alpha)}\times\nabla\times\boldsymbol{v}^{(\alpha)}] \\ & + [\mathbf{v}^{(\alpha)}\times\nabla\times\mathbf{A}^{(\alpha)}], \end{aligned} \tag{48}$$

and, by means of algebraic manipulation, the following expression can be obtained:

$$\nabla\left(\mathbf{A}^{(\alpha)}\cdot\mathbf{v}^{(\alpha)}\right) = -\frac{\partial\mathbf{A}^{(\alpha)}}{\partial t} - \left[\boldsymbol{\omega}\times\mathbf{A}^{(\alpha)}\right] + \left[\mathbf{v}^{(\alpha)}\times\mathbf{B}^{(\alpha)}\right]. \tag{49}$$

Here, $\mathbf{B} = [\nabla\times\mathbf{A}]$. In addition, we notice that the following equality holds:

$$(\mathbf{A}^{(\alpha)}\cdot\nabla)\mathbf{v}^{(\alpha)} = [\boldsymbol{\omega}\times\mathbf{A}^{(\alpha)}], \tag{50}$$

where

$$\mathbf{A}^{(\alpha)} = \sum_{\substack{\beta=1 \\ \beta\neq\alpha}} q^{(\beta)}\frac{\mathbf{v}^{(\beta)}}{r_{\alpha\beta}}, \tag{51}$$

denotes the vector potential actuating on the particle (α) due to every other particle β, and the vorticity is defined by

$$\boldsymbol{\Omega}^{(\boldsymbol{\alpha})} = [\nabla_{\mathbf{r}^{(\alpha)}}\times\mathbf{v}^{(\alpha)}] = 2\boldsymbol{\omega}^{(\boldsymbol{\alpha})}. \tag{52}$$

Therefore, the general equation of dynamics for a physical system (Lagrangian description) follows:

$$\rho\frac{d\mathbf{v}}{dt} = \rho\mathbf{E} + [\mathbf{J}\times\mathbf{B}] - \nabla\phi - \nabla p + \rho[\mathbf{A}\times\boldsymbol{\omega}]. \tag{53}$$

The last term in the RHS of the Eq. 53 is a new term that represents a type of topological spin vector [19], an artifact of non-equilibrium process. We will show that the topological spin vector plays a role in plasma arcs, as well as in magnetocumulative generators [20] and explosive-driven generators to produce high current and energy pulses suitable to excite high power lasers or thermonuclear fusion reactors [21], and suggests a new method for obtaining the helicity transport equation [22].

2.3.1 Application to Plasma Physics: The Plasma Torch

The compression of an electric current by a magnetic field, the z-pinch effect, can be studied on the basis of Eq. 53, which gives the condition for dynamic equilibria. Let us assume a typical geometry for an infinitely long axisymmetric cylindrical arc (Fig. 4) with axial current density $J_z = J_z(r)$. Because the current density is assumed to be constant, Maxwell's equations in the steady state yield the azimuthal

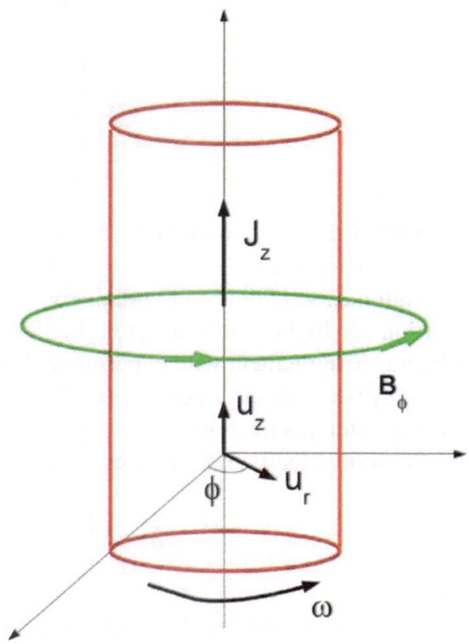

Fig. 4 Geometry and vectorial fields in the Bennett pinch generated by an axial current J_z creating a toroidal field B_ϕ. If, instead, we consider a vacuum arc discharge with radial current J_r and magnetic field B_z, we find a rotating arc with an angular velocity ω

component $B_\theta = \mu_0 J_z r/2$ for $r \leq R$ with R the outer boundary of the cylindrical arc. The vector potential is purely radial and is given by $A_z(r) = -\mu_0 R^2 J_z/4$, for $r < R$, where the Coriolis term plays no role. Based on the above conditions, we can write Eq. 53 in the form:

$$-J_z B_\theta = \frac{dp}{dr}. \tag{54}$$

It follows

$$p(r) = \int_r^R \frac{dp}{dr} dr = \frac{1}{4}\mu_0 J_z^2 (R^2 - r^2), \tag{55}$$

which is a well-known result.

The interaction between vacuum arcs and transverse magnetic fields is used in switching devices for repetitive high current interruption (see, e.g., Refs. [23, 24]). We can instead consider a coaxial configuration with a cathode on-axis with a stabilizing magnetic induction field **B** directed along the axis of symmetry and an arc current density **J** flowing radially (and assuming a "filamentary" current with radius R', $A_r = -\mu_0 R'^2 J_r/4$, with μ_0 representing the permeability of the vacuum). In this case, we may apply Eq. 55 and obtain the pressure differential, from the axis to the wall (at R):

$$\Delta p(r = R) = 2\pi R \left[-J_r B_z + \mid \rho_c \mid \pi R^2 \frac{\mu_0}{4\pi} J_r \omega \right] \Delta\theta. \tag{56}$$

The conditions for retrograde or Amperian rotation are the following:

$$| \rho_c |> \frac{\mu_0}{4\pi}\frac{B_z}{S}\omega, \quad \text{retrograde rotation} \tag{57a}$$

$$| \rho_c |< \frac{\mu_0}{4\pi}\frac{B_z}{S}\omega, \quad \text{amperian rotation.} \tag{57b}$$

Here, S denotes the filamentary current cross-section, with $\rho = \rho_c$. For negative charge carriers ($\rho_c = - | \rho_c |= -en_e$), we obtain an Amperian (clockwise) rotation for high magnetic fields and relatively weak arc currents. We also find a retrograde rotation for higher intensity arcs (higher S) and small transverse magnetic fields, which is in agreement with experimental evidence (e.g., Ref. [23]). Here, we see the interplay between the tendencies of the energy to attain a minimum value, while the entropy attempts to attain a maximum value. From Eq. 56, we obtain an expression for the spot velocity in a transverse magnetic field:

$$v = \omega R = \frac{4\pi}{\mu_0}\frac{1}{| \rho_c |}\left(\frac{\epsilon_0 i}{2\pi\sigma_c} + \frac{B_z R}{S}\right). \tag{58}$$

Here, we have made use of the Bernoulli relation, $\Delta p = \epsilon_0 E^2/2$, and the constitutive equation, $J_r = \sigma_c E$, where σ_c denotes the electrical conductivity of the plasma. Although Eq. 58 is not self-consistent, it shows that the force term $[\mathbf{J} \times \mathbf{B}]$ is not the only player in the process, and that the spot velocity depends linearly on the arc current (see, e.g., Ref. [25]). This approach shows that in out-of-equilibrium systems, a new force term occurs in a way that can suppress the Amperian force under the above referred conditions given by Eq. 57 (see also Ref. [26]).

A better understanding of this phenomenon is crucial, because arc discharges are powerful generators of non-equilibrium atmospheric pressure plasmas. We compare the results predicted by Eq. 58 with experimental data available in the literature in Table 1. The calculations were done using the expression for the electrical conductivity in terms of the microscopic parameters of the plasma, $\sigma_0 = en_e\mu = \varepsilon_0\omega_{pe}^2/\nu_c$, with the electronic mobility given by $\mu = e/m\nu_c$ (nu_c denotes the electron collision frequency) because, for the majority of the data, the transverse magnetic field was below 0.1 T and is not expected to greatly influence the plasma arc electrical conductivity [27] (Table 2).

Table 1 Please write your table caption here

First	Second	Third
Number	Number	Number
Number	Number	Number

Table 2 Comparing Eq. 58 with experimental data for low pressure and atmospheric DC discharges. Electric current $i = 60$ A

	Low pressure	Atmospheric pressure
$T_e(eV)$	1.0[a]	0.87[a]
Electron density (cm^{-3})	$n_e = 2.5 \times 10^{12}$[f]	$n_e = 2 \times 10^{16}$[b]
collision frequency (s^{-1})	$\nu_c = 6.98 \times 10^7$	$\nu_c = 7.2 \times 10^{11}$[c]
Plasma frequency (s^{-1})	$\omega_{pe} = 8.94 \times 10^{10}$	$\omega_{pe} = 8 \times 10^{12}$[d]
Average Speed (m/s)	3.0 ($\sim$2)[e]	10^{-3} (5×10^{-3})[e]

[a] For laboratory discharges, the Coulomb logarithm is $\ln \Lambda \sim 10$, see Ref. [?], for electron temperatures of the order of $T_e \approx 10000$ K [?]
[b] Ref. [?]
[c] The frequency of collision was calculated using the standard expression $\nu_c = 2.91 \times 10^{-6} n_e \ln \Lambda T_e^{-3/2}$, Ref. [?]
[d] We use $\omega_{pe} = 5.65 \times 10^4 \sqrt{n_e}$, (see Ref. [?])
[e] In parenthesis are the experimental data collected for atmospheric pressure, from Ref. [26]; for low pressure, see Ref. [?].
[f] Data interpolated from Ref. [?], assuming $T_e = 0.8$ eV

2.4 Poynting's Theorem

We must now proceed from the notion of the localization of electromagnetic energy in space, and generalize the concept to other forms of energy: $\mathfrak{I}$, kinetic energy; $\mathfrak{P}_g$, total gravitational energy; $\mathfrak{P}_v$, total electrostatic energy; and, $\mathfrak{P}_{rot}$, the total rotational energy.

From Eq. 42, an energy conservative equation can be obtained:

$$\frac{\partial}{\partial t}(\mathfrak{I} + \mathfrak{P}_g + \mathfrak{P}_v + \mathfrak{P}_{rot}) = -\frac{1}{2}\sum_{\alpha} \nabla_{\mathbf{r}^{(\alpha)}} \cdot \left\{ \mathbf{v}^{(\alpha)} \left[-\frac{|\mathbf{J}^{(\alpha)}|^2}{2I^{(\alpha)}} + (\boldsymbol{\omega} \cdot \mathbf{J}^{(\alpha)}) - T\mathfrak{S}^{(\alpha)} \right] \right\}$$
$$+ \frac{1}{2}\sum_{\alpha} \left[-\frac{|\mathbf{J}^{(\alpha)}|^2}{2I^{(\alpha)}} + (\boldsymbol{\omega} \cdot \mathbf{J}^{(\alpha)}) - T\mathfrak{S}^{(\alpha)} \right] (\nabla_{\mathbf{r}^{(\alpha)}} \cdot \mathbf{v}^{(\alpha)}). \quad (59)$$

Finally, we can transform Eq. 59 into a version of *Poynting's Theorem for rotating fluids*:

$$-\frac{\partial}{\partial t}(\mathcal{U}) = \nabla \cdot \mathbf{S} + \mathcal{P}', \quad (60)$$

on which we define a type of *Poynting vector for rotational fluids*, which gives the rate of rotational energy flow:

$$\mathbf{S} = \frac{1}{2}\sum_{\alpha} \mathbf{v}^{(\alpha)} \left[-\frac{|\mathbf{J}^{(\alpha)}|^2}{2I^{(\alpha)}} + (\boldsymbol{\omega}^{(\alpha)} \cdot \mathbf{J}^{(\alpha)}) - T\mathfrak{S}^{(\alpha)} \right]. \quad (61)$$

The *power input* driving the process (source/sink term) is given by

$$\mathcal{P}' \equiv \frac{1}{2}\sum_{\alpha}(\nabla \cdot \mathbf{v}^{(\alpha)})\left[\frac{|\mathbf{J}^{(\alpha)}|^2}{2I^{(\alpha)}} - (\boldsymbol{\omega}\cdot\mathbf{J}^{(\alpha)}) + T\mathfrak{S}^{(\alpha)}\right], \tag{62}$$

The total energy is defined by summing up the different contributions:

$$\mathcal{U} = \mathfrak{I} + \mathfrak{P}_g + \mathfrak{P}_v + \mathfrak{P}_{rot}. \tag{63}$$

Here, the term $T\mathfrak{S}^{(\alpha)}$ represents the thermal energy associated with the species (α) that is equal to $-\Delta F$, the free energy of the physical system. For a system in contact with a reservoir at constant temperature, this is the maximum amount of work extractable from the system; the free energy tends to decrease for a system in thermal contact with a heat reservoir.

2.5 Driving Energy of a Rotating System

In particular, notice that when the angular velocity, ω, is multiplied by Eq. 42, the driving power is obtained. It is worth noting that the presence of the term $\boldsymbol{\omega}\cdot\mathbf{J}$, which plays an analogous role to the slip in electrical induction motors, that is, the lag between the rotor speed and the magnetic field's speed, is provided by the stator's rotational speed. Furthermore, we see that the power input depends on the fluid compressibility $\nabla\cdot\mathbf{v}$. This means that compressibility is a factor that determines the amount of transported angular momentum through the stress-tensor τ_{ij} and may be responsible for a new driving mechanism in addition to the well-known MRI. The driving energy of the rotating system can be expressed in the form:

$$\mathcal{E}_{driv} = \frac{J^2}{2I_p} - (\boldsymbol{\omega}\cdot\mathbf{J}) - \Delta F. \tag{64}$$

Next, we will discuss several examples illustrating the application of the variational method.

2.6 Transport of Angular Momentum in a Hurricane

A hurricane is a natural airborne structure that converts its kinetic and potential energy by means of the transport of angular momentum from the inner core to the outer regions, conveyed either directly by moving matter, or by non-material stresses such as those exerted by electric or magnetic fields [28]. We may apply Eq. 74 to this specific problem, assuming that all of the mechanical and thermal energy is converted into electromagnetic energy U_e, to obtain

$$\mathcal{P}' = \omega \left(\frac{J^2}{2I_p} - \boldsymbol{\omega} \cdot \mathbf{J} - \Delta F \right) = -\frac{\partial U_e}{\partial t} \tag{65}$$

or

$$\mathcal{P}' = \omega \mathcal{E}_{driv}. \tag{66}$$

Let us consider the case of a hurricane in an axisymmetric configuration, with $J = \omega I$. We can safely assume that $\omega^2 I^2 / I_p \gg \omega^2 I + \Delta F$. We can now envision a simple model of a hurricane with a total mass, M, and radius, R, approximated as a solid cylinder with $I = MR^2/2$. Hence, the total power driving the hurricane is given by

$$\mathcal{P}' = \frac{1}{8} \frac{\omega^3}{2} \frac{M^2 R^4}{I_p} \propto \omega^3 \frac{M^2 R^4}{I_p}, \tag{67}$$

or, as a function of the fluid density ρ:

$$\mathcal{P}' = \frac{\pi^2}{4} \omega^3 \frac{R^6 L^2}{I_p} \rho^2. \tag{68}$$

Our result shows the same type of dependency that was demonstrated by Chow and Chey [29], and, in particular, it shows that the intrinsic inertial momentum of the particles constituting the fluid plays a substantial role.

2.7 *Periodic Radiative Heating of the Earth's Atmosphere*

It has been experimentally shown [30] that periodic radiative heating of the Earth's atmosphere transmits angular momentum to it as a result of the Earth-atmosphere coupling through frictional interactions [31]. The images sent by the ESA's Venus Express confirms this fact on Venus (Earth's planetary twin) based on the presence of a "double-eye" atmospheric vortex at the planet's south pole and the presence of high velocity winds whirling westwards around the planet, which is characterized by a four-day period.

Schubert and Withehead's [30] conducted an experiment with the purpose of providing an explanation for the high wind velocities during apparent cloud formation in the upper atmosphere of Venus. In this experiment, a Bunsen flame rotating below a cylindrical annulus filled with liquid mercury induced the rotation of the liquid mercury in a direction counter to that of the rotating flame. The speed of the flame was 1 mm/s and the temperature of the mercury increased from room temperature at a rate of approximately 3 °C/min. After 5 min, a steady-state flow was established, with the mercury rotating in the counter-direction of the flame, with a speed of approximately 4 mm/s. If we consider the liquid to be a spinning body, we can estimate that

$$\frac{d}{dr}\left(\frac{I_p\omega'^2}{2}-\frac{I_p\omega^2}{2}+T\mathfrak{S}\right)>0\Rightarrow I_p\Delta(\omega'+\omega)(\omega'-\omega)+\Delta(T\mathfrak{S})>0. \quad (69)$$

Hence, the following result is obtained:

$$\omega' \gtrsim \pm\omega\sqrt{1-\frac{\Delta F}{2I_p\omega^2}}, \quad (70)$$

and, in the limit $\omega \approx \omega'$,

$$\omega'\Delta\omega \gtrsim -\frac{1}{2}\frac{\Delta T\mathfrak{S}}{I_p}. \quad (71)$$

Here, ω is the angular speed imposed on the system (heat source), ω' is the mass flow induced angular speed due to a sustained source of energy; $\Delta\omega \equiv \omega' - \omega$. We use the following tabulated data: $\mathfrak{S} = 16.6$ J $K^{-1}mol^{-1}$ for mercury at temperature of the experiment and $\rho \approx 13.6 \times 10^{-3}$ kg/m^3. We also consider that the volume of mercury is contained in the channel of the experimental apparatus forming a rim with an average radius of $R = 30$ cm. Using Eq. 71, we estimate that after one minute, the speed must be approximately 3.7 mm/s. It is clear that the sense of rotation and the speed of the wind depend on the latent heat stored in the planetary atmosphere and the temperature difference between the boundaries (through ΔT). Equation 70 is consistent with the results reported by [32], where a maximum mean surface flow is observed corresponding to the angular velocity of the heat source when the convective velocity (in our example, $v_{stam} = \sqrt{\Delta F/2M_\mu}$, with M_μ denoting the molar mass) is attained. Note that, in fact, our v_{stam} represents the limiting speed for the transport of angular momentum. For mercury, this velocity is $v_{stam} \approx 24$ m/s, with $M_\mu = 200.59$ g/mol in the conditions of the moving flame experiment. The existence of a limit to the amplification of the angular speed was also suggested in Ref. [33], which demonstrated the effect of a heating or cooling source in the momentum equilibration. Three possible cases for the clockwise or counter-clockwise motion of a fluid in a rotation frame are observed in Fig. 5.

Equation 59 is not reversible in time due to the presence of the "hyperfine structure" $(\boldsymbol{\omega} \cdot \mathbf{J})$ and the free energy term F. It was shown in Ref. [1] that the power equation can be written under the form of a *Poynting's Theorem* for rotating fluids or extended bodies:

$$-\frac{\partial}{\partial t}(\mathcal{U}) = \nabla\cdot\mathbf{S}+\mathcal{P}', \quad (72)$$

provide we first define the following Poynting vector for rotational fluids and specializing for a single specie, which gives the rate of rotational energy flow:

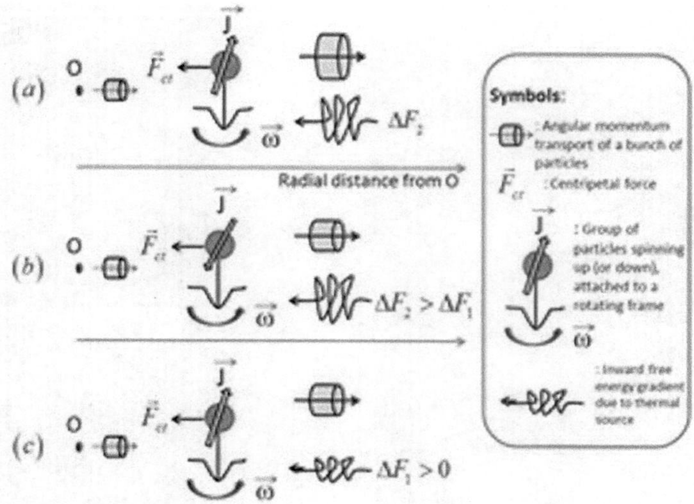

Fig. 5 A group of particles spinning about their axes and revolving around a common axis, Oz, subject to a centripetal force. Three situations typically occur. **a** An outward transport of angular momentum occurs with a larger gradient, while free energy flows to the center of the field. **b** If the inward free energy gradient is dominant relative to the angular momentum gradient, a reversal of the particle' angular momentum may occur. **c** If the angular momentum gradient is of the same order of magnitude as in case **b** but still dominant relative to the inward free energy gradient, the particles may continue spinning in the same direction

$$\mathbf{S} \equiv \frac{1}{2}\left[-\frac{|\mathbf{J}|^2}{2I} + (\boldsymbol{\omega}\cdot\mathbf{J}) + F\right]\mathbf{v}. \tag{73}$$

The *power input* driving the process (source/sink term) is given by

$$\mathcal{P}' \equiv \frac{1}{2}(\boldsymbol{\nabla}\cdot\mathbf{v})\left[\frac{|\mathbf{J}|^2}{2I} - (\boldsymbol{\omega}\cdot\mathbf{J}) - F\right], \tag{74}$$

and the total energy is defined by summing up the different contributions:

$$\mathcal{U} = \mathfrak{T} + \mathfrak{P}_g + \mathfrak{P}_v + \mathfrak{P}_{rot}. \tag{75}$$

Here, $\mathfrak{T}$ is the total kinetic energy, $\mathfrak{P}_g$ is the gravitational energy, $\mathfrak{P}_v$ is the total electrostatic energy, and $\mathfrak{P}_{rot}$ is the total rotational energy (see Ref. [1] for details). Equation 72 shows that it is vortical motion and/or free energy processes that are at the root of non-equilibrium processes and dissipation.

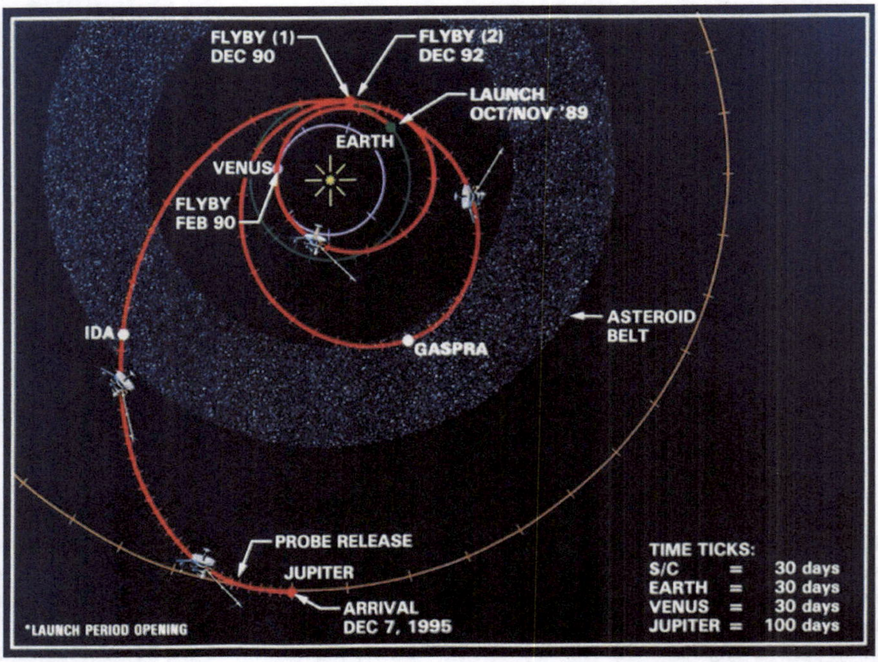

Fig. 6 Trajectory drawing of the Galileo spacecraft's launch from low Earth orbit including Venus (February 1990), two Earth flybies (December 1990 and December 1992), and the asteroids Gaspra and Ida in the asteroid belt. Courtesy: NASA

2.8 *The Flyby Anomaly of the Spacecrafts*

The flyby anomaly is still an unsolved problem in astrodynamics discovered by a research team at the Jet Propulsion Laboratory, lead by John Anderson. The first observation appeared when the trajectory of the Galileo spacecraft approached planet Earth on December 8, 1990, followed by means of Doppler data, showed a difference on its ingoing and outgoing asymptotic velocities, differing in 3.92 mm/s in what was expected theoretically, based on the gravitational orthodox theory (Fig. 6).

Now, we attempt here a simple explanation to the Juno flyby that registered unexpectedly no anomaly.

The equation to solve is

$$\ddot{\mathbf{r}} = -\nabla\phi + \mathbf{f}_{TTC}. \tag{76}$$

We will use the cylindrical coordinate frame $\mathcal{F}_0$ (see Ref. [34]) with basis vectors $(\mathbf{x}_o, \mathbf{y}_o, \mathbf{z}_o)$. In this coordinate frame, the position and velocity are given by

$$\mathbf{r} = r\mathbf{x}_o \tag{77}$$

$$\mathbf{v} = \dot{r}\mathbf{x}_o + r\dot{\theta}\mathbf{y}_o \tag{78}$$

and the angular momentum is given by

$$\mathbf{h} = h\mathbf{z}_o = r^2\dot{\theta}\mathbf{z}_o. \tag{79}$$

The perturbing force in $\mathcal{F}_o$ coordinates is given by

$$\mathbf{f}_{TTC} = f_r\mathbf{x}_o + f_\theta\mathbf{y}_o + f_z\mathbf{z}_o. \tag{80}$$

2.8.1 Spacecraft in Posigrade Direction

$$v_{apx} = V_P - v_\infty \cos(\omega \mp \theta) \tag{81}$$
$$v_{apy} = v_\infty \sin(\omega \mp \theta) \tag{82}$$
$$v_{apz} = 0. \tag{83}$$

with

$$\mathbf{n}' = \sin(\omega \mp \theta)\mathbf{y}_o + \cos(\omega \mp \theta)\mathbf{x}_o \tag{84}$$

and therefore, the term $(\mathbf{v} \cdot \mathbf{n}')$ appearing in Eq. 19 transforms into (we solve now in the frame of the planet, hence $V_P = 0$)

$$(\mathbf{v} \cdot \mathbf{n}') = v_{apx} \cos(\omega \mp \theta) + v_{apy} \sin(\omega \mp \theta) \tag{85}$$
$$= v_\infty \cos^2(\omega \mp \theta) - v_\infty \sin^2(\omega \mp \theta) \tag{86}$$
$$= v_\infty \cos 2(\omega \mp \theta). \tag{87}$$

As we will see next, Eq. 76 is determinant to the issue and, in particular, the dot product $(\mathbf{v} \cdot \mathbf{n})$ (Figs. 7 and 8).

To solve the set of Eq. 76 is not our intention, we just want to solve Eq. in the region of the flyby, assuming that $v_{x_0} \approx 0$ and $R = R_\oplus$, and $\frac{v_{x_0} v_\infty}{v} \sim v_\theta$. The Eq. 76 can be written in the form

$$\frac{dv_\theta}{dt} = \frac{2\omega_\oplus R_\oplus}{c} \frac{v_{x_0} c}{R_\oplus} \sin I + \frac{2\omega_\oplus R_\oplus}{c} \frac{v_{x_0} v_\infty}{R_\oplus} \sin I \tag{88}$$

Here, $K \equiv \frac{2\omega_\oplus}{R_\oplus}$, or

$$\frac{dv_\theta}{dt} = K\frac{v_{x_0} c}{R_\oplus} + K\frac{v_{x_0} v_\infty}{R_\oplus} \sin I \tag{89}$$

Equation 89 shows two forces acting together on the spacecraft. The first term is nearly zero, according to our hypothesis of nearly circular trajectory, while the second term of force actuates as a modulating force and contributes more significantly. We use the approximation of a nearly circular trajectory during the flyby, $dt = R_\oplus d\theta/v$.

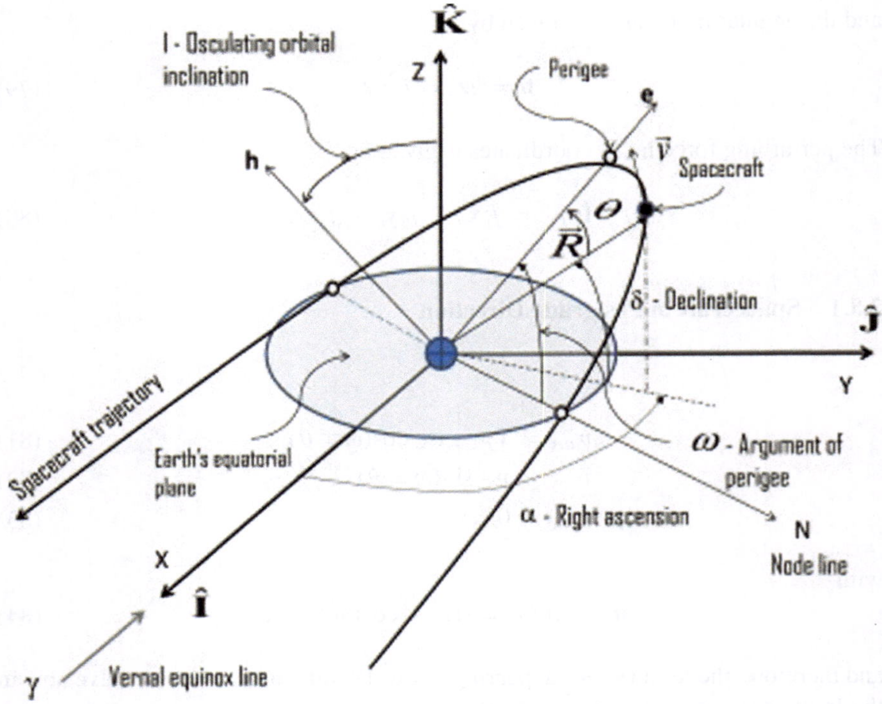

Fig. 7 Spacecraft trajectory in flyby trajectory. Orbital parameters

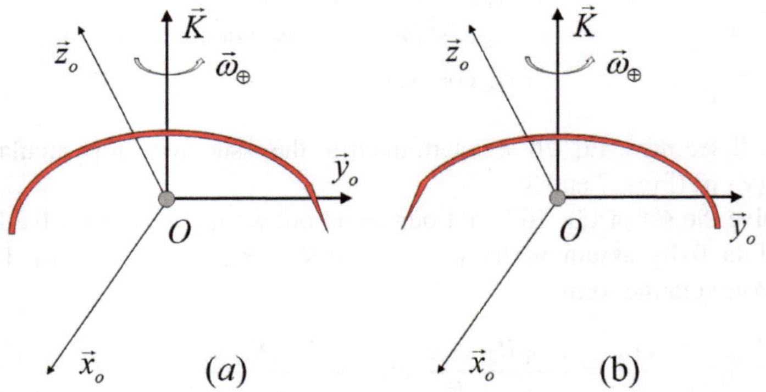

Fig. 8 Spacecraft trajectory in retrograde (**a**) and posigrade (**b**) direction

The equation of motion for the azimuthal velocity becomes

$$\frac{dv_\theta}{d\theta} = \left(\frac{Kcv_r}{v} \sin I + \frac{Kv_\infty v_r}{c} \cos 2(\omega \mp \theta) \sin I \right) \cos(\omega \mp \theta). \tag{90}$$

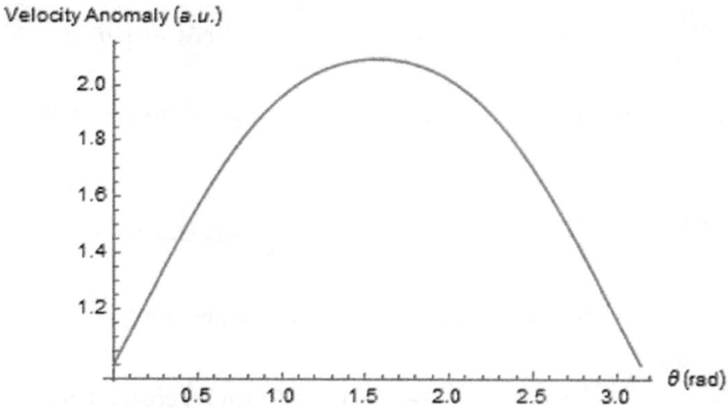

Fig. 9 Azimuthal velocity versus θ in the posigrade direction

The mathematical solution is

$$v_\theta = e^{(c_1 \sin\theta + c_2 \sin(3\theta)} + e^{(c_1 \sin(\theta) + c_2 \sin(3\theta)} \int_1^\theta e^{(-c_1 \sin(\theta') - c_2 \sin(3\theta'))} \cos\theta')d\theta'. \quad (91)$$

The chosen constants are $c_1 = 0.07$ and $c_2 = 0.0233$ that give the best approach to the NEAR spacecraft velocity anomaly. The integral make the approach to $\sin\theta$). Figure 9 shows the velocity anomaly in the posigrade direction. There is an acceleration during half-way but a deceleration that follows, giving an overall null effect. There

2.8.2 Spacecraft in Retrograde Direction

To treat this case, we need the additional components of the velocity:

$$v_{apx} = V_P + v_\infty \cos(\omega \mp \theta) \quad (92)$$

$$v_{apy} = -v_\infty \sin(\omega \mp \theta) \quad (93)$$

$$v_{apz} = 0. \quad (94)$$

and therefore,

$$(\mathbf{v} \cdot \mathbf{n}') = v_{apx} \cos(\omega \mp \theta) + v_{apy} \sin(\omega \mp \theta) \quad (95)$$

$$= V_P \cos(\omega \mp \theta) - v_\infty \cos^2(\omega \mp \theta) + v_\infty \sin^2(\omega \mp \theta) \quad (96)$$

$$= V_P \cos(\omega \mp \theta) + v_\infty \cos 2(\omega \mp \theta) \quad (97)$$

In this case, the equation of azimuthal motion is

$$\frac{dv_\theta}{dt} = 2\omega_\oplus v_\infty \cos(\omega \mp \theta)\sin I + 2\omega_\oplus \frac{v_\infty^2}{c}\cos(\omega \mp \theta)\sin I. \tag{98}$$

Now, we may expect that $v_\infty \sim v_\theta$ and within the range of this assumption the above Eq. becomes

$$\frac{dv_\theta}{d\theta} = K\frac{c}{R_\oplus} v_\infty \cos(\omega \mp \theta)\sin I + \frac{K v_\infty^2}{R_\oplus}\cos(\omega \mp \theta)^2 \sin I. \tag{99}$$

We will make the approach and rewrite the Eq. in the new form

$$\frac{dv_\theta}{d\theta} = \left(\frac{Kcv_r}{v}\sin I + \frac{Kv_\infty v_r}{c}\cos(\omega \mp \theta)\sin I\right)\cos(\omega \mp \theta). \tag{100}$$

We have search the values that give the best similar shape in comparison with the radar data obtained with NEAR spacecraft, and the best guess are, approximately,

$$\frac{Kv_\infty v_r}{c} \sim 0.14 v_\theta \tag{101}$$

$$\frac{Kcv_r}{v_\theta} \sim 1. \tag{102}$$

Considering $v_\infty \sim 6 \times 10^3$ km/s, we obtain $v_r \sim 7$ km/s and $v_\theta \sim 1$ km/s, values that are in the order of the expected magnitude. The mathematical solution is given by

$$v_\theta = \exp[c_3(\omega \mp \theta) + c_4 \sin(\omega \mp \theta)](1 + c_5 \sin(\omega \mp \theta)), \tag{103}$$

with $c_3 = 0.05$, $c_4 = 0.025$, and $c_5 = 0.14$. We may notice that the main difference between the two flyby approaches is the direct dependency on the angle $\omega \mp \theta$ in the retrograde case, which implies a bump in the shape of the figure and an asymmetry on the effect of the TTC.

Figure 10 plots the solution, comparing with the experimental data (see Fig. 3b in Ref. [35]).

2.9 Additional Remarks on the Variational Problem According to the Proposed Reformulation

In 1686, Newton solved the problem of determining the shape of a rotationally symmetric body of least resistance, but the beginning of the calculus of variations is set in the year 1696 with the formulation of the brachistochrone problem by Johann Bernoulli in the *Acta Eruditorum*. Without recurring to minimum-time assumptions, or Euler-Lagrange equations, the basic idea for the application of the concepts of energy and entropy, closely linked to the fundamental equations of thermodynamics,

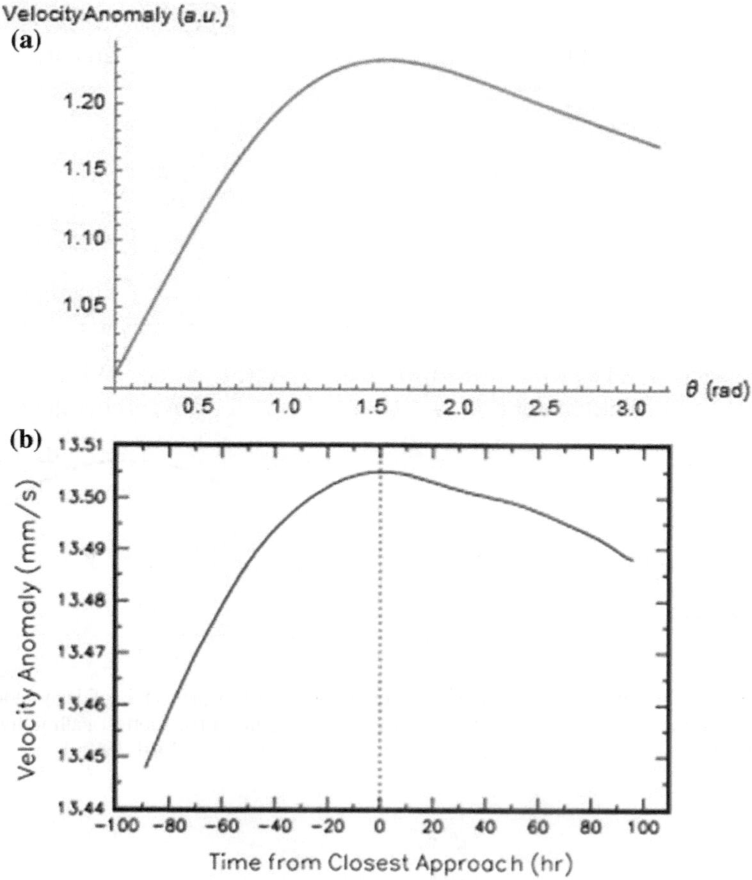

Fig. 10 Anomalous velocity versus θ in retrograde direction. **a** Theoretical prediction; **b** Experimental data from NEAR spacecraft, see Ref

create a unified description of mechanics, in particular yielding a wider frame for variational problems, such as the brachistochrone for uniform (e.g., gravitational) field (Fig. 11).

Following along our research program, let us consider a Cartesian reference frame and a particle with mass m accelerated by gravity from one point to another with no friction and solve the problem using Eq. 14. First, notice that the Cartesian frame is not rotating, hence, $\omega = 0$. Noting that Ω is independent from coordinates (x, y), we may assume $x(0) = -R$, $\frac{dx}{dt} = R\Omega t$ and $\frac{dy}{dt}(0) = 0$, and $y(0) = R$. It gives

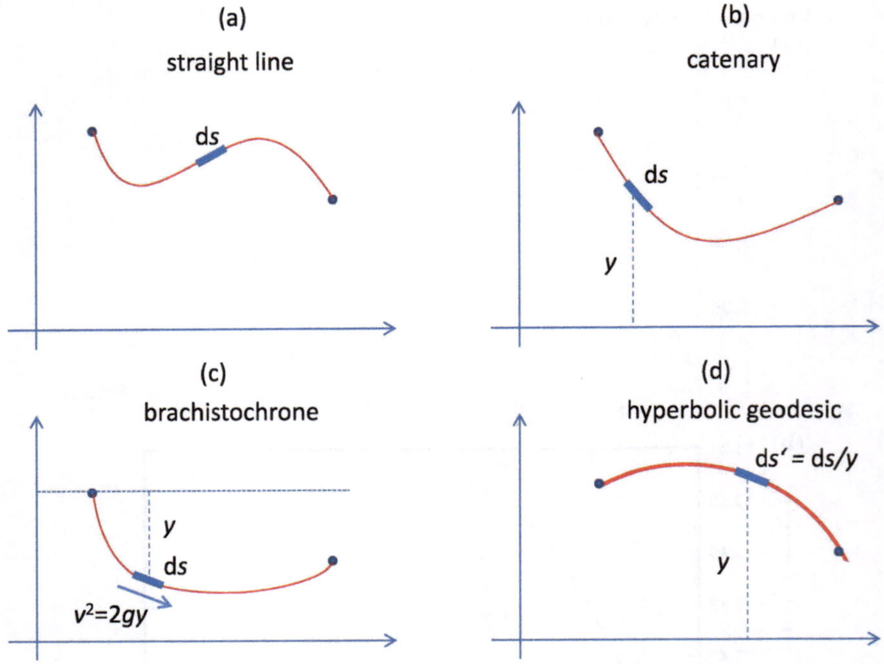

Fig. 11 The four optimization problems: **a** finding the shortest path in Euclidean geometry, **b** the shape of a hanging chain, **c** the path of fastest descent, and **d** the shortest path in hyperbolic geometry. Courtesy from: Raul Rojas, Freie Universität Berlin, arXiv:1401.2660v1

$$\frac{d^2y}{dt^2} = -g - y\Omega^2 \tag{104}$$

$$\frac{d^2x}{dt^2} = -x\Omega^2 \tag{105}$$

since, $\nabla_y(\frac{1}{2}mr^2\Omega^2) = my\Omega^2$ (with similar result for x component), and plugging the identity $dr = \sqrt{dx^2 + dy^2}$. The solutions of the above system of equations are the well-known equations of a cycloid:

$$x(t) = R(\Omega t - \sin \Omega t) \tag{106}$$

$$y(t) = -\frac{g}{\Omega^2}(1 - \cos \Omega t) + R \cos \Omega t. \tag{107}$$

The cycloid equations were straightforwardly obtained because the fundamental equation of dynamics, Eq. 14, retains consistently the two extremal principles of nature, the tendency to minimize energy and maximize entropy. There is no need to make further hypothesis.

The standard solution of one of the earliest problems in the calculus of variations starts with the calculation of the time to travel from a point P_1 to another point P_2:

$$t_{12} = \int_{P_1}^{P_2} \frac{ds}{v}, \tag{108}$$

obtaining the speed at any point by using the equation of conservation of energy, $v = \sqrt{2gy}$. Plugging the arc length $ds = \sqrt{1 + y'^2}dx$, the integral of t_{12} is converted into

$$t_{12} = \int_{P_1}^{P_2} \sqrt{\frac{1 + y'^2}{2gy}} dx = \int_{P_1}^{P_2} f(y, y')dx. \tag{109}$$

Normally, the Euler-Lagrange differential equation is used, and, after some maths, another differential equation is obtained

$$\left[1 + (\frac{dy}{dx})^2\right] y = k^2, \tag{110}$$

where k^2 is a positive constant. The solution is given by the parametric equations, Eqs. 49–50. Our foundational project aims at finding a coherent and possibly true representation of what is physically possible without involving extraneous idealizations or empirical generalizations. In this sense, our formulation seems more natural than the Lagrangian formulation, since it represents a new synthesis of two different propensities in nature (provided by energy and entropy).

3 Direct Conversion of Entropy into Forces

3.1 The Case of Entropic Forces

Emergent phenomena in nature gives rise to surprising order. In particular, it seems that more complex network of interactions between elements result in fewer elements needed for the emergence of phenomena. This conception was adopted by Verlinde [36] (although, much earlier, a more in-depth proposal was advanced, and disregarded [4]), and disproved [37]. Instead, we should look to the concept of entropic forces consistently included in the present formulation. Considering constant temperature, Eq. leads basically to $m\mathbf{a} = m\mathbf{g} - T\nabla S$. But we should remind that irreversible transformations all occur over positive entropy transformations, $\frac{\partial S}{\partial z} > 0$, where z can be a distance, temperature, composition, or volume, for example. Considering that the gravitational force is conservative, this is one of the crucial points against the conception of emergent force.

If a macroscopic system is subject to an entropic force identified as the classical force under thermal equilibrium, it is legitimate to revert the argument and examine whether some microstates provide support for the existence of referenced kinds of forces. This question is illustrate subsequently.

Imagine a particle describing a circular motion at a distance r away from the origin of a central force. The number of configurations in space associated with r is $\Omega = 4\pi r^2$. The entropic force results to be

$$f_r = \frac{T}{2}(\nabla \mathbf{S})_r = \frac{k_B T}{r}. \tag{111}$$

Taking into account the Equipartition Theorem, $\frac{1}{2}k_B T = \frac{1}{2}mv_r^2$, when thermal equilibrium prevails. Hence, the known expression for centrifugal force is found

$$f_r = m\omega^2 r. \tag{112}$$

Notice that the term of mass was introduced via Equipartition theorem. From this the expression for the force becomes (unfortunately, the symbol for period and absolute temperature are identical)

$$f_r = m\left(\frac{4\pi^2}{T^2}\right) r = m\frac{4\pi^2 r^3}{T^2}\frac{1}{r^2}. \tag{113}$$

Recall that according to Kepler law, all the planets on the solar system have the same ratio $\frac{r^3}{T^2} = K$:

$$f_r = 4\pi^2 K \frac{m}{r^2}. \tag{114}$$

Instead, Newton wrote the above equation on the form

$$f_r = \left(\frac{4\pi^2 K}{M_S}\right)\frac{M_S m}{r^2} = G\frac{M_S m}{r^2} \tag{115}$$

defining $G = \frac{4\pi^2 K}{M_S}$ as the gravitational constant. Therefore, the gravitational force is retrieved, but in the present model on the ground of a different viewpoint, giving a new evidence to an old law of physics.

The arguments used above demand further development. First, the number of microstates Ω counts the number of times points on a volume of space in an higher space is projected on a boundary, a surface $\mathbb{S}^2$, forming a two-dimensional information structure, suggesting a projection of information into the cosmological horizon and consistent with the holographic principle. Second, the energy is introduced via a representative scale of the process, the Boltzmann constant k_B, which plays the role of a link between the geometrical representation of space $\mathbb{R}^3$ with the macroscopic physical world.

3.2 In the Realm of Electrochemistry: Electromotive Force and Cell Potential

With the introduction of the concepts of anode, cathode, electrode, electrolyte, and ion, in 1835 by Michael Faraday, it was possible to describe phenomena, and give explanation, in the field of electrochemistry.

The natural inclusion of entropy (or free energy) inside the fundamental equation of dynamics, open the door to treat a more general class of problems, particularly those related to the domain of electrochemistry. Taking Eq. 59 along a closed curve γ, with $\oint_\gamma (\mathbf{F}^{ext} \cdot d\mathbf{s}) = q\mathcal{F}\mathcal{E}$, and taking additionally, $\Delta G = -q\mathcal{F}\mathcal{E}$, according to the laws of electrolysis, and from Maxwell's equations $\Delta S = -(\frac{\partial \Delta G}{\partial T})_p$, then, from Eq. 31, we obtain the Helmholtz formula for the electromotive force of a battery (or cell potential $\mathcal{E}$):

$$\mathcal{E} = U + T\frac{\partial \mathcal{E}}{\partial T}. \tag{116}$$

The consistent integration of dynamics with thermodynamics permits addressing electrochemical processes via the modified Newton equation 59.

4 Conclusion

The usual assumption of point particles when performing calculations in the frame of classical mechanics hinders the correct understanding of motion since in real-world particles or bodies are extended objects, subject to deformation processes that alter their expected behavior. It was shown that from the gradient of rotational energy emerged the well-known pseudo-forces appearing in rotational moving frames. The out-of-equilibrium technique proposed in this Letter may provide a new instrument for a more in-depth analysis for the dynamics of complex systems, in particular the emergence of complex behaviors.

Acknowledgements The author gratefully acknowledge partial financial support by the International Space Science Institute (ISSI) in Bern, Switzerland, as visiting scientist, and express special thanks to Professor Roger-Maurice Bonnet, Dr. Edouard Mendès Pereira, Prof. Rafael Rodrigo, Dr. Maurizio Falanga, and Dr. Rui Ribeiro, for their friendship and support. We acknowledge kind permission to Springer Nature Group and Elsevier, to publish part of the material shown authored by myself and made public via Open Access.

References

1. Pinheiro MJ (2013) A variational method in out-of-equilibrium physical systems. Sci Rep 3. Article number: 3454. https://doi.org/10.1038/srep03454
2. Landau L, Lifchitz E (1967) Physique statistique. Editions MIR, Moscou, p 48

3. Jaynes ET (1957) Information theory and statistical mechanics. Phys Rev 106(4):620–630
4. Pinheiro MJ (2002) Europhys Lett 57:305
5. Pinheiro MJ (2004) Phys Scr 70(2–3):86
6. Peterson MA (1979) Analogy between thermodynamics and mechanics. Am J Phys 47(6):488
7. Pavlov VP, Sergeev VM (2008) Thermodynamics from the differential geometry standpoint. Theor Math Phys 157(1):1484
8. Lavende BH (1974) Principles and representations of nonequilibrium thermodynamics. Phys Rev A 9(2):929
9. Glansdorff P, Prigogine I (1971) Structure, stabilité et fluctuations. Masson Éditeurs, Paris
10. Young RH (1976) Am J Phys 44(6):581–586
11. Truesdell CA (1952) J Appl Math Phys (ZAMP) 3:79. https://doi.org/10.1007/BF02008449
12. Lamb H (1920) Higher mechanics. Cambridge University Press, London
13. Curtiss CF (1956) Kinetic theory of nonspherical molecules. J Chem Phys 24(2):225–241
14. Khinchin AI (1949) Mathematical foundations of statistical mechanics. Dover Publications, New York
15. Liboff RL (1990) Kinetic theory: classical, quantum, and relativistic descriptions. Prentice-Hall, New Jersey
16. Huang K (1987) Statistical mechanics. Wiley, Hoboken
17. Gary SP (1993) Theory of space plasma microinstabilities. Cambridge University Press, Cambridge
18. Stacey WM (2005) Fusion plasma physics. Wiley-VCH, Weinheim
19. Kiehn RM (2007) Topological torsion and topological spin. Ann Fond L de Broglie bf 32:389–408
20. Onoochin V, Phipps TE Jr (2003) On an additional magnetic force present in a system of coaxial solenoids. Eur J Elect Phen 3:256
21. Altgilbers LL et al (2000) Magnetocumulative generators. Springer, New York
22. Steinhauer LC, Ishida A (1997) Phys Rev Lett 79(18):3423
23. Klajn A (1999) Switching vacuum arc in a pulsed transverse magnetic field. IEEE Trans Plasma Sci 27(4):977
24. Flurscheim CH (ed) (1985) Power circuit breaker: theory and, design. Peter Peregrinus, London
25. Boxman RL, Martin PJ, Sanders D (1995) Handbook of vacuum arc science and technology: fundamentals and applications. Noyes Publications, New Jersey
26. James DR (1962) An experimental examination of retrograde motion of an electric arc. PhD dissertation, Wright-Patterson Air Force Base, Ohio
27. Younian W, Tengcai M (1990) Investigation of electric conductivity of plasma in magnetic rotating arc. In: Solonenko OP, Fedorchenko AI (eds) Plasma jets in the development of new materials technology, VSP BV, The Netherlands, pp 27–32
28. Boóirevich V, Freibergs Ya, Shilova EI, Shcherbinin EV (1989) Electrically induced vortical flows. Kluwer Academic Publishsers, Dordrecht
29. Chow KC, Chan KL (2003) Angular momentum transports by moving spiral waves. J Atmos Sci 60:2004–2009
30. Schubert G, Whitehead JA (1969) Moving flame experiment with liquid mercury: possible implications for the Venus atmosphere. Science, New Series 163:71–72
31. Chandrasekar S (1989) Selected papers, vol 4. In: Plasma physics, hydrodynamic and hydromagnetic stability, and applications of the tensor-virial theorem. University of Chicago Press, Chicago
32. Douglas HA, Mason PJ, Hinch EJ (1972) Motion due to a moving internal heat source. J Fluid Mech 54:469–480
33. Stern ME (1971) Generalizations of the rotating flame effect. Tellus XXIII:122–128
34. de Ruiter AH, Dameren C, Forbes JR (2013) Spacecraft dynamics and control: an introduction. Wiley, West Sussex
35. Anderson JD, Campbell JK, Ekelund JE, Ellis J, Jordan JF (2008) Phys Rev Lett 100:091102
36. Verlinde EP (2011) On the origin of gravity and the laws of Newton. JHEP 1104:029 [arXiv:1001.0785 [hep-th]]
37. Kobakhidze A (2011) Once more: gravity is not an entropic force. arXiv:1108.4161

Chapter 4
Minimization of Thermal Conductivity in Nanostructures and Geometric Self-Similar Structures for Thermoelectric Applications

Shiva Kumar Singh and Marcos A. Avila

1 Introduction

The worldwide energy matrix has been shifting toward green and renewable energy sources, and thermoelectric (TE) materials can contribute to the future matrix significantly. For such, there is a pressing demand for the next generation of TE materials which may enable the production of energetically and commercially competitive devices. TE materials are basically those capable of transforming heat into useful electric energy through the Seebeck effect or reversing heat flow by applying a voltage through the Peltier effect. TE devices can harvest waste heat from the Sun, industrial plants, automobiles, and even from our bodies, converting part of the harvested heat into useful electricity [1–3]. In addition, TE devices have many advantages over other energy sources, such as excellent scalability, no moving parts, good reliability, fast response, and low noise [1–3]. A primary goal of research and development in TE materials is the enhancement of their efficiency in heat ↔ electricity conversion, quantified by the material's figure of merit $ZT = \sigma S^2 T/(\kappa_e + \kappa_L)$, a dimensionless parameter, where S, σ, $\kappa_{e,}$ and κ_L are the Seebeck coefficient, electrical conductivity, electronic thermal conductivity, and lattice thermal conductivity, respectively, which are all temperature (T) dependent quantities. ZT should be as large as possible and values close to unity are already considered potentially viable. A good TE material should have a large S that is usually present in semiconductors, high σ similar to metals, and at the same time poor κ_L as in glasses. In its most concise representation, the ideal TE material may be identified as a "phonon-glass, electron-crystal" (PGEC) as proposed by Slack [4]. State-of-the-art TE materials exhibit a peak ZT value close

S. K. Singh (✉) · M. A. Avila
CCNH, Universidade Federal do ABC (UFABC), Santo André, SP 09210-580, Brazil
e-mail: s.kumar@ufabc.edu.br

M. A. Avila
e-mail: avila@ufabc.edu.br

A. Bandyopadhyay and K. Ray (eds.), *Rhythmic Advantages in Big Data and Machine Learning* , Studies in Rhythm Engineering,
https://doi.org/10.1007/978-981-16-5723-8_4

to 2.5. A material with $ZT = 4$ is expected to become competitive with fossil fuel engines in terms of efficiency.

The rational strategies for the enhancement of ZT can be divided into two subgroups: (i) enhancement of the power factor ($S^2\sigma$); (ii) reduction of the lattice thermal conductivity. The main hurdle in improving their performance is that the three physical quantities S, σ, and κ_e are interdependent solid-state quantities derived from the electronic free energy. Therefore, optimization of one often downgrades another. However, decoupling of the electronic and phonon sub-systems is possible through various methods based on theoretical ideas relating to nanostructuring size effects [5–13] and band-structure engineering [7, 14–19]. Theoretically, the κ_L of nanostructured materials can be greatly reduced compared to their bulk counterparts due to the increased scattering of phonons at the nanostructure surfaces. A detailed understanding of the underlying transport processes (electronic and thermal) is crucial for TE materials and their technological applications. In this chapter, we will focus on the manipulation of thermal transport through various mechanisms of nanostructuring and material structure modifications. Thermal conduction in bulk materials and nanostructures can be completely different, since heat conduction on these length scales significantly deviates from diffusive predictions [20–23]. Thermal transport shifts between diffusive, fractional superdiffusive, and ballistic as length scales change from bulk to nanoscale dimensions. Also, phonon confinement is predicted theoretically in superlattice structures [24], which have shown that κ_L depends upon a new kind of Umklapp scattering process, called a mini-Umklapp, associated with the mini-Brillouin zone of the superlattice. We will thus discuss thermal transport (diffusive, fractional superdiffusive and quantum confined) in bulk and nanostructured materials.

The structure of this chapter is as follows: we first present the research focus towards describing thermal transport in bulk and nanostructured materials. Understanding the phonon transport in materials is the basis of controlling their thermal conductivity. We will first discuss the thermal transport in bulk materials and then in nanostructures. The general mechanism of the reduction is explained from the viewpoint of phonon gas kinetics. In the second part, we will discuss self-similar fractal structuring in materials to control the thermal conductivity. With the great progress recently achieved in nanofabrication techniques, self-similar (fractal) structures formed by nanocrystal self-assemblies have been realized [25]. Thus, thermal management using fractal structures may be useful in designing next-generation TE materials and devices.

1.1 Thermal Transport In Bulk and Nanostructures

In this section, we will cover the basic thermal transport mechanism at bulk- and nanoscale in crystalline solid materials. Theoretical research [26–42] has provided great inputs toward understanding this type of transport. Both electrons and phonons propagate heat in crystalline materials, and the total thermal conductivity is a sum of the electronic and lattices thermal conductivity ($\kappa_T = \kappa_e + \kappa_L$). We cannot limit κ_e separately from σ without resorting to violations of the Wiedemann–Franz law, but by minimizing κ_L, ZT has been found to improve significantly. In fact, κ_L is the only independent parameter (in principle) of the TE figure of merit. In crystalline solids, thermal energy combined with the interaction between atoms creates vibration around their equilibrium positions, leading to quantized vibrational waves with various frequencies. The quanta of these waves are termed as phonons, which are characterized by their angular frequency of vibration (ω_{ph}), propagation wavevector (k_{ph}), and longitudinal or transversal polarization. In solid materials, phonons can exist in a wide range of frequencies ω_{ph} determined by the dispersion relation $\omega_{ph}(k_{ph})$. Such phonons are the main factors responsible for heat transport in non-metallic crystalline materials [26]. Various models for phonon dispersion relations $\omega_{ph}(k_{ph})$ have been proposed and we shall discuss some of them here.

Early in 1959, the Callaway model [32] was proposed for calculating κ_L at low temperatures for bulk materials. Callaway assumed an isotropic bulk material with no dispersion in the crystal vibrational spectrum. He further assumed the phonon scattering process could be represented by a frequency-related relaxation time τ_{ph}. Furthermore, identical contributions from both longitudinal and transverse phonons were assumed. Four phonon scattering mechanisms: (i) normal three-phonon processes (N-processes); (ii) Umklapp processes (U-processes); (iii) point impurities; and (iv) boundary scattering were considered to be involved in the heat conduction. Based on the above analysis, Allen [33] revisited the Callaway model and made further corrections and simplifications. The relative roles of the N- and U-processes were reanalyzed on the basis of the Debye-type phonon model. He postulated that Callaway underestimated the suppression of the N-processes in relaxing thermal current. A new equation by elongating the relaxation time was proposed for *thermal conductivity*.

A few years later, a new theoretical model for *lattice thermal conductivity* κ_L of crystalline materials was proposed by Holland [34], which considers the effect of both longitudinal and transverse phonons on heat conduction. Good agreement between experimental data and the calculated κ_L by this model has been found for silicon in the temperature range of 1.7–1300 K [34]. However, Slack in 1964 [36] first pointed out that possible intrinsic troubles with the *Callaway model* might arise from the neglect of dispersion in the phonon spectrum, along with the other corrections considered by Allen and Holland. Dames and Chen [31] considered the scattering of phonons through Born-von Karman dispersion theory [42] and proposed a Born-von Karman-Slack (BvKS) model of thermal conductivity [31]. The BvKS dispersion theory includes a reduced wave vector group velocity close to the first Brillouin zone boundary. In this model, for the simple case of a 1D single-atom harmonic chain with atomic mass M, the phonon dispersion relation is given by [36, 42]:

$$\omega_{ph}(k_{ph}) = 2\left(\sqrt{F/M}\right)\sin\left(\frac{\pi}{2}\frac{k_{ph}}{k_c}\right) \tag{1}$$

where F and k_c are the force constant and cut-off phonon wave vector, respectively.

The heat transport under a thermal gradient is an aggregate behavior of phonons generated within the material at each spatial location, moving along all directions. The extent of heat conduction depends on the three fundamental phonon transport properties; energy of phonons ($\hbar\omega_{ph}$), their group velocity $v_{ph,}$ and phonon mean free path l_{ph}. All three quantities are k_{ph}-dependent phonon transport properties and their statistical average result is a net transport of thermal energy from the hot to the cold end of the material, which is quantified by the lattice thermal conductivity κ_L. Based on the kinetic theory and Boltzmann transport equation and neglecting any dependence of the phonon relaxation times on phonon energy ("*gray* model" approximation), the κ_L of an isotropic bulk material can be expressed as

$$\kappa_L = {}^1\!/_3 C v_{ph} l_{ph} \tag{2}$$

where 'C' is the specific heat of the material. The lattice conductivity κ_L can also be manipulated by engineering these transport properties. Based on the above theoretical concepts, Yang and Dames [26] calculated the κ_L of silicon by six different models. The calculation results of the BvKS model are the most consistent with experimental data [26]. This suggests that the BvKS model is more accurate and superior for studies involving phonon boundary scattering compared to the Holland model, the modified Callaway model, and the *gray* model.

Next, we will discuss how these phonon transport properties can be affected and manipulated in bulk materials.

1.1.1 Engineering Thermal Transport in Bulk Materials

In bulk materials, heat-carrying phonons are scattered via different volumetric processes (also known as internal scattering processes), which include the scattering of phonons by other phonons, electrons, holes, impurity atoms, and isotopes of the constituent atom of the crystal lattice [43]. The strength of internal scattering processes is determined most prominently by l_{ph} and, thus, the κ_L of bulk materials [44, 45]. In these scattering processes, the phonon momentum ($\hbar k_{ph}$) and its ability to carry heat are altered. In isotopically pure bulk materials, thermal transport is mostly diffusive, since phonons are internally scattered by other phonons [46]. During these scattering events, the energy is conserved, while k_{ph} (or momentum $\hbar k_{ph}$) may not be conserved. Those phonon–phonon scattering processes in which momentum is conserved are called N-processes and lead to almost resistance-less thermal transport. While phonon–phonon scattering processes that do not conserve momentum are called U-processes. They are responsible for the resistance to phonon thermal transport and lead to finite values of l_{ph} and κ_L in bulk materials [47].

From the phonon dispersion relation (Eq. 1), it is clear that factors such as the elastic force constants (F), the atomic mass M, and k_{ph} control the phonon frequency. The relaxation time (τ_{ph}), depends on phonon frequency [47] which controls κ_L. The dependence of κ_L on the τ_{ph} is given by rewriting Eq. 2

$$\kappa_L = {}^1\!/_3 C v_s l_{ph} = {}^1\!/_3 C v_s^2 \tau_{ph} \tag{3}$$

Thus these parameters (F, M, and k_{ph}) control the phonon–phonon scattering rate (${\tau_{ph}}^{-1}$), which is another intrinsic bulk material property. ${\tau_{ph}}^{-1}$ is also temperature dependent, since the number of phonons (or phonon occupation numbers) varies with temperature following Bose–Einstein statistics. At low temperatures, ${\tau_{ph}}^{-1}$ is very low due to the small number of phonon states being occupied (small number of active phonons). In this case, l_{ph} is relatively large and phonon transport is primarily controlled by phonon scattering processes occurring at the grain boundaries of the bulk material. Thus, low temperature κ_L follows the evolution of the materials specific heat C which is proportional to T^3 [43, 46]. At high temperatures, when nearly all phonon states are active, ${\tau_{ph}}^{-1}$ becomes high due to a large number of phonons. In this case, l_{ph} is relatively small and thermal conduction is primarily controlled by ${\tau_{ph}}^{-1}$. Thus, with increasing temperature, the thermal conductivity decreases proportionally to T^{-1}.

As discussed above, phonon transport can be manipulated through changes in M or F: materials containing atoms with large M and small F tend to show a low κ_L. Various mechanisms have been explored to reduce κ_L by changing phonon dispersion, through changing M or F. For example, atomic point defects (formed by elemental substitution, removal, or insertion) introduce modifications in F through lattice strains and/or change in atomic mass M, and are effective for scattering short wavelength phonons. Many recent reports on thermoelectric materials [48–53] have used the above-mentioned intrinsic and extrinsic approaches to achieve low κ_L by tuning M or F. For example, when M is constant, anharmonic lattice vibrations correspond to an anharmonicity in F at non-equilibrium atomic positions, which work as *dynamic* strain [48]. This leads to fluctuations (broadening) in ω_{ph} at a given k_{ph}. Such broadening accelerates the change rate of the phonon distribution function, which therefore enables a shorter relaxation time for phonons [49]. Using static lattice strains, remarkable κ_L reduction has been achieved in $Na_{0.03}Eu_{0.03}Sn_{0.02}Pb_{0.92}Te$ and, consequently, a high *ZT* was realized [48]. Fattening in phonon dispersion has been seen in crystalline solids with weakly bonded heavy constituent elements, leading to a shorter relaxation time for phonons and low κ_L [50]. Similarly, both liquid-like ions (weak chemical bonds) and low sound velocity [$v_s \propto \sqrt{F/M}$] indicate a small force constant [51, 52]. In SnTe, alloying Ge (0–30 mol%) induces local structural distortions. The associated ferroelectric lattice instability was shown to induce soft polar phonons, which effectively scatter the heat-carrying acoustic phonons and help achieve ultralow κ_L in cubic SnTe by engineering the instability near room temperature [53].

Phonon scattering rates (${\tau_{ph}}^{-1}$) as a function of phonon frequencies corresponding to various scattering mechanisms (point defects, rattling in cages, dislocations, interfaces, nano-precipitates, etc.) are shown in Fig. 1 [24]. For example, atomic point

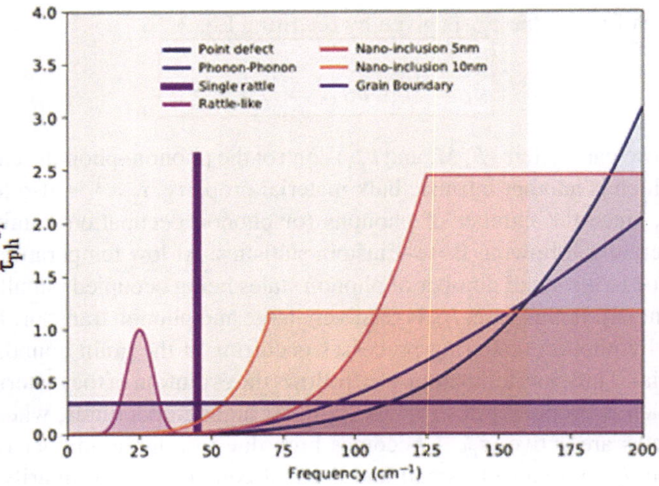

Fig. 1 Schematic of the frequency dependence of phonon scattering rates for various scattering mechanisms [reproduced from Ref. [24] with permission under a Creative Commons Attribution license]

defects (formed by elemental substitution, removal, or insertion) introduce modifications in F through lattice strains and/or change in atomic mass M, and are effective for scattering short wavelength phonons. From Fig. 1, it is also clear that nano-size precipitates or inclusions can significantly scatter medium-wavelength phonons. Mesoscaled grains can scatter all range wavelength phonons, but are more effective for long-wavelength phonons [8, 54, 55]. The dependence of the heat transfer on the microstructure and length scale of the materials will be discussed next.

The Knudsen number, defined as the ratio l_{ph}/L, determines the character of heat transfer at various length scales. For bulk samples, l_{ph} is much smaller than its single grain size L. For these materials when $l_{ph}/L < 10^{-3}$ is satisfied in all three dimensions, phonons are scattered by other phonons before reaching the grain boundary, as shown in Fig. 2a. In this case, the classical diffusive heat transfer occurs as determined by the Fourier law. In the case of nanostructures, L represents their corresponding characteristic sizes, for example, length or diameter of nanotubes, diameter of nanowires, and thickness of films. If the ratio is of the order of the Casimir limit $l_{ph}/L > 10^2$ [56], then almost all phonons transport heat, without scattering, between the boundaries of material (Fig. 2b). This is referred to as a ballistic mechanism of heat transfer.

Between the above limiting cases (i.e., for l_{ph}/L of order 10^{-2} to 10^1), the material is considered to be in a diffusive-ballistic (quasi-ballistic or superdiffusive) regime [57, 58]. In this range, violation of classical theory occurs and thermal conductivity depends on the sample size [56]. The actual heat conduction in these materials is a mixture of diffusive and ballistic phonon transport. In this case, the thermal conduction cannot be explained within the Fourier diffusion-transfer theory and a statistical theory of heat conduction of solids is required [59, 60]. In this context, when the

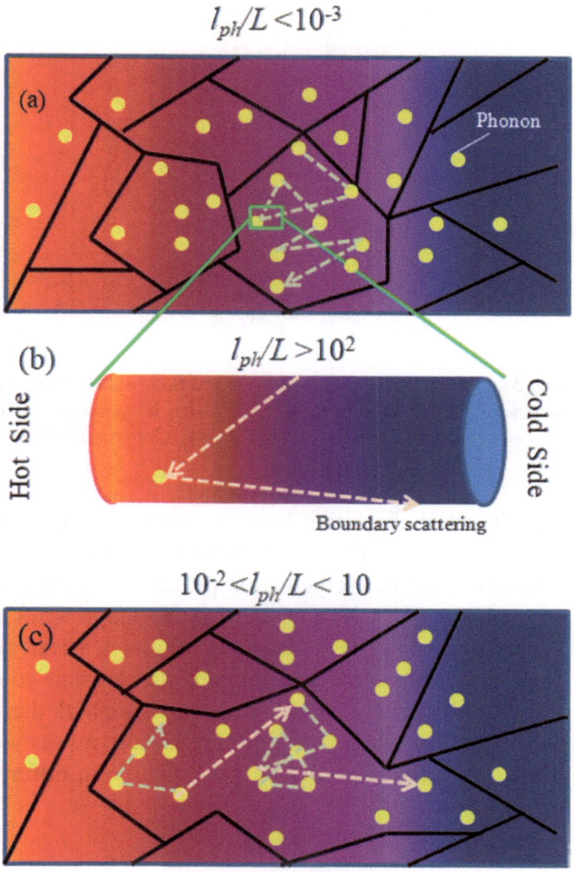

Fig. 2 Schematic of thermal transport at different length scales in **a** bulk grain, **b** thin nanostructures (nanowires/tubes), and **c** thick nanowires, nanotubes and nanostructured bulk materials

thermal conductivity is a function of *L*, heat transfer is often referred to as quasi-ballistic. Such quasi-ballistic dynamics of phonon scattering have been reported in semiconductors [61] and single-walled carbon nanotubes [62, 63].

In the quasi-ballistic regime, heat conduction governed by so-called fractal Lévy transport has been reported in $(ErAs)_x$:InGaAlAs [64]. Figure 3 presents the schematics of quasi-ballistic microscale thermal transport in semiconductor alloys with fractal dimension $1 < D < 2$. The quasi-ballistic heat conduction is found to be virtually independent of the Er content $0.001 < x < 0.1$ and, instead, controlled by alloy scattering of the host matrix. Earlier *ab-initio* BTE analyses on semiconductor alloys have shown that the Fourier theory of diffusion heat transfer breaks down in the quasi-ballistic regime [61]. The random motion of thermal energy is no longer Brownian, but instead governed by Lévy transport. Characteristic random walk patterns with fractal spatial dimension ($1 < D < 2$) induced by fractional superdiffusion processes have already been reported [65]. Lévy parameters provide valuable microscopic metrics about the phonon population [66]. The Lévy parameter can be extracted

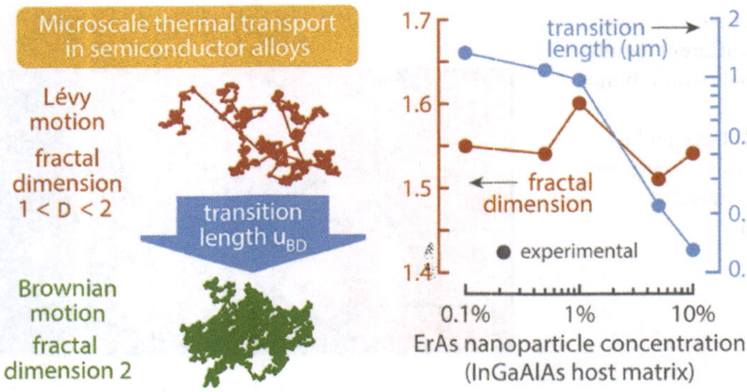

Fig. 3 **a** Schematic of quasi-ballistic and Brownian microscale thermal heat transport. **b** Variation of fractal dimension and ballistic-diffusive length scale (uBD) with nanoparticle concentration [Reprinted with permission from Ref. [64]. Copyright (2015) American Chemical Society]

experimentally by fitting laser thermo reflectance data. The characteristic ballistic-diffusive length scale (u_{BD}), which may be tuned by nanoparticle concentration [64], determines the heat conduction process. The effect on the Lévy characteristics of thermal transport with the inclusion of nanoparticles in an alloy host matrix has been studied in $(ErAs)_x$:InGaAlAs [64]. Although a nearly constant fractal dimension $D\approx$ 1.55 was observed upon varying the nanoparticle concentration from 0.1% to 10%, u_{BD} dropped from ~ 1 µm to ~ 100 nm and bulk thermal conductivity was found to be reduced by nearly threefold. From these results we can conclude that the nanostructuring manipulates thermal conduction in different ways, which we discuss in the next section.

1.1.2 Engineering Thermal Transport in Nanostructures

We now cover in detail the scattering of phonons in solid nanostructured materials. These materials can be roughly classified into four groups according to their dimension [35, 67, 68]: (i) zero-dimensional (0D) nanostructures have all the dimensions within the nanoscale (i.e., all dimensions are smaller than 100 nm). Nanoparticles, nanocrystals, quantum dots, and fullerenes are an example of 0D nanostructures;(ii) One-dimensional (1D) nanostructures, which have only one dimension at the nanoscale, such as nanowires (NWs), nanofibers, and nanotubes; (iii) Two-dimensional (2D) nanostructures, which have two dimensions at the nanoscale, such as graphene nanosheets, and nanofilms; and (iv) Three-dimensional (3D) nanomaterials that are not confined to the nanoscale in any dimension. The latter class can include nanoparticle-embedded bulk systems, dispersions of nanoparticles, nanocomposites, bundles of nanowires, and nanotubes as well as multi-nanolayers [67, 68]. Figure 4 presents a schematic of the classification of these nanostructured materials.

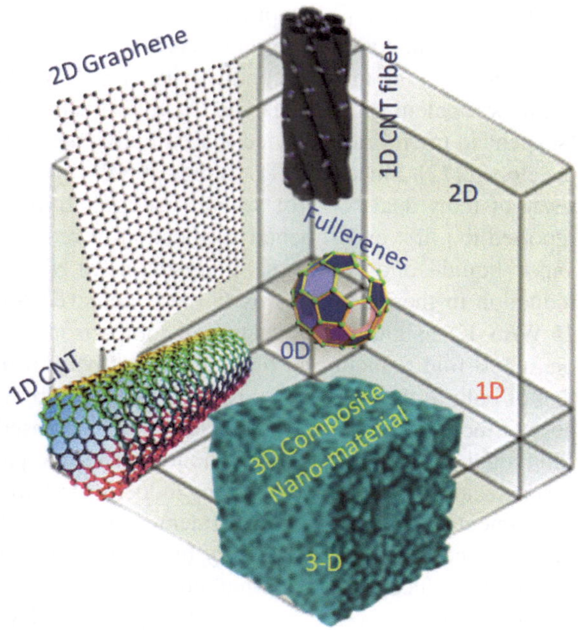

Fig. 4 Schematic representations of nanostructured materials with different dimensions a part of figure is reprinted with permission from Applied Thermal Engineering Vol 130, 2018, Page 1004–1011, Copyright (2018) and Ref. [35] Copyright (2021), with permission from Elsevier

In addition to internal scattering processes in such materials, phonons can also be scattered by the nanostructures of size comparable to l_{ph}. The reduced dimensions of nanostructures increase the possibility of scattering of phonons at the interface, which also has a significant effect on κ_L. The interface scattering critically depends upon the size of the phonon l_{ph} of the material. Significant phonon scattering by interfaces is expected when the phonon l_{ph} is comparable to the characteristic length of interfaces. However, the interface scattering is not very effective for materials with very short l_{ph} and having intrinsically low thermal conductivity [55]. Similarly, in NWs phonons can also be scattered by the nanowire boundaries when l_{ph} is larger than the NW size and phonons can travel far enough to interact with the NW surface (Fig. 2b). This is termed as a boundary scattering process of phonons and, notably, a discrete event occurring only at the surfaces, whereas internal scattering is a volumetric scattering event [43]. Thus, due to several additional scattering processes, κ_L of nanostructure-embedded bulk materials and nanowires is expected to be lower than the corresponding bulk material. Manipulation of both internal scattering and boundary scattering processes of phonons leads to the modulation of thermal transport in nanostructured materials. The interface effects on thermal conductivity of superlattice and periodic thin-film structures was investigated by Chen [69]. The atomic-scale interface roughness was found to be a major cause for the measured reduction in thermal conductivity. Different interface conditions including specular, diffuse, and partially specular, and partially diffuse interfaces were considered. Results obtained from the latter two interface scattering models showed good agreement with experimental data of GaAs/AlAs superlattices.

Walkauskas et al. [70] applied a Boltzmann-transport-based model to predict the reduced thermal conductivity of GaAs NWs with respect to bulk materials. Around the same time, the thermal conductivity of Si NWs with very small diameters (d ~ 5 nm) was calculated by Volz and Chen [71] using molecular dynamics. In 2003, a 'suspended microbridge device'-based individual NW measurement technique was developed [72]. Since then, TE studies of NWs have flourished and *thermal conductivity* of individual NWs of several materials have been evaluated. Li et al. [73] reported the first experimental verification of reduction in thermal conductivity in vapor–liquid–solid (VLS) grown crystalline Si NWs. At 300 K, a dramatic tenfold reduction in thermal conductivities was achieved for NWs with d ~ 30 nm (κ ~ 15 W/m–K) a significant reduction in comparison Si bulk ($\kappa = 156$ W/m–K). Later on, a 100-fold reduction in thermal conductivity of individual rough Si NWs than that of bulk Si was reported at room temperature by Hochbaum et al. [74]. Roughness of the VLS grown Si NW surfaces was increased with the electroless etching. This roughness severely limited phonon mean-free paths with the increased diffuse surface scattering, leading to a strong reduction in lattice thermal conductivity. These experimental findings were in agreement with the theoretical understanding that the frequency of phonon-boundary interactions increased with reducing diameters, leading to reduced lattice thermal conductivity in the NWs. Shi et al. reported the ZT of individual $Bi_{0.54}Te_{0.46}$ NWs [75] and $CrSi_2$ NWs [76]. However, these studies found *no significant* reduction in NW thermal conductivity with respect to their bulk counterparts. The probable reason behind this was the inherent shorter phonon mean-free paths (l_{ph}) of chalcogenide and silicide materials. This indicates that the size effect on thermal transport is not much effective on materials that possess inherently shorter l_{ph} [77–81].

Li et al. [73] also noted that for the thinnest sample (d ~ 22 nm) the change in the thermal conductivity with temperature was distinct from the thicker samples. For thin NWs at low temperatures (below 60 K), thermal conductivity deviated from the T^3 scaling and followed a linear variation with T. However, thick NWs of diameter around 100 nm followed the T^3 scaling. This is expected in thick NWs, as it mimics the heat capacity of the material [47]. The linear variation of thermal conductivity with temperature in thin NWs was speculated to arise due to quantum confinement effects. We will discuss the effect of quantum confinement of phonons briefly in the next section.

1.1.3 Quantum Confinement of Phonons and Thermal Transport

When phonon is confined spatially, it exhibits only certain allowed states with modified phonon density of states and group velocities $v_{ph}(k_{ph})$. Thus the phonon dispersion relation $\omega_{ph}(k_{ph})$ would change (see Eq. 1), however it requires the existence of sufficiently long lph to create standing waves [82]. Thus phonon quantum confinement is expected to play a strong role only in thin NWs because of shortening the lph caused by diffuse scattering of phonons at boundary. However, models are proposed [83–85] to account for reduction in κ_L by including modified density of states and/or group velocities in addition to lph reduction resulted by diffuse scattering of phonons. Another thing to note that, the coherent effects caused by phonon interference, may arise when phonons preserve their phase upon specular scattering with NW surfaces. These coherent effects would lead to modulation in phonon dispersion relation and change in phonon group velocities and density of states [82, 86].

Effects of phonon confinement were first theoretically studied by Ren and Dow [24] on superlattice structures. They have shown that lattice thermal conductivity depends upon a new kind of Umklapp scattering process, called a mini-Umklapp, associated with the mini-Brillouin zone of the superlattice. They considered a superlattice consisting of periodically alternating layers of materials with similar lattice constant but with different mass, leading to a mini-Brillouin zone in the direction of layering. The most significant reduction was seen in large samples with small superlattice periods and with large differences between the superlattice layers.

Later works [87, 88] investigated the effect of confinement of phonons in free-standing quantum wells and predicted a decrease of κ_L by an order of magnitude. Quantum confinement led to a strong modification of phonon group velocities. These modifications increased the phonon relaxation rates significantly and thus reduced κ_L. Hyldgaard et al. [89] investigated the phonon transport in superlattices. Phonon modes were confined due to total internal reflection and significantly reduced the average phonon group velocity. An order-of-magnitude reduction in the ratio of thermal conductivity with phonon relaxation time resulted from the acoustic mismatch. Other reports have also shown a reduction in κ_L due to quantum confinement [90–92].

Discussion and Conclusion

We can conclude that the majority of the lattice thermal conductivity modifications in nanostructured materials result from the presence of internal phonon–phonon scattering and phonon-boundary scattering. Due to the incoherent (diffuse) scattering of phonons at the nanostructure boundaries, l_{ph} are shortened and κ_L is reduced in comparison to bulk. Diffuse scattering of phonons at the interface of the core and shell layers of heterogeneous structures such as core–shell NWs [93] is one of the reasons for the suppression in thermal conductivities. Recent developments show that these structures have another potential avenue to influence the overall scattering. Above 300 K, lattice thermal conductivity in Ge-Si core–shell NWs is found to be lower than Ge NWs of similar diameters [94]. TE measurements of these core–shell-structured NWs suggest higher TE efficiencies than bulk Ge [95]. Molecular dynamics simulations suggested that thermal conductivity could be manipulated by modifying the vibrational density of states in the core–shell NWs [96, 97]. For Si-Ge core–shell NWs with a Si-core of square cross-Sect. (2.3 × 2.3 nm), the thermal conductivity was found to lie in the range of 3-9 W/m–K depending on the Ge shell thickness.

Equiatomic Bi-Sb alloy nanoparticles and with core–shell configuration (Fig. 5) were simulated by Shishulin et al. [98]. The dependences of phase composition and κ_L associated with the shape of nanoparticles were studied. The shape of a particle is simulated by a coefficient corresponding to the extent of deviation of the particle

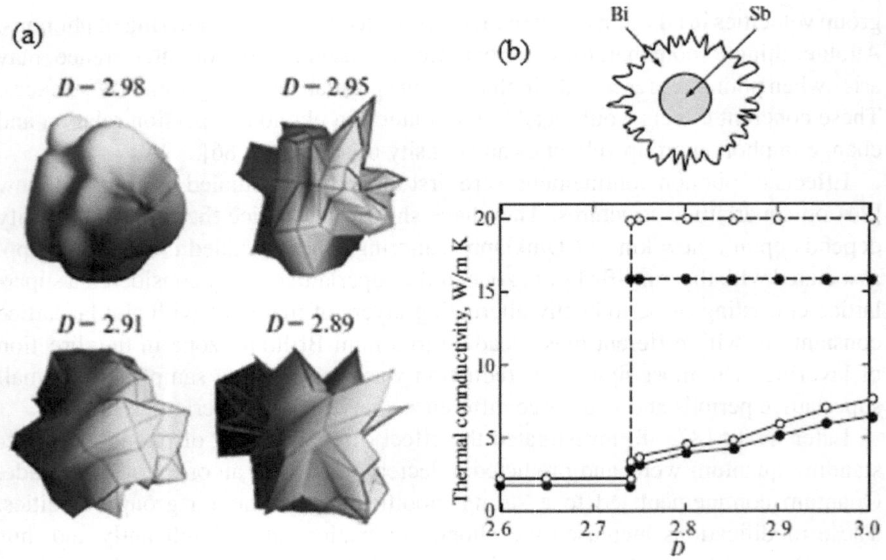

Fig. 5 **a** Structures of different fractal dimensions of Bi-Sb core–shell nanoparticles. **b** Jumpwise decrease in thermal conductivity for D < 2.74. Reprinted with permission of Ref. [98] and Springer Nature, Copyright (2019)

shape from spherical, i.e., in fractal dimension D ranging from 3 to 2.74 (Fig. 5a). They reported that for the configuration in which the core phase was formed by an Sb-based solid solution, a decrease in D from 3.00 to 2.74 reduced the lattice thermal conductivity of the shell phase by approximately 56% for two crystallographic directions (Fig. 5b). However, the thermal conductivity of the core phase was almost unchanged. For $D < 2.74$, the homogeneous state with 50 % Sb became a thermodynamic equilibrium state, which leads to a jumpwise decrease in thermal conductivity (Fig. 5b). The jump magnitude was found to be 71% as compared to the thermal conductivity of the shell phase for $D = 3.00$ and approximately 88% as compared to the thermal conductivity of the core phase. This indicates that fractal structures have another potential avenue to influence the overall scattering and lattice thermal conductivity. In the next section, we will discuss fractal self-similar structures and the possibility to manipulate thermal conductivity in such structures.

1.2 Fractal Structure and Thermal Conductivity

In this section, we will discuss the effect of introducing fractal self-similar structures in materials on their thermal conductivity. Let us first introduce a fractal entity. The measurement of an object/quantity is carried out in its particular dimension. Ordered objects such as points, lines, surfaces, and cubes/cuboids can be described by Euclidean geometry in general, using integer dimensions 0, 1, 2, and 3, respectively [99]. However, not all objects/quantities are regular and many of them are irregular and disordered in nature. These irregular entities with non-integral dimensions/values are called fractals. The value of π is an apt example of a fractal quantity. If D is the fractal dimension of an object, then the measure of a fractal object $M\ (L)$ is related to the length scale L by the following scaling law form [100] $M\ (L) \sim L^{D}$, where M can be any quantity such as the length of a line, the area of a surface, the volume of a three-dimensional object, the mass of an object, and so on.

Fractal structure formation in solid materials usually happens under conditions of high energy dissipation [101]. Under thermodynamic nonequilibrium conditions, atoms and molecules organize themselves into nanosize particles or clusters. This eventually leads to fractal aggregates with characteristic features of structural hierarchy, self-similar character, scale invariance, low density, and the presence of nano-dimensional porosity [101]. In fractal self-similar geometry, there are two kinds of self-similar sets: exactly self-similar fractals and statistically self-similar fractals. The Sierpinski carpet (SC), Sierpinski gasket, and Koch-curve are examples of self-similar fractals. Exact self-similar geometric fractals are rare in nature. However, real porous structures could be considered as statistically self-similar fractal structures. Several studies have shown that most of the natural porous structures and some synthetic porous materials are statistically self-similar fractal structures [102–105]. These self-similar fractal structures can be scaled with fractal scaling laws over multiple scales [102–105]. Two kinds of fractal models have been used, in which fractal scaling laws can be applied to characterize the geometrical and morphological structures, either for pore or solid phases in porous structures [105]. In the fractal-pore model the pore phase is fractal while the solid phase is non-fractal, and vice-versa in the case of fractal-solid model.

Several studies have shown a decrease in effective thermal conductivity of porous structures [99, 106]. The proposed thermal conductivity model is expressed as a function of porosity (related to stage n of Sierpinski carpet) [106]. The reduction in κ_L of porous materials is calculated by Tarkhanyan and Niarchos considering that κ_L varies according to Eq. 2. Four different models of thermoelectric materials are examined, including a model with randomly distributed hollow pores of varying diameters from nano-to micro-scale sizes [107]. In their model of porous thermoelectric materials with an all-scale hierarchical disorder, they have considered randomly distributed pore sizes from nano- to micro-meter scales. A model of a porous medium in the presence of two different groups of non-overlapping spherical hollow pores with various size scales was taken to calculate effective thermal conductivity. In this model, the material is composed of nanoscale pores with volume fraction ϕ_1 and randomly distributed pore diameters with average diameter d_1, plus microscale pores with volume fraction ϕ_2 and average diameter d_2. They assumed that the pores were distributed randomly (Fig. 6a) and that pore size distribution in each group is described by a probability density function.

In their earlier work [108] they showed that in "gray medium" approximation and assuming completely diffuse phonon scattering on pore/medium interfaces, the ratio of the effective κ_L for a porous medium ($\kappa_L{}^p$) to that in bulk medium with zero

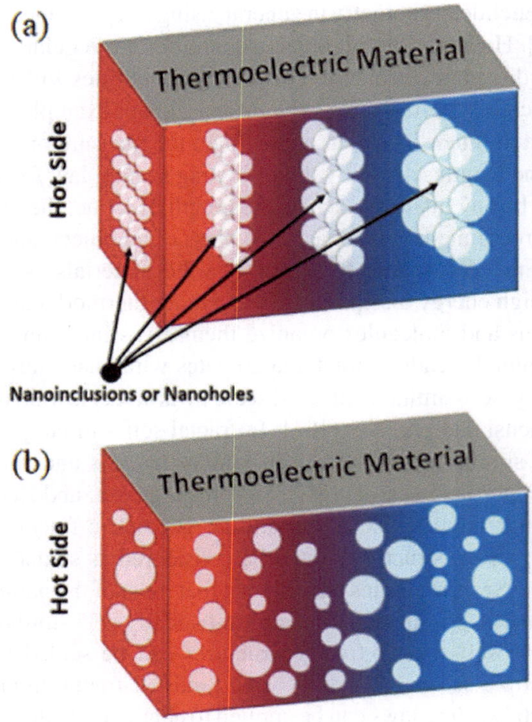

Fig. 6 **a** Model of porous composite with graded inhomogeneous porosity **b** Model of all-scale hierarchical fractal in porous thermoelctric material (Published with permission of Ref. [107] under a Creative Commons Attribution license

porosity ($\kappa_L{}^b$) is given by

$$\frac{\kappa_L^p}{\kappa_L^b} = \frac{1-\phi}{l_b - l_p} \tag{4}$$

where ϕ is the net porosity, l_b and l_p are the phonon mean-free paths in the bulk and porous medium, respectively. They have reported a decrease in thermal conductivity with the introduction of pores, which further decreases with pore size reduction. For example, the reduction is around twofold at $d_1 = 198$ nm and more than fourfold at $d_1 = 32$ nm. They have also shown for *a model of porous composite with graded inhomogeneous porosity* a kind of multi-scale hierarchical architecture structure (Fig. 6b) further reduction in thermal conductivity (see Table 1). With multi-scale hierarchical nano- and micro-sized inclusions, reduction in thermal conductivity has already been reported for thermoelectric materials [8]. These nano- and micro-size inclusions in materials represent a kind of *solid fractal model* where the pore is replaced with a solid phase. Further exploration of multi-scale hierarchical nanostructuring from the viewpoint of fractal (and/or self-similar) structures is needed.

With the great progress in the development of nanofabrication methods, nanocrystal self-assembly have become available as a means to make self-similar structures [25]. Also, a whole series of defect-free molecular fractals, specifically

Table 1 Theoretical calculations [107] show reduction in lattice thermal conductivity of porous material in comparison to bulk

	Porous medium with all-scale hierarchical disorder using $\phi = 35\%$, $d_2 = 0.8\ \mu m$ (fixed) and $l_p = 300$ nm	Porous composite with graded inhomogeneous porosity
$< d_1 >$ (nm)	$\kappa_L{}^p / \kappa_L{}^b$	$\kappa_L{}^p / \kappa_L{}^b$
198	0.488	0.433
81	0.383	0.314
68	0.358	0.288
32	0.246	0.183

Sierpiński triangles of aromatic-bromo compounds, were designed and fabricated on Ag(111) surfaces[109]. Thus, it will be possible to fabricate more sophisticated self-similar fractal structures in future. Theoretical calculations have indicated a reduction in thermal conductivity of several materials with such structures [110–113]. Using the molecular dynamics simulation method, the thermal transport properties of graphene with a self-similar SC structure have been investigated by Kang et al. [110]. The thermal conductivity of graphene decreased from 157.62 W/m–K to 19.60 W/m–K as the fractal number increased from 1 to 3. Han et al. [113] systematically studied thermal transport in isotopically doped WS_2 monolayers modified with Rectangle Carpet (RC) and SC structures using molecular dynamics simulations (Fig. 7). The effects of fractal number and temperature on κ_L were evaluated. For

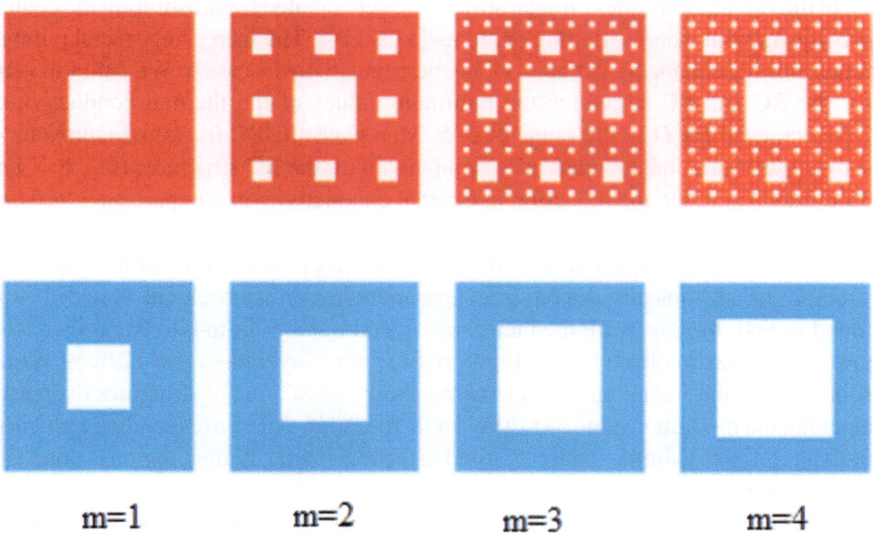

Fig. 7 Schematic representation of Sierpinski carpet and rectangle carpet of self similar fractal of different dimensions [Published with permission of Ref. 113] under a Creative Commons Attribution license

each fractal number, the SC fractal structures have shown lower thermal conductivities in comparison with the RC structures. Furthermore, an initial decreasing and then increasing trend of κ_L is observed with an increasing fractal number. Zhang et al. [112] introduced the SC fractal structure into Si/Ge nanocomposites and investigated the effect of SC fractal structures on their thermal conductivity. They found that the SC fractal structures could reduce κ_L due to the enhanced phonon scattering at the Si/Ge interfaces.

The fabrication of a self-similar structure is done with subdivision of a shape into smaller copies of itself then removing one or more copies, and proceeding recursively. For example, in the SC fractal structure, the construction of the SC begins with a square, then this square is further subdivided into smaller squares [100]. The SC fractal structure can tune the bandgap feature in phononic crystals to dominate the elastic wave propagation. The SC fractal structure can be generated as follows: the first step is to cut a square into nine identical sub-squares by trisecting each side of this square, and then the central sub-square is removed. Thus formed fractal structures have a fractal number (f_m) of 1 and are termed SC1. Afterward, the same trisection-divide-remove procedure is applied to the remaining 8 sub-squares to generate the SC2 structure. By applying this process successively the SC3, SC4 structures, and so on can be generated (Fig. 7). The self-similar fractal dimension (D) of the SC fractal structure is calculated by [$D = \log(b)/\log(a)$]. Here a is the number of parts of the evenly divided square side, and b is the ratio between the remaining area of the square being divided and the area of the small central square of the $f_m{}^{\text{th}}$ SC fractal structure, which is being removed. For example, the dimension of the Sierpinski Carpet SC1is given by $D = \log(8)/\log(3) = 1.893$.

In the calculations, these fractal structures were created by substituting the central sub-square with isotopic doping of atoms [113, 114]. The increasing fractal number represents the enhancing isotopic doping percentage. In the case of WS_2 monolayers, for the RC and SC structures the minimum values of the thermal conductivities were achieved for D in the range of 3–5. Moreover, the SC fractal structures have shown lower thermal conductivities compared with the RC structures (Fig. 8a). For each fractal number, the reduction in thermal conductivity in comparison to pristine (monolayer without fractal insertion) structures is around 20.80% and 29.94% for the RC and SC fractal, respectively. Similarly, the thermal conductivity of the SC fractal structure of (isotopically doped) graphene monolayers has been calculated. It was found to first decrease with the increasing fractal number from 0 to 6 and then tend toward stabilization from 6 to 8 (Fig. 8b). SC0 monolayer has $\kappa_L = 1228.59$ W/m–K and, when the fractal number increases, the κ_L of SC fractal structures decreases reaching the minimum value 585.20 W/m–K for $D = 6$. The corresponding reduction ratio is 52.37%. A similar declining trend in the low fractal number region is observed in other works.

Han et al.[111] further evaluated the thermal conductivity of these structures in the temperature range of 200 to 600 Kas depicted in Fig. 8c. With the increasing temperature, the thermal conductivities of pristine RC3 and SC3 WS_2 monolayers decrease monotonically. The corresponding reduction ratios are 49.39%, 34.13%, and 29.18%, respectively. Due to the increasing temperature, more high-frequency phonons are activated, and the Umklapp phonon scattering is enhanced. The thermal conductivity dependence with temperature is fitted using $\kappa \sim T^{-\beta}$, where the β exponents are 0.621, 0.397, and 0.313 for pristine, RC3, and SC3 WS_2 monolayers, respectively. The smaller value of the β exponent in the SC3 WS_2 monolayer indicates that some phonon scattering processes in the SC3 WS_2 monolayer possess a weaker dependence on temperature. In other words, in the SC WS_2 monolayer, because of the isotopic dopants, stronger phonon scattering is induced at lower temperatures and thus lower thermal conductivity is realized. This should lead to higher *ZT* in a large temperature range and thus higher ZT_{avg}.

Chakraborty et al. [115] calculated *ZT* for a Sierpinski Gasket (SPG) triangle made of 15 atoms fractal lattice, which is connected with two heat reservoirs kept at different temperatures as shown in Fig. 8d. Thermoelectric properties were calculated using Landauer's prescription in the framework of tight-binding model. Two kinds of hopping integrals, horizontal (t_x) and non-horizontal (t_y) were considered to examine the asymmetry effect on its thermoelectric behavior (for the details of integral function see Ref. [115]). A remarkably large value of $ZT > 30$ has been

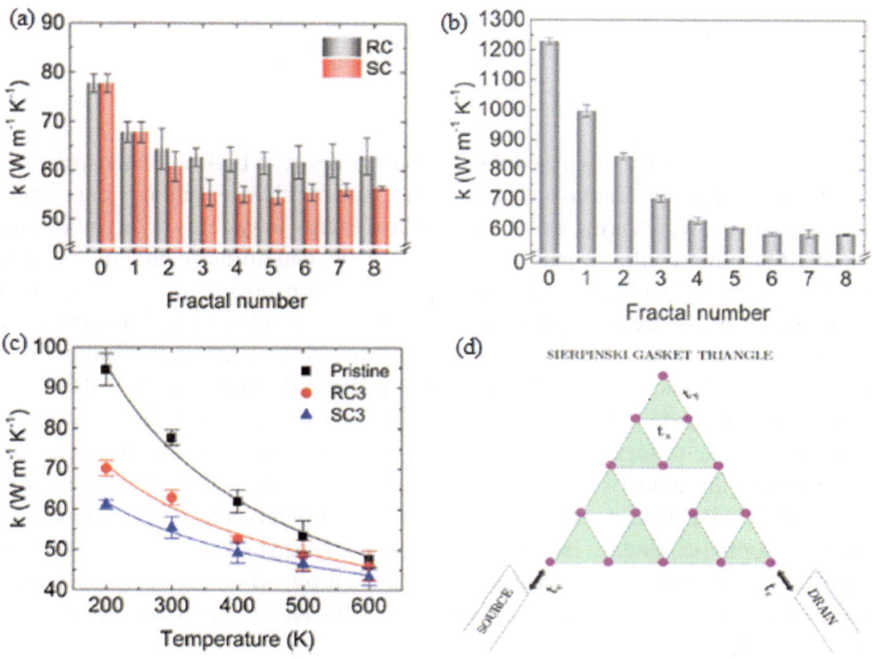

Fig. 8 **a** SC fractal structures have shown lower calculated thermal conductivities compared with the RC structures for each fractal number Ref. [113], **b** Calculated thermal conductivity of the graphene SC fractal structure monolayer fractal number from 0 to 8, **c** Calculated thermal conductivities of pristine, RC3 and SC3 WS2 monolayers in the temperature range 200 to 600 K Ref. [111], **d** Schematic of Sierpinski Gasket (SG) triangle made of 15 atoms connected with heat reservoirs Ref. [115]. Figures reprinted under a Creative Commons Attribution license with permission from Refs. [111, 113, 115]

reported for the $t_x > t_y$ asymmetric case. *ZT* results close to 2 for the symmetric case and its value is enhanced to 5 for $t_x < t_y$.

Conclusion

Overall, the introduction of fractal structures in materials is expected to provide an alternative to tuning the thermal conductivity and thus the thermoelectric properties. In bulk materials, synthetic porous structures are examples of statistically self-similar fractal structures. Silica aerogels, which can be prepared with porosities ranging typically from 80 to 99%, can have fractal structures extending over more than two orders of magnitude in length. This has been inferred from small-angle neutron scattering experiments [114]. Theoretical studies have shown minimization of thermal conductivity in self-similar fractal structured materials. Earlier, an almost tenfold higher thermo-emf of nanotubular fractal deposits as compared to that of polycrystalline carbon was reported [101]. However, rigorous experimental studies on more compounds and especially thermoelectric compounds are still awaited. Hopefully, advancements in nanofabrication and chemical methods may lead to the successful fabrication of fractal structures in other 2D materials in the future.

2 Conclusion

In this chapter, we have discussed the thermal transport in bulk and nanostructured materials. Altering the length-scale of material leads to changes in the phonon transport mechanism, which is the main carrier of heat energy in systems of thermoelectric interest. Quantum-confined, ballistic, quasi-ballistic, and diffusive phonon transport has been observed with changing length scales from nanostructures to bulk. With the presence of different length-scale mean free paths, hierarchical modifications in the material structure have led to drastic suppression of lattice thermal conductivity. Multi-scale inclusions with nano- and micro-sized phases, which is also a type of solid fractal modeling where a nano- and micro-size pore is filled with a solid phase, may come to serve the purpose. This is an open field and further studies are required from this point of view. Also, porous structures which are examples of statistically self-similar fractal structures have shown thermal conductivity suppression. However, for these systems, the electronic properties need to be optimized in order to improve their thermoelectric performance. Recent progress in nanofabrication techniques may lead to the fabrication of self-similar fractal structures for experimental studies and devices.

Acknowledgements This work is partially supported by FAPESP, grant numbers 2017/10581-1 and 2018/17384-0. We thank the authors of references [24, 35, 64, 98, 107, 110, 111, 113, and 115] for giving their consent to use figures from the corresponding papers.

References

1. Bell LE (2008) Science 321:1457. Vining CB (2009) An inconvenient truth about thermoelectrics. Nat Mater 8:83
2. Twaha S et al (2016) Renew Sustain Energy Rev 65:698
3. Snyder GJ et al (2008) Complex thermoelectric materials. Nat Mater 7:105
4. Slack GA (1995) CRC handbook of thermoelectrics. In: Rowe DM (ed), pp 407–440. CRC, Boca Raton
5. Hsu KF, Loo S, Guo F, Chen W, Dyck JS, Uher C, Hogan T, Polychroniadis E, Kanatzidis MG (2004) Cubic AgPbmSbTe2+m: bulk thermoelectric materials with high figure of merit. Science 303:818–821
6. Poudel B, Hao Q, Ma Y, Lan Y, Minnich A, Yu B, Yan X, Wang D, Muto A, Vashaee D et al (2008) High-thermoelectric performance of nanostructured bismuth antimony telluride bulk alloys. Science 320:634–638
7. Biswas K, He J, Zhang Q, Wang G, Uher C, Dravid VP, Kanatzidis MG (2011) Strained endotaxial nanostructures with high thermoelectric figure of merit. Nat Chem 3:160–166
8. Biswas K, He J, Blum ID, Wu C-I, Hogan TP, Seidman DN, Dravid VP, Kanatzidis MG (2012) High-performance bulk thermoelectrics with all-scale hierarchical architectures. Nature 489:414–418
9. Vineis CJ, Shakouri A, Majumdar A, Kanatzidis MG (2010) Nanostructured thermoelectrics: big efficiency gains from small features. Adv Mater 22:3970−3980
10. Hicks L, Dresselhaus M (1993) Effect of quantum-well structures on the thermoelectric figure of merit. Phys Rev B Condens Matter Mater Phys 47:12727−12731
11. Dresselhaus MS, Chen G, Tang MY, Yang R, Lee H, Wang D, Ren Z, Fleurial JP, Gogna P (2007) New directions for low-dimensional thermoelectric materials. Adv Mater 19:1043-1053
12. Hicks L, Dresselhaus M (1993) Thermoelectric figure of merit of a one-dimensional conductor. Phys Rev B Condens Matter Mater Phys 47:16631−16634
13. Kim SI, Lee KH, Mun HA, Kim HS, Hwang SW, Roh JW, Yang DJ, Shin WH, Li XS, Lee YH et al (2015) Dense dislocation arrays embedded in grain boundaries for high- performance bulk thermoelectrics. Science 348:109–114
14. Heremans JP, Jovovic V, Toberer ES, Saramat A, Kurosaki K, Charoenphakdee A, Yamanaka S, Snyder GJ (2008) Enhancement of thermoelectric efficiency in PbTe by distortion of the electronic density of states. Science 321:554–557
15. Liu W, Tan X, Yin K, Liu H, Tang X, Shi J, Zhang Q, Uher C (2012) Convergence of conduction bands as a means of enhancing thermoelectric performance of N-Type $Mg_2Si_{1-x}Sn_x$ solid solutions. Phys Rev Lett 108:166601
16. Pei Y, Shi X, LaLonde A, Wang H, Chen L, Snyder GJ (2011) Convergence of electronic bands for high performance bulk thermoelectrics. Nature 473:66–69
17. Pei Y, Wang H, Snyder G (2012) Band engineering of thermoelectric materials. Adv Mater 24:6125–6135
18. Zhao L-D, He J, Hao S, Wu C-I, Hogan TP, Wolverton C, Dravid VP, Kanatzidis MG (2012) Raising the thermoelectric performance of P-Type PbS with endotaxial nanostructuring and valence-band offset engineering Using CdS and ZnS. J Am Chem Soc 134:16327–16336
19. Zhao L-D, Hao S, Lo S-H, Wu C-I, Zhou X, Lee Y, Li H, Biswas K, Hogan TP, Uher C et al (2013) High thermoelectric performance via hierarchical compositionally alloyed nanostructures. J Am Chem Soc 135:7364–7370
20. Liang L, Li B (2006) Phys Rev B 73:153303
21. Cai W, Moore AL, Zhu Y, Li X, Chen S, Shi L, Ruoff RS (2010) Nano Lett 10:1645–1651
22. Minnich AJ, Chen G, Mansoor S, Yilbas BS (2011) Phys Rev B 84:235207
23. Minnich A (2012) J Phys Rev Lett 109:205901
24. Ren SY, Dow JD (1982) Thermal conductivity of superlattices. Phys Rev B 25:3750–3755
25. Shang J, Wang Y, Chen M, Dai J, Zhou X, Kuttner J, Hilt G, Shao X, Gottfried J, Mand, Wu K (2015) Assembling molecular Sierpiński triangle fractals. Nat Chem 7:389

26. Yang F, Dames C (2013) Mean free path spectra as a tool to understand thermal conductivity in bulk and nanostructures. Phys Rev B 87:035437
27. Chen G (2005) Nanoscale energy transport and conversion: a parallel treatment of electrons, molecules, phonons, and photons. Oxford University Press, New York
28. Nakagawa J, Kage Y, Hori T, Shiomi J, Nomura M (2015) Crystal structure dependent thermal conductivity in two-dimensional phononic crystal nanostructures. Appl Phys Lett 107:023104
29. Wang Z, Alaniz JE, Jang W, Garay JE, Dames C (2011) Thermal conductivity of nanocrystalline silicon: importance of grain size and frequency-dependent mean free paths. Nano Lett 11:2206–2213
30. Hu Y, Zeng L, Minnich AJ, Dresselhaus MS, Chen G (2015) Spectral mapping of thermal conductivity through nanoscale ballistic transport. Nat Nanotechnol 10:701–706
31. Dames C, Chen G (2005) Thermoelectrics handbook: macro to nano. Taylor & Francis, New York
32. Callaway J (1959) Model for lattice thermal conductivity at low temperatures. Phys Rev 113:1046–1051
33. Allen PB (2013) Improved callaway model for lattice thermal conductivity. Phys Rev B 88: 144302
34. Holland MG (1963) Analysis of lattice thermal conductivity. Phys Rev 132:2461–2471
35. Qiu L, Zhu Z, Feng Y, Michaelides EE, Żyła G, Jing D, Zhang X, Norris PM, Markides CN, Mahian O (2020) A review of recent advances in thermophysical properties at the nanoscale: from solid state to colloids. Phys Rep 843:1–81
36. Slack GA, Galginaitis S (1964) Thermal conductivity and phonon scattering by magnetic impurities in CdTe. Phys Rev 133:A253–A268
37. Cuffe J, Eliason JK, Maznev AA, Collins KC, Johnson JA, Shchepetov A, Prunnila M, Ahopelto J, Sotomayor Torres CM, Chen G, Nelson KA (2015) Reconstructing phonon mean-free-path contributions to thermal conductivity using nanoscale membranes. Phys Rev B 91:245423
38. Balandin AA, Ghosh S, Bao W, Calizo I, Teweldebrhan D, Miao F, Lau CN (2008) Superior thermal conductivity of single layer graphene. Nano Lett 8:902–907
39. Larkin JM, McGaughey AJH (2014) Thermal conductivity accumulation in amorphous silica and amorphous silicon. Phys Rev B 89:144303
40. Feldman JL, Kluge MD, Allen PB, Wooten F (1993) Thermal conductivity and localization in glasses: numerical study of a model of amorphous silicon. Phys Rev B 48:12589–12602
41. Feldman JL, Allen PB, Bickham SR (1999) Numerical study of low-frequency vibrations in amorphous silicon. Phys Rev B 59:3551–3559
42. BornVon M, Karman T (1912) Vibrations in space gratings (molecular frequencies). Z Phys 13:297–309
43. Malhotra A, Maldovan M (2019) Phononic pathways towards rational design of nanowire heat conduction Nanotechnology 30:372002
44. Slack GA (1973) Nonmetallic crystals with high thermal conductivity. J Phys Chem Solids 34:321–335
45. Glassbrenner CJ, Slack GA (1964) Thermal conductivityof silicon and germanium from 3 K to the melting point. Phys Rev 134:A1058–A1069
46. Ashcroft NW, Mermin ND (1976) Solid state phys. Saunders College Publishing, Philadelphia
47. Ziman JM (1960) Electrons and phonons. Oxford University Press, London. Ziman JM (1960) Electrons and phonons, The theory of transport phenomena in solids. Clarendon Press
48. Wu Y, Chen Z, Nan P, Xiong F, Lin S, Zhang X, Chen Y, Chen L, Ge B, Pei Y (2019) Lattice Strain Advances Thermoelectrics. Joule 3:1276–1288
49. Cowley RA (1968) Anharmonic crystals. Rep Prog Phys 31:123
50. Lin S, Li W, Li S, Zhang X, Chen Z, Xu Y, Chen Y, Pei Y (2017) High thermoelectric performance of Ag_9GaSe_6 enabled by low cutoff frequency of acoustic phonons. Joule 1:816–830
51. Morelli DT, Jovovic V, Heremans JP (2008) Intrinsically minimal thermal conductivity in cubic I-V-VI2 semiconductors. Phys Rev Lett 101:035901.

52. Liu H, Shi X, Xu F, Zhang L, Zhang W, Chen L, Li Q, Uher C, Day T, Snyder GJ (2012) Copper ion liquid-like thermoelectrics. Nat Mater 11:422–425
53. Banik A et al (2019) Engineering ferroelectric instability to achieve ultralow thermal conductivity and high thermoelectric performance in $Sn_{1-x}Ge_xTe$, Energy. Environ Sci 12:589
54. Yang J et al (2016) On the tuning of electrical and thermal transport in thermoelectrics: an integrated theory–experiment perspective. NpjComput Mater 2:15015
55. Qiu B, Bao H, Zhang G, Wu Y, Ruan X (2012) Molecular dynamics simulations of lattice thermal conductivity and spectral phonon mean free Path of PbTe: bulk and nanostructures. Comput Mater Sci 53:278–285
56. Khvesyuk VI, Skryabin AS (2017) Heat conduction in nanostructures. High Temperature 55(3):434–456
57. Alvarez FX, Jou D (2008) J Appl Phys 103:094321
58. Shiomi J (2016) APL Mater 4:104504
59. Peierls RE (1929) Ann. Phys. (New York) 395:1055
60. Klemens P (1958) Solid State Phys 7:1
61. Vermeersch B, Carrete J, Mingo N, Shakouri A (2015) Superdiffusive heat conduction in semiconductor alloys - I. Theoretical foundations. Phys Rev B 91:085202
62. Shiomi J, Maruyama S (2006) Phys Rev B 73(20):205420
63. Shiomi J, Maruyama S (2008) Jpn J Appl Phys, Part 1 47:2005
64. Mohammed AMS, Koh YR, Vermeersch B, Lu H, Burke PG, Gossard AC, Shakouri A (2015) Fractal lévy heat transport in nanoparticle embedded semiconductor alloys. Nano Lett 15:4269−4273
65. Metzler R, Klafter J (2000) Phys Rep 339:1−77
66. Vermeersch B, Mohammed AMS, Pernot G, Koh YR, Shakouri A (2014) Superdiffusive heat conduction in semiconductor alloys II. Truncated L'evy formalism for experimental analysis. Phys Rev B 91:085203
67. Eighth Nanoforum Report: Nanometrology. Nanoforum 2006, pp 13–14
68. Pokropivny VV, Skorokhod VV (2007) Classification of nanostructures by dimensionality and concept of surface forms engineering in nanomaterial science. Mater Sci Eng C 27(5–8):990–993
69. Chen G (1997) Size and interface effects on thermal conductivity of superlattices and periodic thin-film structures. J Heat Transfer 119(2):220–229
70. Walkauskas SG, Broido DA, Kempa K, Reinecke TL (1999) Lattice thermal conductivity of wires. J Appl Phys 85:2579–2582
71. Volz SG, Chen G (1999) Molecular dynamics simulation of thermal conductivity of silicon nanowires Appl. Phys Lett 75:2056–2058
72. Shi L, Li D, Yu C, Jang W, Kim D, Yao Z, Kim P, Majumdar A (2003) Measuring thermal and thermoelectric properties of one-dimensional nanostructures using a microfabricated device. J Heat Transfer 125:881
73. Li D et al (2003) Thermal conductivity of individual silicon nanowires Appl. Phys Lett 83:2934
74. Hochbaum AI, Chen R, Delgado RD, Liang W, Garnett EC, Najarian M, Majumdar A, Yang P (2008) Enhanced thermoelectricperformance of rough silicon nanowires. Nature 451:163–167
75. Zhou J, Jin C, Seol JH, Li X, Shi L (2005) Thermoelectric properties of individual electrodeposited bismuth telluride nanowires. Appl Phys Lett 87:133109
76. Zhou F, Szczech J, Pettes MT, Moore AL, Jin S, Shi L (2007) Determination of transport properties in chromium disilicide nanowires via combined thermoelectric and structural characterizations. Nano Lett 7:1649–1654
77. Boukai AI, Bunimovich Y, Tahir-Kheli J, Yu J-K, Goddard WA, Heath JR (2008) Silicon nanowires as efficient thermoelectric materials. Nature 451:168–171
78. Chen CL, Chen YY, Lin SJ, Ho JC, Lee PC, Chen CD, Harutyunyan SR (2010) Fabrication and characterization of electrodeposited bismuth telluride films and nanowires. J Phys Chem C 114:3385–3389
79. Shin HS, Jeon SG, Yu J, Kim YS, Park HM, Song JY (2014) Twin-driven thermoelectric figure-of-merit enhancement of Bi2Te3 nanowires. Nanoscale 6:6158–6165

80. Zhou F, Moore AL, Pettes MT, Lee Y, Seol JH, Ye QL, Rabenberg L, Shi L (2010) Effect of growth base pressure on the thermoelectric properties of indium antimonide nanowires. J Phys D Appl Phys 43:025406
81. Lee SH, Shim W, Jang SY, Roh JW, Kim P, Park J, Lee W (2011) Thermoelectric properties of individual single-crystalline PbTe nanowires grown by a vapor transport method. Nanotechnology 22:295707
82. Maldovan M (2015) Phonon wave interference and thermal bandgap materials. Nat Mater 14:667–674
83. Balandin A (2001) Phonon heat conduction in a semiconductor nanowire. J Appl Phys 89:2932–2938
84. Mingo N, Yang L, Li D, Majumdar A (2003) Predicting the thermal conductivity of Si and Ge nanowires. Nano Lett 3:1713–1716
85. Mingo N (2003) Calculation of Si nanowire thermal conductivity using complete phonon dispersion relations. Phys Rev B 68:113308
86. Maldovan M (2013) Sound and heat revolutions in phononics. Nature 503:209–217
87. Balandin A, Wang KL (1998) Effect of phonon confinement on the thermoelectric figure of merit of quantum wells. J Appl Phys 84:6149–6153
88. Balandin A, Wang KL (1998) Significant decrease of the lattice thermal conductivity due to phonon confinement in a free-standing semiconductor quantum well. Phys Rev B 58:1544–1549
89. Hyldgaard P, Mahan G (1997) Phonon superlattice transport. Phys Rev B 56:10754–10757
90. Tamura S-I, Tanaka Y, Maris HJ (1999) Phonon group velocity and thermal conduction in superlattices. Phys Rev B 60:2627–2630
91. Khitun A, Wang K, Chen G (2000) Thermoelectric figure of merit enhancement in a quantum dot superlattice. Nanotechnology 11:327–331
92. Venkatasubramanian R, Siivola E, Colpitts T, O'Quinn B (2001) Thin-film thermoelectric devices with high room-temperature figures of merit. Nature 413:597
93. Prasher R (2006) Thermal conductivity of tubular and core/shell nanowires. Appl Phys Lett 89:063121
94. Wingert MC et al (2011) Thermal conductivity of Ge and Ge-Si core–shell nanowires in the phonon confinement regime. Nano Lett 11:5507–13
95. Moon J, Kim J-H, Chen ZCY, Xiang J, Chen R (2013) Gate-modulated thermoelectric power factor of hole gas in Ge–Si core–shell nanowires. Nano Lett 13:1196–202
96. Hu M, Giapis KP, Goicochea JV, Zhang X, Poulikakos D (2011) Significant reduction of thermal conductivity in Si/Ge core–shell nanowires. Nano Lett 11:618–23
97. Chen J, Zhang G, Li B (2012) Impacts of atomistic coating on thermal conductivity of germanium nanowires. Nano Lett 12:2826–2832
98. Shishulin AV, Fedoseev VB, Shishulina AV (2019) Phonon thermal conductivity and phase equilibria of fractal Bi–Sb nanoparticles. Techn Phys 64(4):512–517
99. Kou J, Liu Y, Wu F, Fan J, Lu H, Xu Y (2009) Fractal analysis of effective thermal conductivity for three-phase unsaturated porous media. J Appl Phy 106:054905
100. Mandelbrot BB (1982) The fractural geometry of nature. W. H. Freeman, New York
101. Zolotukhin IV, Golev IM, Popov AA, Ievlev VP (2002) Techn Phys Lett 28(8):678–679
102. Perfect E, Kay BD (1995) Applications of fractals in soil and tillage research: a review. Soil Till Res 36:1–20
103. Adler PM, Thovert JF (1998) Real porous media: local geometry and macroscopic properties. Appl Mech Rev 51:537–585
104. Yu BM (2008) Analysis of flow in fractal porous media. Appl Mech Rev 61:050801
105. Xu P (2015) A discussion on fractal models for transport physics of porous media. Fractals 23:1530001
106. Ma Y, Yu B, Zhang D, Zou M (2004) Fractal geometry model for effective thermal conductivity of three-phase porous media. J Appl Phys 95:6426
107. Tarkhanyan RH, Niarchos DG (2014) Reduction of thermal conductivity in porous "gray" materials. APL Mater 2:076107

108. Tarkhanyan RH, Niarchos DG (2013) Reduction in lattice thermal conductivity of porous materials due to inhomogeneous porosity. Int J Thermal Sci 67:107e112
109. Shang J, Wang Y, Chen M, Dai J, Zhou X, Kuttner J, Hilt G, Shao X, Gottfried JM, Wu K (2015) Nat Chem 7:389
110. Kang Y, Duan F, Shangguan S, Zhang Y, Zhou T, Si B (2018) Molecules 23:3294
111. Han D, Ding W, Wang X, Cheng L (2019) Tunable thermal transport in a WS_2 monolayer with isotopic doping and fractal structure. Nanoscale 11:19763–19771
112. Zhang CB, Cheng QK, Chen YP (2014) Acta Phys Sin 63:236601
113. Han D, Fan H, Wang X, Cheng L (2020) Atomistic simulations of phonon behaviors in isotopically doped graphene with Sierpinski carpet fractal structure. Mater Res Express 7:035020
114. Vacher R, Courtens E (1989) Phonon-fracton crossover and self-similarity in real fractals. Physicascripta T29:239–243
115. Chakraborty S, Maiti SK (2020) Fractal lattice as an efficient thermoelectric device. J Phys Conf Ser 1579:012004

Chapter 5
Hypercomputation of the Brain by Superluminal Particles

Takaaki Musha

1 Introduction

It is widely believed that Moore's law for microprocessor performance will fail to hold in the next decade due to the brick wall arising from fundamental physical limitations of the computational process. Richard P. Feynman discussed the possibility of a quantum computer in which quantum computational energy cost versus speed is limited by energy dissipation during computation by taking an example of reversible computing [1]. According to his idea, the computational speed is limited by minimum energy required to transport a bit of information irreversibly between two devices, which prevents the speeding up of quantum computation. In this chapter, the author studies the energy limit of the computer system utilizing evanescent photon (quantum tunneling photon), which is considered to be a superluminal particle called a tachyon. Recami claimed in his paper [2] that tunneling photons which travel in evanescent mode (i.e. evanescent photon) can move with superluminal group speed inside the barrier. Chu and Wong at AT&T Bell Labs measured superluminal velocities for light traveling through the absorbing material [3]. Furthermore, Steinberg, Kwait and Chiao measured the tunneling time for visible light through the optical filter consisting of the multilayer coating about 10^{-6} m thick [4]. Experimental results obtained by Steinberg and co-workers have shown that the photons seemed to have traveled at 1.7 times the speed of light. Recent optical experiments at Princeton NEC have verified that superluminal pulse propagation can occur in transparent media [5]. These results indicate that the process of tunneling in quantum physics is superluminal, as claimed by Recami [2].

T. Musha (✉)
Advanced Science-Technology Research Organization, Yokohama, Japan

Foundation of Physics Research Center (FoPRC), Cosenza, Italy

A. Bandyopadhyay and K. Ray (eds.), *Rhythmic Advantages in Big Data and Machine Learning* , Studies in Rhythm Engineering,
https://doi.org/10.1007/978-981-16-5723-8_5

In mathematics and computer science, an accelerated Turing machine is a hypothetical computational model related to Turing machines which can perform a countable infinite number of computational steps, within a finite time. It is also called a Zeno machine, a concept proposed independently by Russdel, Blake and Weyl [6, 7], which performs its first computational step in one unit of time and each subsequent step in half the time of the step before, allowing an infinite number of steps to be completed within a finite interval of time. However, such a machine cannot be physically realized, from the standpoint of the Heisenberg uncertainty principle, because the energy to perform the computation will be exponentially increasing as the computational step is accelerating. Thus the Zeno machine is considered a mere mathematical concept with no possibility to realize it in the physical world. Contrary to this conclusion, the author can show the possibility to realize a real Zeno machine, by utilizing superluminal particles instead of subliminal particles, superluminal particles are related to evanescent photons. He studies the possibility to realize a high performance computing system by utilizing superluminal tunneling photon and he has also studied the possibility that the microtubular structure of neurons in a human brain is functioning as a quantum computational system that can attain higher efficient computation compared with conventional silicon processors.

Starting from standpoint of a model of the brain based on superluminal tunneling photons, the author has also described theoretically the possibility of a brain-like computer that would be more powerful than Turing machines, would allow non-Turing computation, and that may hold the key to the origin of human consciousness itself. According to the brain model proposed, it has been shown that microtubules in the biological brain have the possibility to achieve large quantum bit computation at room temperature which is superior in performance to conventional processors.

2 Signal Processing by Using Tunneling Photons

Recently it has been discovered by the research team at the University of Maryland that a signal processing device based on single photon tunneling can be realized [8]. They conducted the experiment for the transmission of light through nanometer-scale pinholes in a gold film covered by a nonlinear dielectric, which revealed that transmittance of a nanopore or nanopore array at one wavelength could be controlled by illumination with a second, different, wavelength. This opens the door to optical signal processing devices, such as all-optical switches realized on a microscopic scale which can manipulate single electrons, atoms or photons. If the atoms can be localized at distances smaller than the radiation wavelength, they can be coherently coupled by photons and an entangling quantum logic gate can be realized.

By applying this technology, the computer gate consisting of a large number of small, interconnected electromagnetic ion traps created for both memory and logical processing by manipulating atoms with the tunneling photon, may be realized as shown in Fig. 1.

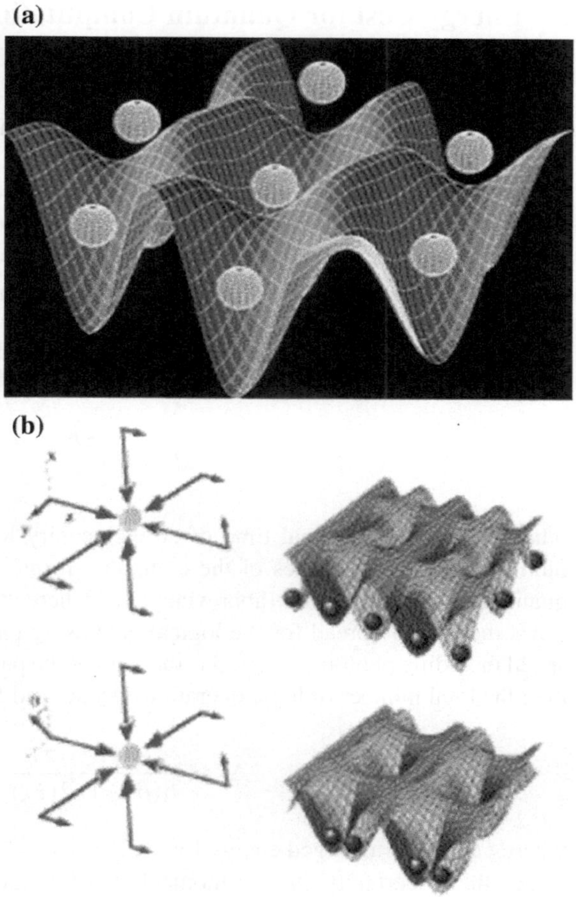

Fig. 1 Computer gate consisting of tunneling photons

The possibility of nano-switching by using evanescent photons in optical near-field was also proposed by Kawazoe and his co-workers [9].

By utilizing this technology, it is considered that the computation utilizing quantum tunneling photons has the possibility to achieve much faster computational speed and the influence of energy cost due to the uncertainty principle which prevents speeding up the quantum computation can be overcome as suggested in the following section.

3 Energy Cost for Quantum Computation UtilizingTunneling Photons

Hereafter we assume that tunneling photons which travel in an evanescent mode can move with a superluminal group speed, as confirmed by some experimenters.

We consider the computer system which consists of quantum gates utilizing quantum tunneling photons to perform logical operations. Benioff showed that the computation speed was close to the limit by the time-energy uncertainty principle [10]. Margolus and Levitin extended this result to a system with an averaged energy $\langle E \rangle$, which takes time at least $\langle E \rangle = \pi\hbar/(2\Delta t)$ to perform logical operations [11].

Summing over all logical gates of operations, the total number of logic operations per second is no more than

$$\Delta t \approx N\frac{\pi\hbar}{2\langle E \rangle}, \tag{1}$$

where Δt is a operational time of an elementary logical operation and N is the number of consisting gates of the computer. From Eq. (1), energy spread for the quantum tunneling photon (abbreviated QTP, hereafter) gate becomes $1/\beta(\beta - 1)$ times the energy spread for the logical gate using particles moving at sub-luminal speed including photons, where β is the ratio of the particle speed vs. the light speed, then the total number of logic operations per second for QTP gates can be given by

$$\Delta t \approx \frac{N}{\beta(\beta - 1)}\frac{\pi\hbar}{2\langle E_* \rangle}, \tag{2}$$

where $\langle E_* \rangle$ is an averaged energy for QTP gates.

As the uncertainty in the momentum of tunneling photons moving at the superluminal speed can be given by

$$\Delta p = \frac{m_* v}{\sqrt{v^2/c^2 - 1}} - \frac{\hbar\omega}{c}, \tag{3}$$

where m_* is an absolute value of the mass for the tunneling photon moving at superluminal speed and ω is an angular frequency of the photon, the velocity of the tunneling photon can be estimated as

$$v \approx c\left(1 + \frac{1}{\sqrt{\omega\Delta t}}\right), \tag{4}$$

from uncertanity principle for superluminal particles and $E = \hbar\omega$ [12].

If we denote the tunneling distance by d, the time for a photon tunneling across the barrier can be roughly estimated to be $\Delta t = d/v$. Then the velocity of the tunneling photon can be given by

$$v \approx c\left(1 + \frac{c}{2\omega d} + \sqrt{\frac{c}{\omega d} + \frac{c^2}{4\omega^2 d^2}}\right). \tag{5}$$

From which, the ratio of the minimum energy required for computation by QTP gates and the conventional computation can be given, as follows, by equating Eqs. (1) and (2);

$$R = \frac{\langle E_* \rangle}{\langle E \rangle} \approx \frac{1}{\beta(\beta - 1)}, \tag{6}$$

where

$$\beta \approx 1 + \frac{c}{2\omega d} + \sqrt{\frac{c}{\omega d} + \frac{c^2}{4\omega^2 d^2}}. \tag{7}$$

By the Higgs mechanism in quantum field theory, the penetration depth of tunneling photons can be estimated by the formula shown as [13]

$$r_0 = \frac{\hbar}{Mc}, \tag{8}$$

where M is the effective mass, which yields $M \approx 10$ eV.

From the above considerations, the penetration depth of the tunneling photon is estimated to be 1.5×10^{-8} m. Then the ratio of minimum energy required for the computation by QTP gates to the conventional computation processes can be estimated as shown in Fig. 2, when we let the tunneling distance of the barrier be $d \approx 10$ nm [14].

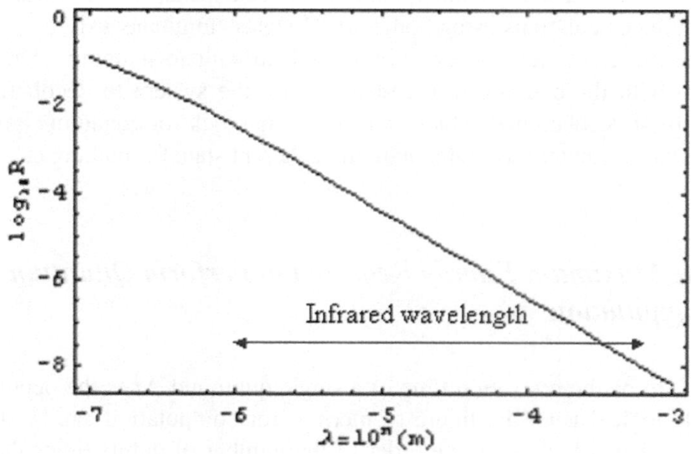

Fig. 2 Energy ratio of minimum energy required for the computation by QTP gates

In this figure, the horizontal axis is for the wavelength of tunneling photons and the vertical axis is for the ratio of their energy required for the computation.

If the wavelength of the tunneling photon is in a far infrared region ($\lambda = 10^5$ nm), the energy cost of computation for the computer, which consists of QTP gates, reaches to 10^{-6} times smaller than that of conventional computer systems.

In recent years, many studies on the quantum computation were conducted and it was recognized that the computational speed by quantum computing was much higher than that of conventional silicon processors. But the energy cost due to the uncertainty principle which prevents speeding-up of the quantum computation was not considered.

From the theoretical analysis of the energy limit of the quantum computer system which utilizes tunneling photons, it can be shown that energy loss of computation by utilizing superluminal tunneling photons is much lower than that of conventional silicon processors. Moreover, this superluminal effect can eventually speed up computers significantly because it can compensate interconnect delays inside logic gates which can never be fully eliminated from any real electronic components and bringing overall transmission rate closer to the ultimate speed limit [15], which actually boost the speed of a signal traveling on an electromagnetic wave for achieving high performance computers.

4 Decoherence Problem of Quantum States to Conduct Quantum Computation

The problem for conducting quantum computation is decoherence of quantum states. The qubit calculations of the quantum computer are performed while the quantum wave function is in a state of superposition between states, which is what allows it to perform the calculations using both 0 and 1 states simultaneously.

But this coherent state is very fragile and difficult to maintain. The slightest interaction with the external world would cause the system to decohere. This is the problem of decoherence, which is a stumbling block for quantum computation. Therefore the computer has to maintain the coherent state for making calculations.

4.1 The Maximum Energy Required to Perform Quantum Computation

If we let T to be the relaxation time of a single qubit and Δt is the operation time of a single logical gate, the figure of merit γ for computation can be defined as $\gamma = T/\Delta t$ [16], which is on the order of the number of qubits times the number of gate operations. As a superposition state of the L-qubits system would cause decoherence approximately 2^L times faster than a superposition state of one qubit

[17], then the relaxation time of the L-qubits system can be roughly estimated to be 2^{-L} times the relaxation time of a single qubit computation. Thus the minimum energy required to perform quantum computation for the L-qubits system can be given by

$$E_0 \approx \frac{\hbar \nu_G L}{T} 2^L, \tag{9}$$

where ν_G is the number of gate operations.

Similar to this equation, the minimum energy required to perform quantum computation utilizing superluminal particle can be estimated as [18]

$$E_0' \approx \frac{\hbar \nu_G L}{\beta(\beta - 1)T} 2^L. \tag{10}$$

Supposing that $E_0 = E_0'$, an increase of qubit size to perform computation by superluminal evanescent photon compared with the conventional computation can be given by

$$\Delta L \approx \frac{\log_2[\beta(\beta - 1)]}{1 + 1/L \log 2}, \tag{11}$$

when satisfying $\Delta L < L$.

From which, we can estimate an increase of qubit size of the quantum computation utilizing superluminal evanescent photons compared with the conventional computer system.

4.2 Decoherence Time of Superluminal Computation

Consider the Klein-Fock-Gordon equation for a scalar field with an imaginary mass $m = im_0$ given by [19]

$$\left(\frac{\partial^2}{\partial t^2} - \nabla^2 - m^2\right)\varphi = 0, \tag{12}$$

where $m = m_0 c/\hbar$.

From which, we can obtain an elementary solution shown by

$$\varphi_k(x) = \frac{1}{(2\pi)^{3/2}} \exp[i(\mathbf{k} \cdot \mathbf{x} - \omega t)] \equiv \frac{1}{(2\pi)^{3/2}} \exp(ikx), \tag{13}$$

where $\omega = \sqrt{k^2 - m_0^2}$.

Thus it can be seen that tachyons cannot be localized in space from the superposition of solutions given by Feinberg shown as [19]

$$\phi(x) = \int \varphi_k(x) f(k) d^3k \neq \delta^3(x), \tag{14}$$

where $f(k)$ is a function of bounded support, φ_k is the set of elementary solutions of the Klein-Fock-Gordon equation with an imaginary mass and $\delta(x)$ is the Dirac's delta function. Such a quantum superposition cannot be made to vanish outside a sphere of finite radius, but rather has a finite tail which is made to decrease with an arbitrary power of x for large values, by choosing the weight function $f(k)$ to have a zero of suitable order at $|k| = m_0$.

If we consider a quantum computer system utilizing superluminal photons, computational gates could be entangled with each other, according to the property of non-locality, to exert influences outside the extent of the processor.

According to Zurek, the density matrix $\rho(x, x')$ of the particle in the position representation evolves by the master equation shown as [20]

$$\dot{\rho} = -\frac{i}{\hbar}[H, \rho] - \gamma(x - x')\left(\frac{\partial}{\partial x} - \frac{\partial}{\partial x'}\right)\rho - \frac{2m\gamma k_B T}{\hbar^2}(x - x')^2\rho, \tag{15}$$

where H is the particle's Hamiltonian, γ is the relaxation rate, k_B is the Boltzmann constant, and T is the temperature of the field.

Zurek has shown that the off-diagonal peaks of the density matrix, $\rho(x, x') = \phi(x)\phi^*(x')$ will decay at the rate

$$\frac{d}{dt}(\rho^{\pm}) \approx 2\gamma m k_B T/\hbar^2(\Delta x)^2\rho^{\pm} = \rho^{+}/\tau_D. \tag{16}$$

From which, quantum coherence will disappear on a time scale shown by

$$\tau_D \cong \tau_R\left(\frac{\hbar}{\Delta x\sqrt{2mk_B T}}\right)^2. \tag{17}$$

The decoherence time for a quantum computation utilizing superluminal particles, has been shown by the author using the formula [18];

$$\tau'_D/\tau_D \approx [\beta(\beta - 1)]^2, \tag{18}$$

where τ'_D is the decoherence time for the superluminal particle.

In the following formula β is the ratio of the velocity of a superluminal particle relative to the speed of light

$$\beta \approx 1 + \frac{c}{2\omega d} + \sqrt{\frac{c}{\omega d} + \frac{c^2}{4\omega^2 d^2}}, \tag{19}$$

where ω is an angular frequency of the particle and d is the size of the tunneling barrier.

When we utilize an evanescent wave with frequencies in the infra-red spectrum, whose wavelength is given by $\lambda = 100\ \mu\text{m}$, we have $\tau'_D/\tau_D = 1.3 \times 10^{12}$ for the case when we let $d = 15\,\text{nm}$.

Thus the decoherence time utilizing evanescent photons in a quantum processor is given by

$$\tau'_D \approx \tau_R \left(\frac{\hbar}{\Delta x \sqrt{2mk_B T}} \right)^2 [\beta(\beta - 1)]^2 \approx 9.82 \times 10^2 \tau_R(\text{s}), \tag{20}$$

where we suppose the effective mass of an evanescent photon to be $m = 1.78 \times 10^{-35}$ Kg, $T = 300\,^\circ\text{K}$ and $\Delta x = 1.0\,\text{cm}$.

For the case when the error rate is small enough, it may be possible to use quantum error correction, which corrects errors due to decoherence, thereby allowing the total calculation time to be longer than the decoherence time.

If we suppose that the required error rate in each gate is 10^{-4}, this implies that each gate must be able to perform its task in one 10,000th of the decoherence time of the system. We suppose that the relaxation time was an order of macroscopic time satisfying $\tau_R \gg 1$ s, each gate can perform its task within 0.1 s and it is enough for each gate in a quantum computer system to complete its computation task.

5 Possibility of Quantum Computation Inside Microtubules Inside the Brain

From the high performance capabilities of quantum computation, there are many researches that can explain the higher performance of the human brain, including consciousness. As proposed by Feynman the optical computing network is the most realizable quantum mechanical computer among many possibilities such as a superconducting computer, and Hameroff suggested that microtubules in the brain were acting as waveguides for photons and as holographic processors [21].

The cytoskeleton of biological cells, including neurons of the brain, is made up of microtubules as shown in Fig. 3.

Each microtubule is a hollow cylindrical tube of tubulin proteins as shown in Fig. 4, which outer core diameter is 25 nm.

Microtubules are comprised of subunits of the protein, named tubulin [22]. Proteins contain hydrophobic (water repellent) pockets and these pockets contain atoms with electrons called π electrons, which means electrons in the reactive outer part (outer shell) of the atom that are not bonded to other atoms. The tubulin protein subunits of the microtubules have hydrophobic pockets within two nanometers of one another, which is close enough for the π electrons of the tubulin to become quantum entangled.

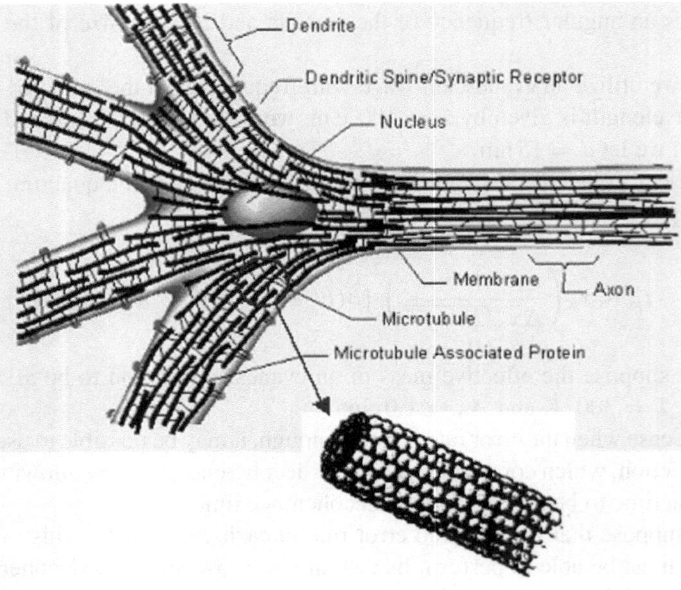

Fig. 3 Tubulin structure and microtubules

Fig. 4 Schematic diagram of the microtubule structure

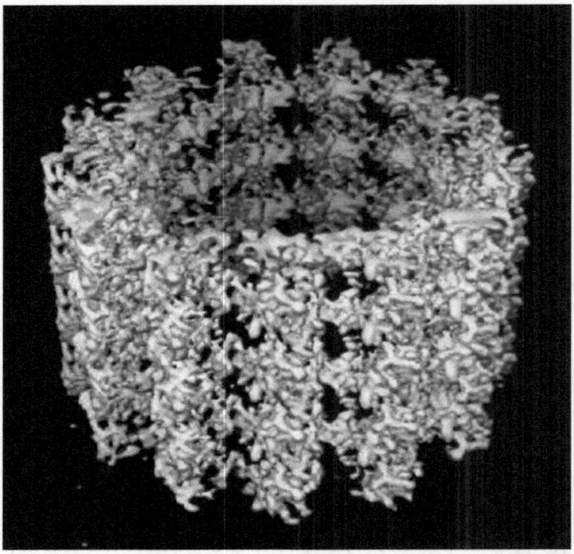

Each tubulin molecule can function as a switch between two conformations which exist in quantum superposition of both conformational states.

The dimensions of centrioles are close to the wavelengths of light in the infrared and visible spectrum, such that they may act as phase coherent, resonant waveguides

[23] (as shown in Fig. 5) which can be mapped onto the spin of material and they may play the role of retrieval, coupling/entangling of spin states of materials [24].

Thus each tubulin molecule can function as a switch between two conformations which exist in quantum superpositions of both conformational states.

According to Jibu et al. [25, 26], microtubule quantum states link to those of other neurons by quantum coherent photons tunneling through membranes in biological systems and the cytoskeletal protein conformational states are entangled by these photons which form coherent domains in their interaction with the local electromagnetic field.

Hameroff and Tuszynski [27] have proposed that microtubule subunit tubulins undergo coherent excitations, which leads to the automatic sequence where quantum coherence superposition is emerged in certain tubulins and consciousness is occurred as the process shown in Fig. 6.

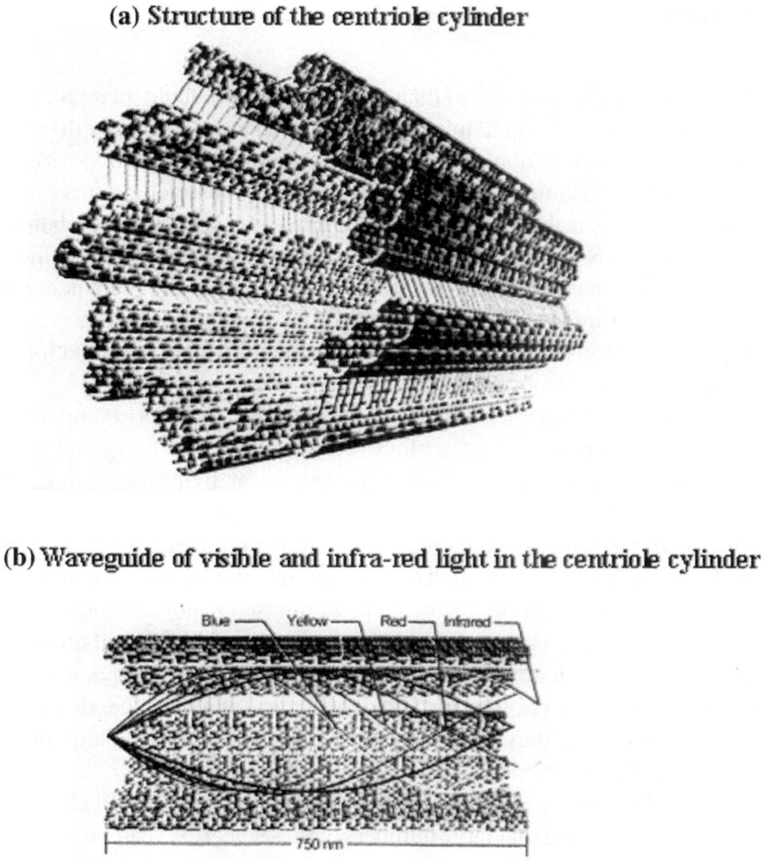

Fig. 5 Structure of the centriole cylinder and waveguide of visible and infra-red light in the centriole cylinder

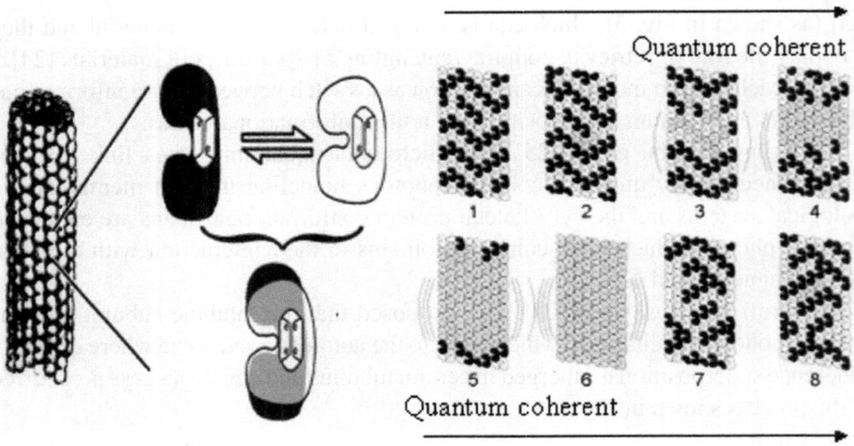

Fig. 6 Quantum computation conducted inside the microtubule

According to their hypothesis of quantum brain, microtubule quantum states link to those of other neurons by quantum coherent photon tunneling through membranes in biological systems, functioning in a way that resembles an ion trap computer.

On the other hand, Hameroff and Penrose [28] have constructed a theory, in which human consciousness is the result of quantum gravity effects in microtubules, which they dubbed Orch-OR (orchestrated object reduction) that involves a specific form of quantum computation conducted at the level of synapses among brain neurons. They have suggested that microtubules in brain neurons function as quantum computers with tubulin proteins in macrotubules acting as a quantum bit of computation through two-state q-bits formed from tubulin monomers.

Hameroff proposed that microtubules were suitable candidates to support quantum processing [29]. Microtubules are comprised of subunits of the protein, tubulins. Proteins constitute much of the driving machinery of living organisms. Proteins contain hydrophobic (water repellent) pockets. These pockets contain atoms with electrons called π electrons, which means electrons in the reactive outer part (outer shell) of the atom that are not bonded to other atoms. The tubulin protein subunits of the microtubules have hydrophobic pockets within two nanometers of one another. Hameroff claims that this is close enough for the π electrons of the tubulin to become quantum entangled. Quantum entanglement is a state in which quantum particles can alter one another's properties instantaneously and at a distance, in a way which would not be possible, if they were large scale objects obeying the laws of classical as opposed to quantum physics.

In the case of the electrons in the tubulin subunits of the microtubules.

Hameroff has proposed that large numbers of these electrons can become involved in a state known as a Bose–Einstein condensate. These occur when large numbers of quantum particles become locked in phase and exist as a single quantum object. These are quantum features at a macroscopic scale, and Hameroff suggests that through a feature of this kind quantum activity, which is usually at a very tiny scale, could be boosted to be a large scale influence in the brain.

Hameroff has proposed that condensates in microtubules in one neuron can link with other neurons via gap junctions. In addition to the synaptic connections between brain cells, gap junctions are a different category of connections, where the gap between the cells is sufficiently small for quantum objects to cross it by means of a process known as quantum tunneling. Hameroff proposes that this tunneling allows a quantum object, such as the Bose–Einstein condensates mentioned above, to cross into other neurons, and thus extend across a large area of the brain as a single quantum object. He further postulates that the action of this large-scale quantum feature is the source of the gamma (40 Hz) synchronization observed in the brain, and sometimes viewed as a correlate of consciousness. In support of the much more limited theory that gap junctions are related to the gamma oscillation, Hameroff quotes a number of studies from recent years.

The Orch OR theory combines Penrose's hypothesis with respect to the Gödel theorem with Hameroff's hypothesis with respect to microtubules. Together, Penrose and Hameroff have proposed that when condensates in the brain undergo an objective reduction of their wave function, that collapse connects to non-computational decision taking/experience embedded in the geometry of fundamental space–time.

The theory further proposes that the microtubules both influence and are influenced by the conventional activity at the synapses between neurons. The Orch in Orch OR stands for orchestrated to give the full name of the theory Orchestrated Objective Reduction. Orchestration refers to the hypothetical process by which connective proteins, known as microtubule-associated proteins (MAPs) influence or orchestrate the quantum processing of the microtubules.

It is perhaps necessary to recap the various objections to this complicated theory. Penrose's take on the Gödel theorem is rejected by many philosophers, logicians and artificial intelligence (robotics) researchers. His proposal for objective reduction is distinct from anything else in physics. The main objection to the Hameroff side of the theory is that any quantum feature in the environment of the brain would undergo wave function collapse (reduction) as a result of interaction with the environment, far too quickly for it to have any influence on neural processes.

The interiors of neurons alternate between liquid (solution: sol) states and quasi-solid (gelatinous: gel) states. In the gel state, water molecules which are electrical dipoles, are ordered, or orientated in the same direction, along the outer edge of the microtubule tubulin subunits. Hameroff et al. proposed that this ordered water could screen any quantum coherence within the tubulin of the microtubules from the environment of the rest of the brain. The tubulins also have a tail extending out from the microtubules, which is negatively charged, and therefore attracts positively charged ions. It is suggested that this could provide further screening. Furthermore, it was suggested that the microtubules could be pumped into a coherent state by biochemical energy.

Finally, it is suggested that the configuration of the microtubule lattice might be suitable for quantum error correction, a means of holding together quantum coherence in the face of environmental interaction. There is little existing evidence to directly support the Orch OR theory, although a paper in the journal Nature in 2007, claiming evidence for quantum coherence in the photosynthetic systems of plants has been seen as a possible indicator for quantum coherence in biological tissue.

Jibu et al. [25] also proposed that the conscious process in the brain is related to the macroscopic condensates of massive evanescent photons generated by the Higgs mechanism, arising from dynamical effects of electromagnetic interaction among electric dipoles in biological systems, which was studied by Del Guidance et al. [30].

They claimed that human consciousness could be understood as arising from those creation-annihilation dynamics of a finite number of evanescent (tunneling) photons in the brain. They also considered that each microtubule was a coherent optical encoder in a dense microscopic optical computing network in the cytoplasm of each brain cell, acting as a holographic information processor. Vitiello [31] proposed that consciousness mechanisms in the brain were achieved by the condensation of collective modes (called Nambu-Goldstone bosons) in the vacuum.

Faber et al. [32] proposed a model of the mind based on the idea that neuron microtubules could perform computations and they estimated the favorable condition for storage and information processing, which was found at temperatures close to the human body.

6 Difficulties of the Orch OR Model Proposed by Penrose

The wave or superposition form of the quanta is referred to as being quantum coherent. Interaction with the environment results in decoherence otherwise known as wave function collapse. It has been questioned as to how such quantum coherence could avoid rapid decoherence in the working conditions of the brain.

With reference to this question, a paper by the physicist Max Tegmark refuting the Orch OR model and published in the journal Physical Review E, is widely quoted [33].

It was also found by Unruh [34] that the time required in quantum computation must be less than the thermal time scale $\hbar/k_B T$, which yields 2.6×10^{-14} s at room temperature ($T = 300\,^\circ$K).

This value is too small for the conventional silicon processors to conduct a single gate operation.

Tegmark developed a model for time to decoherence, and from this calculated that microtubule quantum states would persist for only 10–13 s at brain temperatures, far too brief to be relevant to neural processing [33].

Tegmark published a refutation of the Orch-OR model in his paper [33] that the time scale of neuron firing and excitations in microtubules was slower than the decoherence time by at least a factor of $1/10^{10}$. According to his paper, it is reasonably unlikely that the brain functions as a quantum computer at room temperature.

In their reply to his paper, also published in Physical Reviews E, the physicists, Scott Hagan and Jack Tuszynski and Hameroff [29] claimed that Tegmark did not address the Orch OR model, but instead a model of his own construction. This involved superpositions of quanta separated by 24 nm (billionths of a meter) rather than the much smaller separations stipulated for the Orch OR model.

As a result, Hameroff's group claimed a decoherence time seven orders of magnitude greater than Tegmark's, but still well short of the 25 ms required if the quantum processing in the theory was to be linked to the 40 Hz gamma synchrony, as Orch OR suggested.

Hagen, Hameroff, and Tuszynski have claimed that Tegmark based his calculations on a model which was different from the Orch-Or model and Tegmark's assumptions should be amended [35]. Subsequently to it, Hameroff and Tuszynski proposed an idea [36] that microtubule subunit tubulins underwent coherent excitations, which led to the automatic sequence, where quantum coherence superposition emerged in certain tublins and consciousness occurred in the brain.

But some researches suggest that the brain requires quantum computing for perception, which means that the human brain must work as a quantum processing system.

Hameroff proposed the idea in his paper [29] that the biological information processing could be occurred by computer-like transfer and resonance among subunits of cytoskeletal proteins in microtubules. But the need to maintain macroscopic quantum coherence in a warm wet brain is certainly a serious problem for the Penrose-Hameroff model.

But the need to maintain macroscopic quantum coherence in a warm wet brain is certainly a serious problem for the Penrose-Hameroff model.

7 High Performance Computation in the Brain Utilizing Tunneling Photons

Contrary to researches mentioned above, we postulate that the human brain is a quantum computer system operated by superluminal evanescent photons.

Figure 7 shows the table of processing speed per CPU of the PowerPC. From this data, the ratio of energy loss/processing speed can be roughly estimated as 2.9 J/s/G-Flops.

Supposing that the computational speed of the human brain is 2×10^{16} operations/s, the energy of which required for computation by the PowerPC can be estimated as 5.8×10^7 J/s from Fig. 7. As the human brain consumes an estimated energy of 500 kcal per day (=24.2 J/s), the energy ratio of the human brain and the silicon processor become $R = 4.2 \times 10^{-7}$, which is similar to the calculation obtained for QTP gates from $R = \langle E_* \rangle / \langle E \rangle \approx 1/\beta(\beta - 1)$ [37].

Hence, it is seen that the human brain consumes much less energy than the conventional silicon processors, and it is considered that the human brain is an efficient computational system utilizing superluminal evanescent photons (Fig. 8).

Power PC	Performance (1CPU)	Watt (1CPU)
PPC603	0. 4GFLOPS	6. 0W
PPC750	0. 6GFLOPS	8. 0W
PPC7410	3. 2GFLOPS	8. 3W
PPC7410	3. 7GFLOPS	10. 4W
PPC7447	6. 4GFLOPS	18. 5W

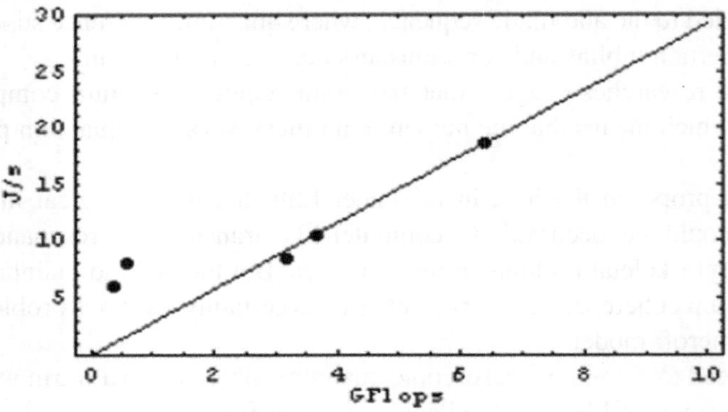

Fig. 7 Power loss inside the CPU of Power PC

Fig. 8 The human brain consumes an estimated energy of about 500 kcal per day

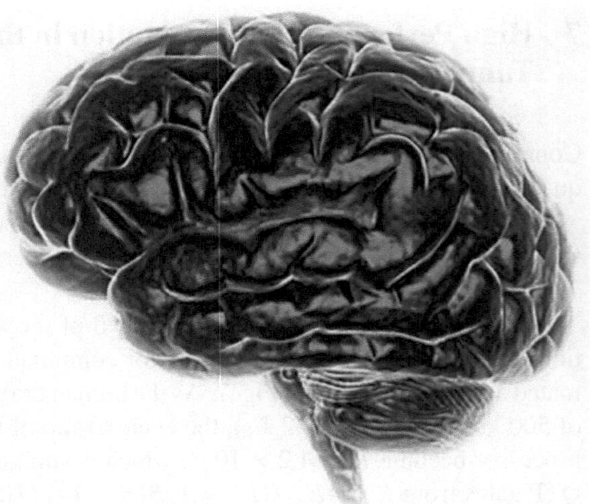

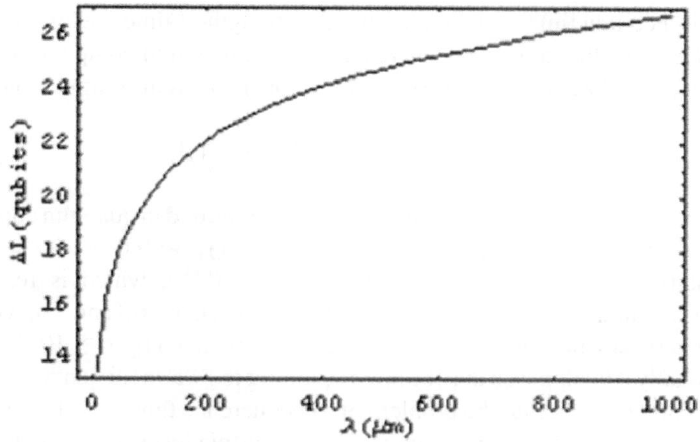

Fig. 9 ΔL and the wavelength of superluminal evanescent light

From Eqs. (7) and (11), it can be seen that the biological brain has the possibility to perform high efficient computation up to 20 qubits more than conventional silicon processors with the same energy dissipation when satisfying $\Delta L < L$, as shown in Fig. 9 for the infrared light frequency, the wavelength of which is $\lambda = 100\,\mu\text{m}$, where photons would propagate losslessly in a microtubule in the infra-red spectrum region [38], if we let $d \approx 15$ nm, which is the same order as the extracellular space between the brain cells.

This calculation result as shown in Fig. 9 suggests that the human brain can perform computational processes more efficiently in a high temperature environment than the conventional silicon processors.

Tegmarkcalculated that microtubules in the warm and wet brain would cause decoherence on the order of $\tau \approx 10^{-13}$ s, which is slowerby a factor at least 10^{10} than the time scale of neuron firing, $\tau \approx 10^{-3} \sim 10^{-4}$ s, from

$$\tau \approx \frac{D^2\sqrt{mk_BT}}{Ngq_e^2}, \tag{21}$$

where D is the tublin diameter, m is the mass of an ion, k_B is the Boltzmann constant, $g = 1/4\pi\varepsilon_0$, q_e is an ionic charge and $N = Q/q_e$.

Contrary to his calculation, it can be shown that a quantum processor utilizing superluminal evanescent photons can perform computation to satisfy the time scale required for quantum computation in microtubules based on the following considerations.

The decoherence time of the quantum system [39] can be given by

$$\tau_D = \frac{\hbar^2}{\tau_0}\frac{1}{(\Delta E)^2}, \tag{22}$$

where τ_0 is a certain time-scale to measure the strength of time uncertainties.

From Eq. (18), the ratio of decoherence times of quantum computation utilizing evanescent (tunneling) photons and the conventional quantum computation becomes

$$\tau'_D/\tau_D \approx (E_0/E'_0)^2 = [\beta(\beta-1)]^2. \tag{23}$$

When we let $\tau_D = 2.6\times10^{-14}$, which is the time required in quantum computation at the room temperature, we have $\tau'_D = \tau_D \times [\beta(\beta-1)]^2 \approx 0.03$ s for $\lambda = 100\,\mu$m.

This satisfies the decoherence time, $10^{-5} \sim 10^{-4}$ s, which is required for conducting quantum computation estimated by Hagen, Hameroff and Tsuzynski [35], and which also satisfies the time scale of neuron firing given by $\tau \approx 10^{-3} \sim 10^{-4}$ s. Conventionally it is supposed that the brain seems far too warm for quantum computation apparently running into the problem of a decoherence time, which would persist for only 10^{-13} s at brain temperatures. But recent evidence shows that quantum processes in biological molecules are actually enhanced at higher temperatures [40].

Thus it is considered that coherent quantum states may be preserved in microtubules at room temperature by superluminal photons, which attain high efficient computation compared with the conventional silicon processors without the mechanism of quantum gravity proposed by Penrose [41].

The warm and wet inner environment of the brain does not allow any long-time entanglement and superposition of two functional units from the conventional physical mechanism, as pointed by the Tegmark's calculation result and thus we must search for other mechanism, for example, via superluminal photons called tachyons.

If this mechanism is true for the brain function, we can obtain the possibility to realize much more efficient computer systems like a human brain by utilizing evanescent photons.

In conclusion, on the basis of the theorem that the evanescent photon is a superluminal particle, the possibility of high performance computation in biological systems has been studied. From the theoretical analysis presented in this chapter, it is shown that the biological brain has the possibility to achieve large quantum bits computation at the room temperature compared with the conventional processors. Hence it is considered that the human brain can attain high efficient computation compared with the silicon processors.

However, still remain the questions: how to determine a qubit in the brain, how it is related to a functioning neuron and how to determine the difference between memory registers and processing units in the brain. These questions must be clarified by further researches.

8 Holographic Memory in Human Biological Systems

Daniel Pollen and Michael Trachtenberg [42] proposed the holographic brain theory [43] to help explain the existence of photographic memories in some people. They suggested that such individuals had more vivid memories because they somehow could access a very large region of their memory holograms. S. R. Hameroff suggested in his paper that a centriole cylinder composed of microtubules functions as a waveguide for the evanescent photons for quantum signal processing [29]. Georgiev also proposed the idea that consciousness can be the result of quantum computation via applied laser-like pulses in quantum gates within the brain cortex [44]. Subsequently Georgiev concluded that this mechanism cannot be used for manipulation of the qubits inside the microtubule cavities, or centrioles, because the photons wavelength is two orders of magnitude longer than the size of these centrioles [45]; super radiant photons in the microtubule cavities could have wavelength of $\lambda = 100\,\text{nm}$ or more suggested by Smith [46], incompatible with the length of a moderate-sized microtubule cavity, which is about 1 nm. Therefore super radiant emissions could not be used to signal qubits in a fashion similar to standing wave lasers in an ion trap computation.

As an alternative mechanism, the author proposes that the cylinder formed by the microtubule cavity may have a negative refractive index, similar to a metamaterial [47]. Typically metamaterials are artificial materials, engineered to have a negative refractive index, a property that is not normally found in nature [48]. They usually gain their properties from structure rather than composition, using microscopic inhomogeneities to create an effective macroscopic behavior.

Negative refractive index materials appear to permit the creation of super lenses which can have a spatial resolution below that of the wavelength [49]. If the inner medium of the cylinder of microtubules possesses characteristics of a negative refractive index, the generation of evanescent photons is enhanced, and they propagate without loss inside the neurons according to the properties of a metamaterial.

8.1 Holographic Memory Based on Evanescent Superluminal Photons in the Microtubule

Due to the high performance of quantum computation, there is considerable research into how quantum computing could explain the performance behaviors of human brains, including consciousness itself. As Feynman proposed, the optical computing network is the most realizable quantum mechanical computer and Hameroff has suggested that microtubule cavities in the brain can act as waveguides for photons and as holographic processors [29].

The cytoskeleton of biological cells, including neurons of the brain, are made up of microtubules. Microtubules are comprised of subunits of the protein, named tubulin. Each microtubule is a hollow cylindrical tube of tubulin proteins as shown in Fig. 10, whose outer core diameter is 25 nm. Each tubulin molecule within each microtubule can function as a switch between two conformations that exist in a quantum superposition of both conformational states. The dimensions of centrioles are close to the wavelengths of light in the infrared and visible spectrum such that they may act as phase coherent, resonant waveguides [38].

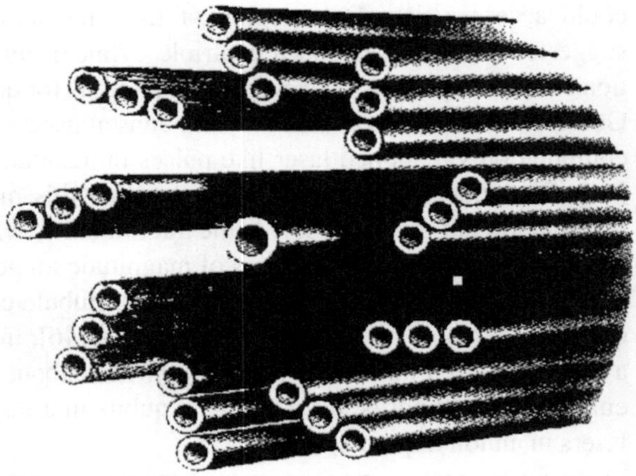

Fig. 10 Structure of microtubules forming a centriole cavity

According to Jibu et al. [25, 26], microtubule quantum states link to those of other neurons by quantum coherent photons tunneling through the membranes in biological systems and the cytoskeletal protein conformational states are entangled by these photons which form coherent domains in their interaction with the local electromagnetic field. They claimed that human consciousness could be understood as arising from those creation-annihilation dynamics of a finite number of evanescent (tunneling) photons in the brain.

Recami [50, 51] claimed in his papers that tunneling photons traveling in an evanescent mode can move with superluminal group speed, which can be shown as follows.

The evanescent photon generated in a quantum domain satisfies the following Klein-Fock-Gordon equation given by

$$\left[-\frac{1}{c^2}\frac{\partial^2}{\partial t^2}+\nabla^2-\frac{m_0^2c^2}{\hbar^2}\right]\phi(x,t)=0, \tag{24}$$

where c is the light speed, m_0 is an absolute value of the proper mass of the evanescent photon and $\hbar$ is the Planck's constant divided by 2π.

This equation gives the solution for the photon traveling in an evanescent mode as

$$\phi(x,t)=A_0\exp\left[-\frac{Et+px}{\hbar}\right], \tag{25}$$

which corresponds to the elementary particle with an imaginary mass im_0 that travels at a superluminal speed satisfying

$$E^2=p^2c^2-m_0^2c^4, \tag{26}$$

where E is the energy of the superluminal particle and p is its momentum. Hence, it can be seen that tunneling photons traveling in an evanescent mode can move at a superluminal speed. From the assumption that the evanescent photon is a superluminal particle, the author has shown that a microtubule in a biological brain can achieve quantum bit (qubit) computations on large data sets, which would account for the high performance of the computations as compared with the conventional processors. It therefore seems highly plausible that macroscopic quantum ordered dynamical systems of evanescent photons in the brain could play an essential role in realizing long-range biological order in living systems. Ziolkowski pointed out in his paper that superluminal pulse propagation which permits consequent superluminal exchange without a violation in causality is possible in electromagnetic metamaterials [52].

Metamaterials are artificial materials engineered to have properties that may not be found in nature. They are assemblies of multiple individual elements fashioned from conventional microscopic materials such as metals or plastics, but the materials are usually arranged in periodic patterns as shown in Fig. 11.

Metamaterials gain their properties not from their composition, but from their exactingly designed structures. Their precise shape, geometry, size, orientation and arrangement can affect the waves of light or sound in an unconventional manner, creating material properties which are unachievable with conventional materials. These metamaterials achieve desired effects by incorporating structural elements of sub-wavelength sizes, i.e. features that are actually smaller than the wavelength of the waves.

Negative refractive index materials appear to permit the creation of super lenses that can have a spatial resolution below that of the wavelength. If the inner medium of a cylinder of microtubules possesses the characteristics of a negative refractive

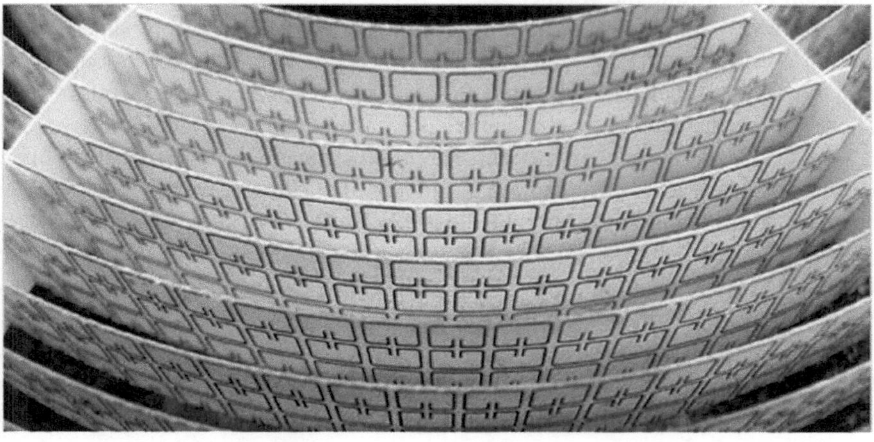

Fig. 11 Structure of the electromagnetic meta-material (left-handed metamaterial configuration consisting of copper split-ring resonators and wires)

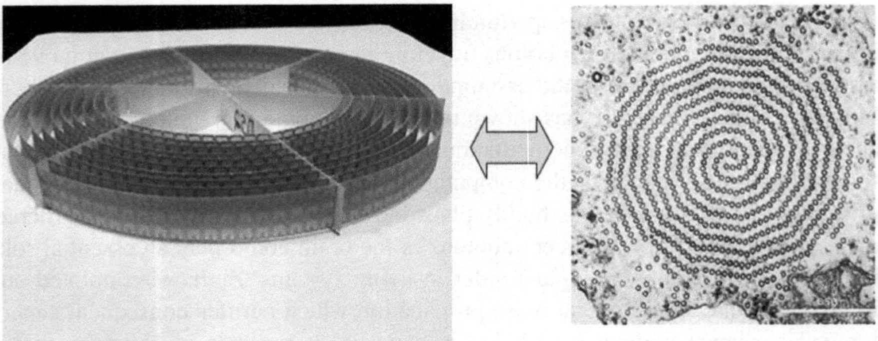

Fig. 12 Similarity between the metamaterial (left figure) and the microtubule (right figure)

index, the generation of evanescent photons is enhanced, and they propagate without loss inside the neurons according to the properties of a metamaterial.

If the inner medium of a cylinder of microtubules possesses the characteristics of a metamaterial with negative refractive index, tunneling photons will propagate losslessly inside the neurons in a way that is not restricted by wavelength, and therefore infrared photons could be used for the manipulation of qubits in the brain (Fig. 12).

8.2 Mechanism of Holographic Memory of the Brain

Similar to Prof. Hameroff's idea, Dr. Georgiev presented an idea that consciousness could be the result of quantum computation via short laser-like pulses controlling quantum gates within the brain cortex.

However, he later rejected this theory because the wavelength of super radiant photon emission in the infrared spectrum is two orders of magnitude longer than any sized microtubule cavity [45]. Thus if one proposes this mechanism, infrared photon cannot account for manipulation of the quantum qubits inside the microtubule cavity. To avoid this problem, we suggest the substance in the microtubule cylinder has the characteristics of a metamaterial composed of sub-wavelength structures. As evanescent waves inside the microtubule cavity waveguide can propagate below the cutoff frequency defined by $f_c = c/2d$, where d is a diameter of the waveguide, thus we can model the evanescent wave according to the theory of evanescent wave holography [53] shown as follows;

The basic idea of evanescent wave holography is to use an evanescent wave as a reference wave and as a readout wave [54]. Suppose that the field distribution of a wave propagating in a microtubule is periodic inside the core, and evanescent outside this region given by

$$U_e(x, z) = U_e \exp(ik_e x) \exp(-k_e z), \tag{27}$$

and

$$U_p(x, z) = U_p \exp[i(k_p x + k_p z)], \tag{28}$$

where U_e is an evanescent wave and U_p is a propagating wave from outside of the microtubule waveguide as shown in Fig. 13.

As shown in Fig. 13, U_p interferes with U_e and this interference is recorded. This interference in the storage material will produce a proportional variation in the index of refraction, which will be periodic along the length of each microtubule. From Eqs. (27) and (28), the intensity of the wave can be given by

$$I = \frac{1}{2}\left\{U_p^2 + U_e^2 \exp(-2k_{ez}z)\right\} \\ + 2\mathrm{Re}\left\{U_e^* U_p \exp(i[k_{px} - k_{ex})x + k_{pz}z]) \exp(-k_{ez}z)\right\}. \tag{29}$$

We suggest the cylinder of the microtubule consists of a storage material, whose thickness is much smaller compared to the penetration depth of evanescent waves. The response of the recoding medium to this intensity distribution is that the dielectric constant ε is a function of the local intensity given by $\varepsilon = f(I \cdot \tau)$, where τ is an exposure time.

Then the total field, consisting of an illuminated wave and a diffracted wave, can be described by

$$\nabla^2 U + \left(\frac{\omega}{c}\right)^2 \varepsilon(r) U = 0, \tag{30}$$

and thus the wavelet scattered by each volume element of the hologram can be obtained using Green's function and integrated over the volume of the hologram,

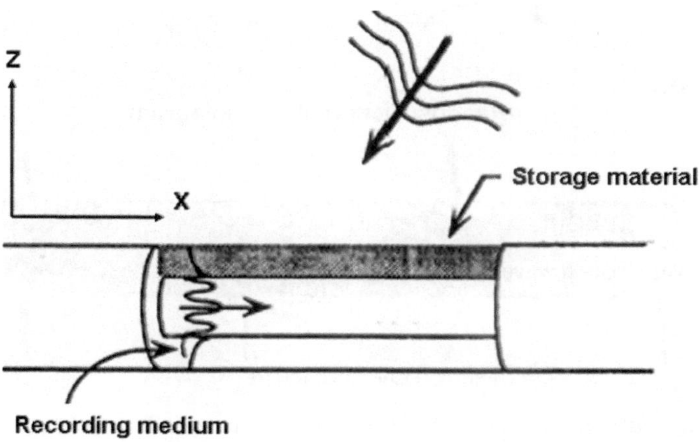

Fig. 13 Recording of evanescent wave hologram in the micro tubulewaveguide

according to the integral equation for hologram reconstruction, given by

$$U(\vec{r}) \approx \iint U_0(\vec{r}') \exp(-i\beta\vec{n} \cdot \vec{r}/|\vec{r}|)\varepsilon(\vec{r}) \cdot d\vec{r}, \tag{31}$$

where U_0 is an illuminating wave, U is a wave including the information of object wavefield and β is a parameter satisfying $[\nabla^2 + \beta^2]G(r', r) = -\delta(|r' - r|)$.

This means that the guided readout wave is propagating in the same direction as the reference wave during recording. The readout wave is diffracted by the hologram structure and the object field can be reconstructed.

This is the same mechanism of holographic memory of the brain, which was proposed by Pribram et al. [43].

Figure 14a shows a schematic diagram of the reconstruction of stored memory by the propagating evanescent wave inside the microtubule structure. As shown in this figure, an illumination with the reference wave U_r reconstructs the objective wave

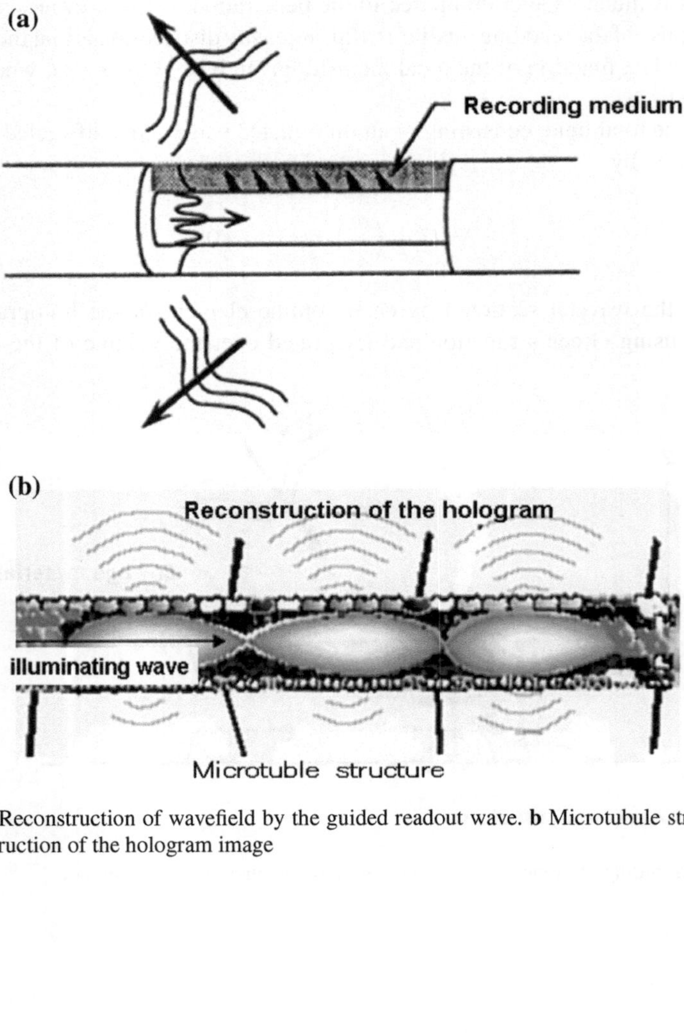

Fig. 14 **a** Reconstruction of wavefield by the guided readout wave. **b** Microtubule structure and the reconstruction of the hologram image

field U_o by interaction with the term $U_r U_o$ in the dielectric structure of the physical substrate for memory inside the microtubule structure.

By this mechanism, the evanescent photon propagating inside the microtubule can manipulate the storage and provide for the retrieval of stored data, or memory, as shown in Fig. 14b. The holographic mechanism would also explain how our brains could store so many memories in so little space. If the microtubules store memories using evanescent wave holography, this suggests that impairment of consciousness such as Alzheimer's disease may be due in part to the loss of metamaterial characteristics of the substructure inside the microtubule within the brain.

Furthermore, a new computer hardware system could be constructed using a similar technique for data storage and retrieval, with memories which rely on interfering waves of photons passing through multiplexed holographic gratings similar to the brain's neural structure; i.e. an artificial holographic brain system could now be constructed using this same mechanism.

A holographic theory of human memory has now been proposed which is based on the use of evanescent superluminal photons for recording and retrieving holographically stored qubits. We have now theoretically established the microtubule structure inside the biological brain is capable of storing memories as holograms, using microtubule substrate as the storage material, if the interior of the microtubule cavities has the characteristics of a metamaterial composed of sub-wavelength structures.

9 Holonomic Model of the Brain Function

Thus microtubules have both the ability to store memories and conduct quantum computation as shown in Fig. 15. As the microtubules are not just restricted to inside the brain, but distributed over all of cells within the human body, we can consider consciousness to be distributed throughout the human body.

David Bohm believed that objective reality does not exist and, despite its apparent solidity, the Universe is at heart a phantasm, a gigantic and splendidly detailed hologram. Bohm was involved in the early development of the holonomic model of brain function, a model for human cognition which is drastically different from conventionally accepted ideas. Bohm developed the theory that the brain operates in a manner

Fig. 15 Microtubule functions, quantum computation and the memory

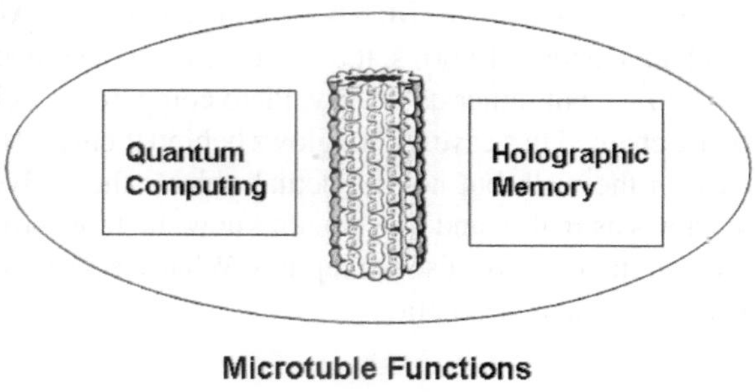

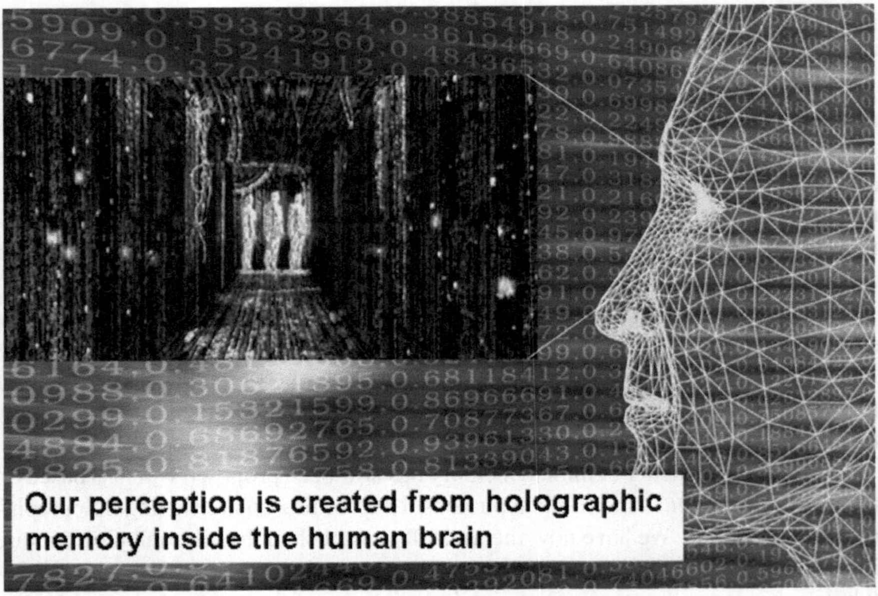

Fig. 16 Is our world the hologram created by our brains?

similar to a hologram, in accordance with quantum mathematical principles and the characteristics of wave patterns [55].

To understand why Bohm makes this startling assertion, one must first understand that a hologram is a three-dimensional photograph made with the aid of a laser. To make a hologram, the object to be photographed is first bathed in the light of a laser beam. Then a second laser beam is bounced off the reflected light of the first and the resulting interference pattern (the area where the two laser beams conflate) is captured on film. When the film is developed, it looks like a meaningless swirl of light and dark lines. But as soon as the developed film is illuminated by another laser beam, a three-dimensional image of the original object appears.

This is similar to the plot of the 1999 Holywood film called "Matrix". This film depicts a future world in which reality, as perceived by most people, is actually a simulated reality called "the Matrix", created by sentient machines to subdue the human population (Fig. 16).

The premise of "The Matrix" is related to Plato's Allegory of the Cave. According to Plato's theory of Forms, the true essence of an object is not what we perceive with our senses, but rather its quality. Plato compares people uneducated in this theory to being chained in a cave. A fire glows behind them and they see the shadows of objects cast on the wall, but not the actual objects themselves. These people perceive the shadows as reality and thus do not know the true form of the objects, and therefore, are confined to this false perception. What we observe and accept as reality, may be just such a false perception.

10 Hypercomputingby Superluminal Particles

Supposing that there exist superluminal particles, it is possible to make an infinite computation within a finite time.

An accelerated Turing machine named a Zeno machine is a hypothetical computational model which can perform the countable infinite number of computational steps within a finite time. However, it cannot be physically realized from the standpoint of the Heisenberg uncertainty principle, because the energy to perform the computation will be exponentially increased when the computational step is accelerated.

In mathematics and computer science, an accelerated Turing machine is a hypothetical computational model related to Turing machines which can perform the countable infinite number of computational steps within a finite time. It is also called a Zeno machine which concept was proposed independently by B. Russdel, R. Blake and H. Weyl. As mentioned in the Introduction, it performs its first computational step in one unit of time and each subsequent step in half the time of the step before, allowing an infinite number of steps can be completed within a finite interval of time [56, 57]. However, this machine cannot be physically realized from the standpoint of the Heisenberg uncertainty principle, $\Delta E \Delta t \approx \hbar$, because the energy to perform the computation will be exponentially increased when the computational step is accelerated. Thus it is considered that the Zeno machine is a mere mathematical concept and there is no possibility to realize it in a physical world. Contrary to this conclusion, the author shows the possibility to realize it by utilizing superluminal particles instead of subliminal particles, including photons.

10.1 Computational Time Required to Perform Infinite Steps of Computation

Feynmann defined the required energy per step for the computation as shown in Fig. 17, given by [1]

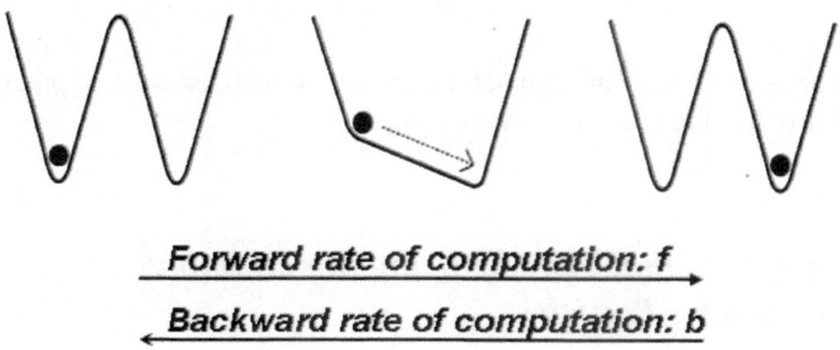

Fig. 17 Computational steps necessary to perform reversible computation

$$\text{energy per step} = k_B T \frac{f-b}{(f+b)/2}, \tag{32}$$

where k_B is Boltzmann's constant, T is a temperature, f is a forward rate of computation and b is backward rate.

Supposing that there is no energy supply and parameters f and b are fixed during the computation, we can consider the infinite computational steps given by

$$E_1 = kE_0, E_2 = kE_1, \ldots E_n = kE_{n-1}, \tag{33}$$

where we let the initial energy of computation be $E_0 = k_B T, k = 2(f-b)/(f+b)$, and E_n is the energy for the n-th step computation.

From the above we have $E_n = k^n E_0$, and then the energy loss for each computational step becomes

$$\begin{aligned} \Delta E_1 &= E_0 - E_1 = (1-k)E_0 \\ \Delta E_2 &= E_1 - E_2 = (1-k)kE_0 \\ \Delta E_n &= E_{n-1} - E_n = (1-k)k^{n-1}E_0. \end{aligned} \tag{34}$$

According to the paper by Lloyd [58], it is required for the quantum system with average energy ΔE to take time at least Δt to evolve to an orthogonal state given by

$$\Delta t = \frac{\pi \hbar}{2\Delta E}, \tag{35}$$

from which, the total energy for the infinite steps yields E_0 if setting $E = \Delta E_i$ into Eq. (35), then the total time for the computation with infinite steps becomes

$$T = \sum_{n=1}^{\infty} \Delta t_n = \frac{\pi \hbar}{2E_0} \sum_{n=1}^{\infty} \frac{1}{(1-k)k^{n-1}}. \tag{36}$$

As the infinite sum of Eq. (36) diverges to infinity when satisfying $0 < k < 1$, the Feynman model of computation requires infinite time to complete the calculation (see Fig. 18).

Hence an accelerated Turing machine cannot be realized for computers utilizing ordinary particles due to the uncertainty principle.

10.2 Computational Time by Using Superluminal Elementary Particles

Recami claimed in his paper [51] that tunneling photons which travel in evanescent mode can move with superluminal group speed inside the barrier. Chu and Wong

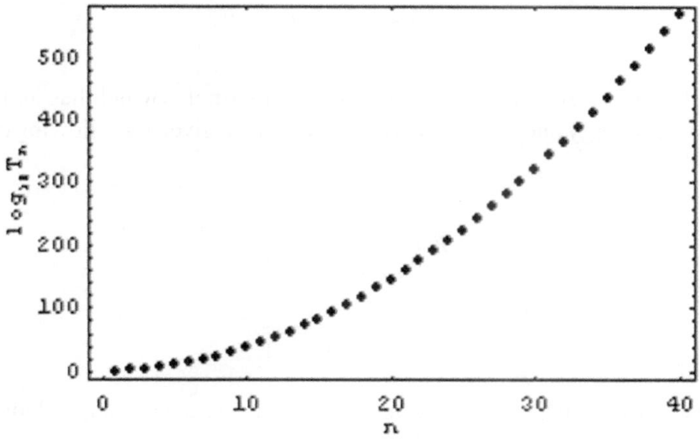

Fig. 18 Time required to conduct computation at each step (for the case, $k = 1/2$ and $\gamma = 1.0$)

at AT&T Bell Labs measured superluminal velocities for light traveling through the absorbing material [3]. Furthermore, Steinberg, Kwait and Chiao measured the tunneling time for visible light through the optical filter consisting of the multilayer coating about 10^{-6} m thick [4]. Experimental results obtained by Steinberg and co-workers have shown that the photons seemed to have traveled at 1.7 times the speed of light. Recent optical experiments at Princeton NEC have verified that superluminal pulse propagation can occur in transparent media [59]. These results indicate that the process of tunneling in quantum physics is superluminal, as claimed by E. Recami [51].

From the relativistic equation of energy and momentum of the moving particle, given by

$$E = \frac{m_0 c^2}{\sqrt{1 - v^2/c^2}}, \tag{37}$$

and

$$p = \frac{m_0 v}{\sqrt{1 - v^2/c^2}}, \tag{38}$$

the relation between energy and momentum can be shown as $p/v = E/c^2$.

From which, we have [60]

$$\frac{v\Delta p - p\Delta v}{v^2} = \frac{\Delta E}{c^2}. \tag{39}$$

Supposing that the approximation $\Delta v/v^2 \approx 0$ holds, Eq. (39) can be simplified as

$$\Delta p \approx \frac{v}{c^2}\Delta E. \tag{40}$$

This relation is also valid for the superluminal particle (which has an imaginary mass im_*), the energy and the momentum of which are given by following equations, respectively.

$$E = \frac{m_* c^2}{\sqrt{v^2/c^2 - 1}}, \tag{41}$$

$$p = \frac{m_* v}{\sqrt{v^2/c^2 - 1}}. \tag{42}$$

According to the paper by Park and Park [61], the uncertainty relation for the superluminal particle can be given by

$$\Delta p \cdot \Delta t \approx \frac{\hbar}{v - v'}, \tag{43}$$

where v and v' are the velocities of a superluminal particle after and before the measurement. By substituting Eq. (40) into (43), we obtain the uncertainty relation for superluminal particles given by

$$\Delta E \cdot \Delta t \approx \frac{\hbar}{\beta(\beta - 1)}, \tag{44}$$

when we let $v' = c$ and $\beta = v/c$.

Instead of subluminal particles including photons, the time required for the quantum system utilizing superluminal particles becomes [60]

$$T = \sum_{n=1}^{\infty} \Delta t_i = \frac{\pi \hbar}{2E_0} \sum_{n=1}^{\infty} \frac{1}{\beta_n(\beta_n - 1)(1 - k)k^{n-1}}, \tag{45}$$

from the uncertainty principle for superluminal particles given by Eq. (44), where β can be given by

$$\beta_n = \sqrt{1 + \frac{m_*^2 c^4}{E_n^2}} = \sqrt{1 + \frac{m_*^2 c^4}{k^{2n} E_0^2}}, \tag{46}$$

which is derived from Eq. (42).

From Eqs. (45) and (46), it is seen that the computation time can be accelerated as shown in Fig. 19.

By the numerical calculation, it can be shown that the infinite sum of Eq. (45) converges to a certain value satisfying $0 < k < 1$, as shown in Fig. 20.

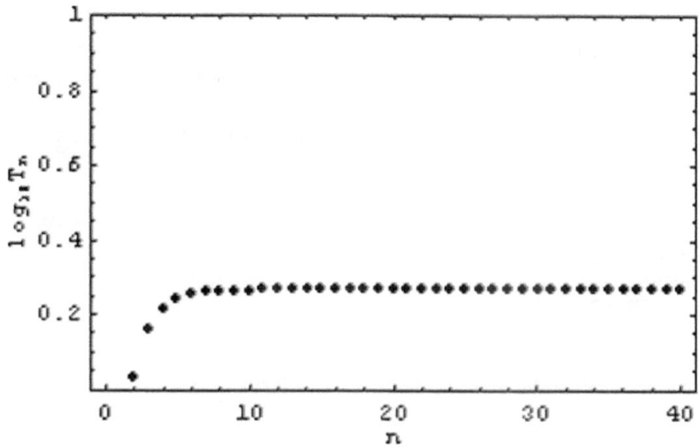

Fig. 19 Time required to conduct computation at each step by using superluminal particles (for the case, $k = 1/2$ and $\gamma = 1.0$)

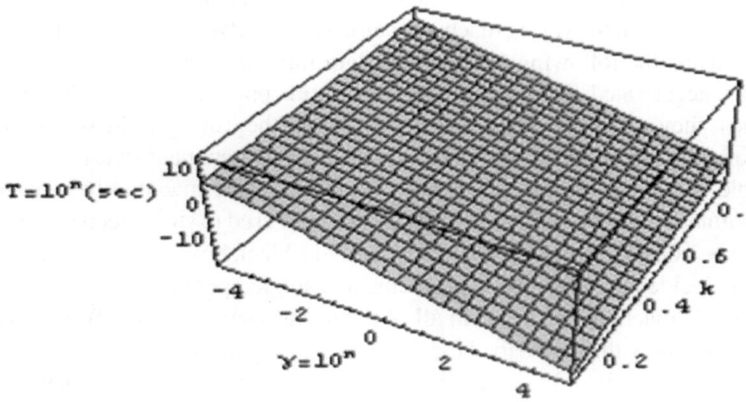

Fig. 20 Computational required time for the superluminal particles

In this figure, the horizontal line is for the parameter $\gamma = m_* c^2 / E_0$ and the vertical line is for the time to complete infinite step calculations.

From these calculation results, an accelerated Turing machine can be realized by utilizing superluminal particles, instead of subluminal particles, for the Feynman's computational model.

Thus, contrary to the conclusion obtained relatively to the Feynman's model of computation when using ordinary particles, it can be seen that superluminal particles permits the realization of an accelerated Turing machine.

It is known that an accelerated Turing machines allow us to be computed some functions which are not Turing-computable, such as the halting problem [62],

```
begin program
    write 0 on the first position of the output tape;
begin loop
        simulate 1 successive step of the given Turing
machine on the given input;
if the Turing machine has halted then
            write 1 on the first position of the out-
put tape and break out of loop;
end loop
end program
```

Fig. 21 Pseudocode algorithm to solve the Halting problem

described as "given a description of an arbitrary computer program, decide whether the program finishes running or continues to run forever".

This is equivalent to the problem of deciding, given a program and an input, whether the program will eventually halt when run with that input, or will run forever.

Halting problem for Turing machines can easily be solved by an accelerated Turing machine using the following pseudocode algorithm (see Fig. 21).

As an accelerated Turing machines are more powerful than ordinary Turing machines, they can perform computation beyond the Turing limit which is called hypercomputation, such as to decide any arithmetic statement that is infinite time decidable. From this result, we can construct an oracle machine [63] by using a superluminal particle, which is an abstract machine used to study decision problems.

It can be conceived as a Turing machine with a black box, called an oracle, which is able to decide certain decision problems in a single operation.

An oracle machine can perform all of the usual operations of a Turing machine, and can also query the oracle to obtain a solution to any instance of the computational problem for that oracle. For example, if the problem is a decision problem for a set A of natural numbers, the oracle machine supplies the oracle with a natural number, and the oracle responds with "yes" or "no" stating whether that number is an element of A.

Given a device that tells you in advance whether a given computer program would halt, or go on running forever, you would be able to prove or disprove any theorem whatsoever about integers: the Goldbach conjecture, Fermat's Last Theorem, and the famous Riemann hypothesis because it is equivalent to the following problem:

$$\forall n > 0, \left(\sum_{k \le \delta(n)} \frac{1}{k} - \frac{n^2}{2} \right)^2 < 36n^3, \ \text{where}\ \delta(n) = \sum_{d|n} \frac{1}{d}. \tag{47}$$

You would simply show this "Oracle" a program that would loop through all the integers, testing every possible set of values and only halting if it came to a set that violated the conjecture.

By using the superluminal particle, there is a possibility to realize the hyper-computational system. As an accelerated Turing machines are more powerful than ordinary Turing machines, they can perform computation beyond the Turing limit, which is called hypercomputation, such as to decide any arithmetic statement that is infinite time decidable.

11 Human Intuition from the Standpoint of Superluminal Hyper-Computation

According to the stochastic electrodynamics, it can be shown that a tachyon field can be spontaneously created from the sea of omnipresent zero-point-energy field pervading all of the Universe. If the brain has the possibility to correlate with the outer tachyon field created from other living organisms and cosmic zero-point field, the totality of them, we may call it a universal brain or a cosmic consciousness, can exert an influence on the individual mind and this can explain the extraordinary ability of human biological brains.

To further interpret this result, we consider S. Berkovich suggestion of a "cloud computing paradigm", in which is given an elegant constructive solution to the problem of the organization of mind. Within his article, he defines a situation where individual brains are not stand-alone computers but collective users who have shared access to portions of a holographic memory of the Universe [64]. He proposed that the cosmic microwave background (CMB) radiation has nothing at all to do with the residual radiation leftover from the Big Bang.

Instead, he claimed that CMB radiation is nothing but noise from writing operations in the holographic memory of the Universe.

Such holographic write operations would require some type of universal clocking rate for these operations. Since the virtual superluminal particle pairs are created and annihilated in the vacuum within a short, finite period of time according to the uncertainty principle, we could logically consider this duration as the clock rate for these operations.

There are some papers on the hypothesis that the human mind consists of evanescent tunneling photons, which have a property of superluminal particles called tachyons [65]. If the human consciousness consists of superluminal particles, as claimed by Prof. Dutheil, the superiority of the human brain to conventional silicon processors may be explained.

He proposed a new set of hypothesis based on superluminal consciousness:

- The brain is nothing more than a simple computer that transmit information.
- Consciousness, or the mind, is composed of a field of tachyons or superluminal matter, located on the other side of the light barrier in superluminal space–time.

Professor Dutheil presented his hypothesis in his book, "L'HommeSuperlumineux" [66], proposing that consciousness is a field of

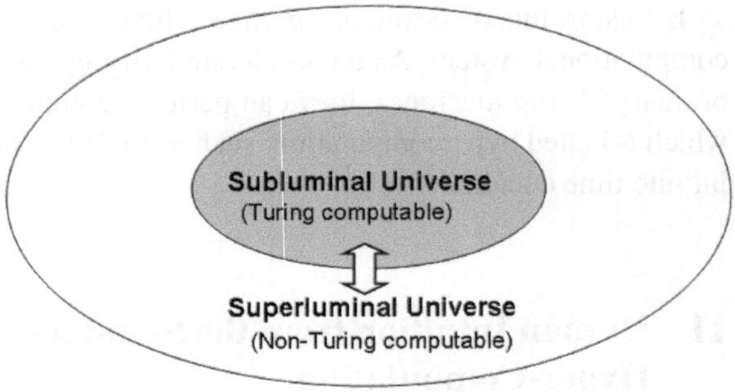

Fig. 22 Structure of the Universe from the standpoint of superluminal particles

superluminal matter belonging to the true fundamental Universe, as shown in Fig. 22, and that our world is merely a subluminal holographic projection of it.

As shown in this chapter, an accelerated Turing machine can be realized by using superluminal particles instead of subluminal particles including photons, and it is considered that the tachyon field outside the human brain acts as an oracle, and this enables us to know the mathematical problem, whether it is true or false.

From the theoretical analysis, it is seen that an accelerated Turing machine can be realized by using superluminal particles from the standpoint of quantum mechanics. Thus an extraordinary capability of human brains such as intuition compared with the ordinary silicon processors might be explained if they are composed of superluminal photons, because they have a capability to function beyond the ordinary Turing machines.

Figure 23 shows the similarity between the computer system and the human brain.

According to the hypothesis proposed by Prof. Dutheil, the brain is nothing more than a simple computer that transmits information.

Human brain receives the program to recognize the world from the Cosmic brain as the terminal computers receive their commands from the central processor, and thus it can be considered that we can recognize the world as the three-dimensional reality.

Figure 24 shows the structures of the brain cell and the Universe.

It shows the similarity of their structures. From which, it is considered that there is a possibility that the Universe is a huge brain operated by superluminal particles.

12 Discussions and Conclusion

In this chapter, it has been shown that the biological brain possibly utilizes large quantum bit computation at room temperature and thus is not synonymous with conventional processors. Due to the non-locality characteristic of superluminal photons, the organism's coherence goes beyond the coherence of the physical biological system. This non-local coherence is what gives the human brain such high performance when compared to silicon digital processors.

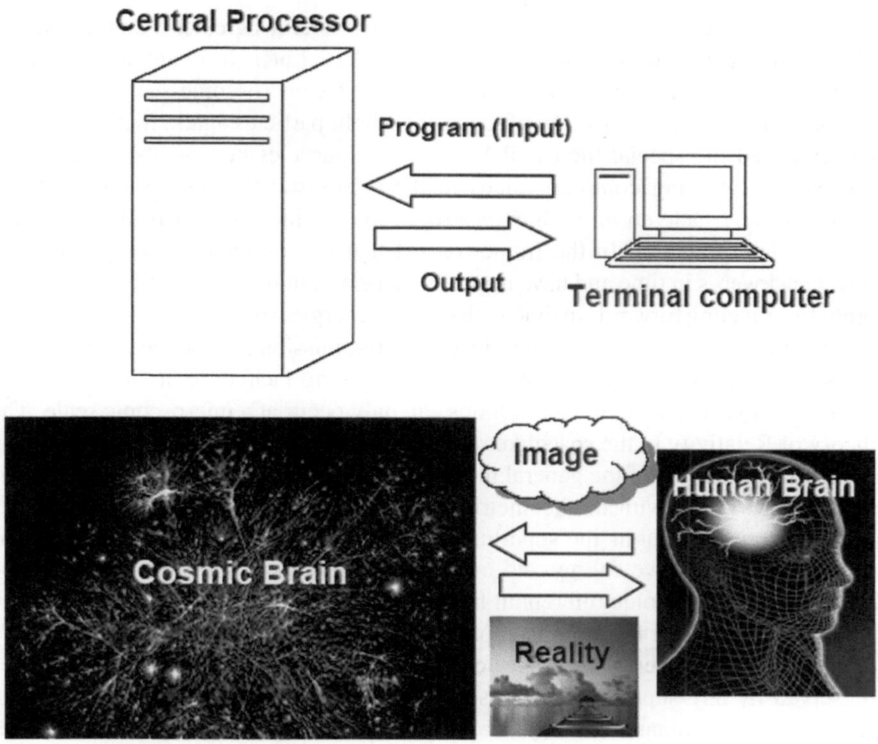

Fig. 23 Similarity between the computer system and thehumanbrain

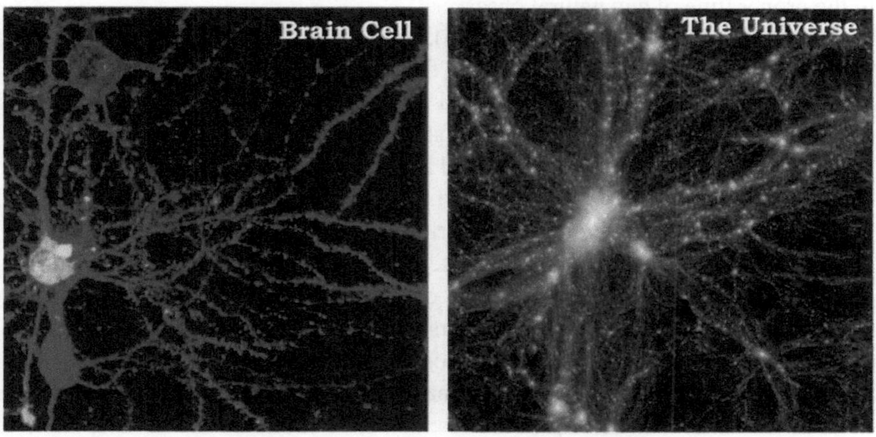

Fig. 24 Similar structure of the brain cell and the Universe. (Is our Universe a huge brain?)

Very early, A. Einstein showed particles with velocities greater than the velocity of light in vacuum may produce causal anomalies. Later, in quantum mechanics CPT transformations have allowed causal loops at a microscopic scale. So there is reason to analyze the possibility of faster-than-light particles again. Meta-Relativity has extended the special theory of Relativity to particles beyond the light barrier (tachyons), by using complex values in the relativistic formula. Meta-Relativity assigns to any tachyon an imaginary proper mass which has no readily apparent physical interpretation. In the framework of that theory, tachyons may appear to travel backwards in time and have negative energies, but they still must also be interpreted as traveling forwards in time with positive energies (reinterpretation principle). Meta-Relativity allows a tachyon reflection or re-emission to produce a causal loop, but some authors reject this objection by postulating tachyon emission cannot be systematically repeated. So causal loops can only occur at a microscopic scale. The theory of Relativity in the spacelike region has been developed by R. Dutheil using the tensor formalism of the general theory of Relativity. He defined tachyonic referential frames (TRF) with another metric tensor and showed it leads to another Lorentz group of transformations-the superluminal Lorentz group. In this theory, tachyons always have a positive energy and a real proper mass, but their behavior must be described with tachyonic referential frames.

R. Dutheil argued from the isomorphism of both Lorentz groups that Zeeman's theorem can be proven to be respected by tachyons; so a sequence order is always preserved by any superluminal transformation. In his present communication, he showed time coordinates of tachyonic referential frames do not preserve causal order and do not make sense for natural observers. Nevertheless, he showed the causal order is preserved within the superluminal proper time of tachyons, which can be related to the proper time of any natural observer.

As regards to Ramanujan (see Fig. 25), who was an Indian mathematical genius, he found many mysterious mathematical formulas related to Number Theory shown in the right part of Fig. 25. He often said that "an equation for me has no meaning, unless it represents a thought of God", that is a Universal mind.

Another case is of the twins, John and Michael, mentioned by Olver Sacks in his book, "The Man Who Mistook His Wife for a Hat". The twins, John and Michael, were idiot savants who exhibited a mysterious human ability with primes using some unknown, unconscious algorithm. They seemed to have a peculiar passion and grasp of numbers even though they could not do simple mathematical calculations, and lacked even the most rudimentary powers of arithmetic. In front of Dr. Sacks, they exhibited an extraordinary ability to recognize an eight-digit number told to them, was a prime number, after some unimaginable internal process of testing. After some time, the twins were able to produce twenty-figure primes, which is difficult even for computers if one uses Eratosthenes's sieve or any other algorithm. There is no simple method of calculating primes. He supposed that they visualized prime patterns instead of utilizing calculation, but the riddle of how they visualized the primes and used them for communication to each other remains unanswered.

According to stochastic electrodynamics, it can be shown a tachyon field can be created from the sea of omnipresent zero-point-energy field in the Universe. If the

Fig. 25 Ramanujan: the man who can communicate with the Universalmind

brain can possibly correlate with a universal tachyon field created from all other living organisms and the cosmic zero-point field, the totality of which we may call a universal brain or a cosmic consciousness, then this cosmic consciousness could be exerting an influence upon an individual mind. This could be the explanation for the extraordinary abilities of the human mind, such as the enigmas of Ramanujan and the John and Michael twins.

Thus it is considered that the true mechanism of the human electromagnetic field may be similar to that of the Universe as a whole, according to the statement, "as above, so below".

Assuming a computing device utilizing superluminal particles could actually be built, what could you do with such a superluminal computer? Such a device would fall into a class of processing machine known as a hypercomputer. These are hypothetical devices more powerful than Turing machines, that allow non-Turing computations. They were first discussed by Alan Turing in the 1930s.

In theory, a hypercomputer can compute certain kinds of otherwise non-computable functions. That sounds handy but even though there are uncountable non-computable functions, it's actually quite hard to come up with an example of one that might seem useful. O. Finkel proved that some basic questions on automata reading infinite words depended on the model of the axiomatic system ZFC [67]. One of possibilities is;

"There is a model of a ZFC set theory, in which the complement $L(A)$ has a cardinal $\aleph_1$ with $\aleph_0 < \aleph_1 < 2^{\aleph_0}$", which means there is a possibility of computation with a higher hierarchy than ordinary Turing machines.

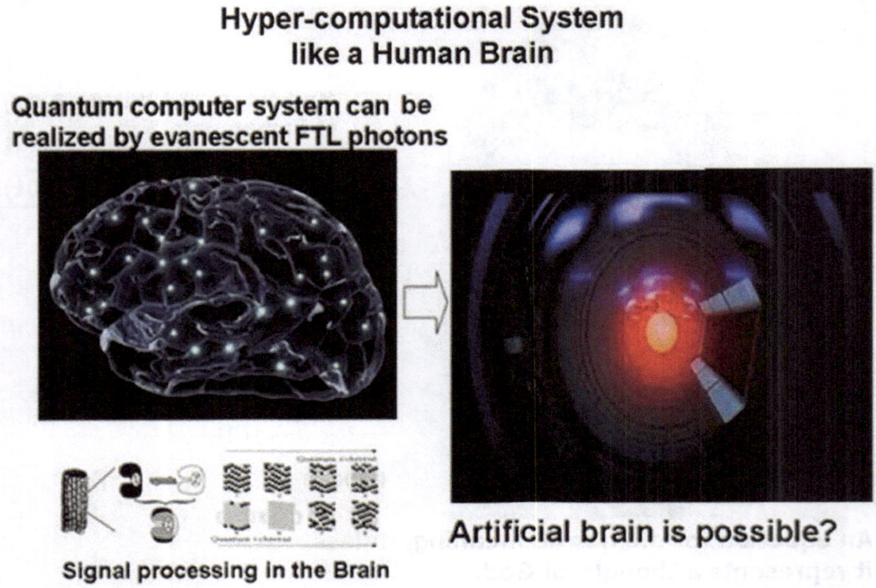

Fig. 26 Is there a possibility to realize the hypercomputer like a brain?

We can suppose it is possible higher order computation might be conducted in the world of a superluminal Universe. It was shown by the hypothesis of Prof. Dutheil, that human consciousness may be composed of a field of tachyons or superluminal matter located on the other side of the light barrier, in superluminal space–time, and would have the capability to conduct infinite steps of computation within a finite time.

By applying quantum mechanics, the author has shown it may be possible to realize a hypercomputational system capable offunctioning beyond an ordinary Turing machine, by information processing conducted inside microtubule structures of neurons, utilizing concepts of superluminal particles.

If such a hypercomputer as shown in Fig. 26 can be realized, it will change theworld completely as we pass the singularity point of computer technology.

References

1. Feynman RP (1999) Feynman lectures on computation. Penguin Books, London
2. Recami E (2001) A bird's-eye view of the experimental status-of-the-art for superluminal motions. Found Phys 32:1119–1135
3. Brown J (1995) Faster than the speed of light. New Scientist 146:26–30
4. Steinberg AM, Kwait PG, Chiao RY (1993) Measurement of the single-photon tunneling time. Phys Rev Lett 71(5):708–711
5. Wang LJ, Kuzmich A, Dogariu A (2000) Gain-assisted superluminal light propagation. Nature 406:277–279

6. Ben-Amram AM (2005) The church-Turing thesis and its look alikes. SIGACT News 3(13):113–116
7. Copeland M, Jack B (2002) Accelerating turing machines. Mind Mach 12:281–301
8. Whipple CT (2002) Nano switching in optical near-field, eye on technology, oe magazine. http://www.oemagazine.com/fromTheMagazine/mar02/pdf/eyeontech.pdf
9. Smolyyaninov II, Zayats AV, Gungor A, Davis CC (2004) Single photon tunneling. http://arXiv.org/PS-cache/cond-mat/pdf/0110/0110252.pdf
10. Benioff P (1982) Quantum mechanical models of Turing machines that dissipates no energy. Phys Rev Lett 48(23):1581–1585
11. Margolus N, Levitin LB (1998) The maximum speed of dynamical evolution. Physica D 120:188–195
12. Musha T (2006) A study on the possibility of high performance computation using quantum tunneling photons. Int J Simul Process Model 2(1/2):63–66
13. Jibu M, Pribram KH, Yasue K (1996) From conscious experience to memory storage and retrieval: the role of quantum brain dynamics and Boson condensation of evanescent photons. Int J Mod Phys B 10(13 & 14):1753–1754
14. Musha T (2005) Superluminal effect for quantum computation that utilizes tunneling photons. Phys Essays 18(4):525–529
15. Chiao RY, Hickmann JM, Soli D (2001) Faster-than-light effects and negative group delays in optics and electronics, and their applications. http://arXiv.org/abs/cs/0103014
16. Haroche S, Raymond J-M (1996) Quantum computing: dream or nightmare? Phys Today 49(8):51–52
17. Gea-Banacloche J (2005) Fundamental limit to quantum computation: the energy cost of precise quantum logic. Math Sci (Tokyo, SaiensuCo. Ltd.) 508:47–57
18. Musha T (2009) Possibility of high performance quantum computation by superluminal evanescent photons in living systems. BioSystems 96:242–245
19. Feinberg G (1967) Possibility of faster-than-light particles. Phys Rev 159:1089–1101
20. Zurek WH (2002) Decoherence and transtion from quantum to classical-revisited. Los Alamos Sci 27:86–109
21. Jibu M, Hagan S, Hameroff SR, Pribram KH, Yasue K (1994) Quantum optical coherence in cytoskeletal microtubules: implications for brain function. Biosystems 32(3):195–209
22. Satinover J (2001) The quantum brain. Wiley, New York, USA
23. Albrecht-Buehler G (1992) Rudimentary form of cellular "vision." Proc Natl Acad Sci USA 89(17):8288–8292
24. Hameroff S, Tuszynski J (2004) Quantum states in proteins and protein assemblies: the essence of life? In: SPIE conference, Grand Canary Island, May 2004
25. Jibu M, Yasue K (1997) What is mind?-Quantum field theory of evanescent photons in brain as quantum theory of consciousness. Informatica 21:471–490
26. Jibu M, Yasue K, Hagan S (1997) Evanescent (tunneling) photon and cellular vision. Biosystems 42:65–73
27. Hameroff SR, Tuzynsli J (2004) Quantum states in proteins and protein assemblies: the essence of life? In: SPIE conference, Grand Canary Island, May 2004
28. Hameroff SR, Penrose R (1996) Conscious events as orchestrared space-time entanglement, selections. J Conscious Stud 3:36–53
29. Hameroff SR (1982) Information processing in microtubules. J Theor Biol 98:549–561
30. Del Guidance E, Doglia S, Milani M (1986) Electromagnetic field and spontaneous symmetry breaking biological matter. Nucl Phys B 257:185–199
31. Vitiello G (1995) Dissipation and memory capacity in the quantum brain model. Int J Mod Phys B 9(8):973–989
32. Faber J, Portugal R, Rosa LP (2006) Information processing in brain microtubles. Biosystems 83:1–9
33. Tegmark M (2000) Importance of quantum coherence in brain processes. Physical Reviews E 61:4194–4206
34. Unruh WG (1995) Maintaining coherence in quantum computers. Phys Rev A 51(2):992–997

35. Hagan S, Hameroff SR, Tuszynski J (2002) Quantum computation in brain microtubles: decoherence and biological feasibility. Phys Rev E 65:061901-1-11
36. Hameroff SR, Tuzynsli JJ (2003) Search for quantum and classical mode of information processing in microtubles (bioenergitic organization in living systems). World Scientific, Singapore
37. Musha T (2006) A study on the possibility of high performance computation using quantum tunneling photons. Int. J. Simulations and Process Modeling 2(1/2):63–66
38. Hameroff SR (2004) A new theory of the origin of cancer: quantum coherent entanglement, centrioles, mitosis, and differentiation. Biosystems 77(1–3):119–136
39. Diosi L (2005) Instrict time uncertainties and decoherence: comparison of 4 models. Braz J Phys 35(2A):260–265
40. Ouyang M, Awschalom DD (2003) Coherence spin transfer between moleculary bridged quantum dots. Science 301:1074–1078
41. Penrose R (1999) The large, the small and the human mind. Cambrigde University Press, Cambridge
42. Talbot M (1991) The Holographic Universe, Harper Perennial, New York, NY, USA, pp 23–24
43. Pribram KH, Nuwer M, Baron RJ (1974) The Holographic hypothesis in memory structure in brain function and perception. In: Atkinson RC, Krantz SH, Luce RC, Suppes P (eds) Contemporary development in mathematical psychology. W.H. Freeman & Co., San Francisco, USA, pp 416–467
44. Georgiev DD (2004) Bose-Einstein condensation of tunneling photons in the brain cortex as a mechanism of conscious action. http://cogprint/3539/01/tunneling.pdf (31 Mar. 2004)
45. Georgiev DD (2004) Quantum computation in the neuronal microtubules: quantum gates, ordered water and superradiance. arxiv.org/abs/quant-ph/0211080 (16 Apr. 2004)
46. Smith T (2003) Quantum consciousness. Water, light speed, and microtubules. http://www.innerx.net/personal/tsmith/QuanCon2.html (31 Mar. 2003)
47. Veselago VG (1968) The electrodynamics of substances with simultaneously negative values of and Soviet. Phys Uspekhi 10:509–514
48. Ung B (2009) Metamaterials: a metareview. www.polymtl.ca/doc/art_2_2.pddf (23 Nov. 2009)
49. Veselago V, Braginsky L, Shklover V, Hafner C (2006) Negative refractive index material. J Comput Theor Nanosci 3:1–30
50. Recami E (2001) A bird's-eye view of the experimental status-of-the-art for superluminal motions. Found Phys 31:1119–1135
51. Recami E (2004) Superluminal tunneling through successive barriers: does QM predict infinite group velocities? J Mod Opt 51(6–7):913–923
52. Ziolkowski RW (2001) Superluminal transmission of information through an electromagnetic material. Phys Rev E 63(4):046604–046617
53. Claus RO (1990) Noncontact measurement of high temperature using optical fiber sensors. Final report, NAG-1–831 (NASA-CR-186975) Virginia Polytechnic Inst. And State Univ., VA, USA, pp 40–50
54. Musha T (2012) Holographic view of the brain memory based on evanescent superluminal photons. Information 3:344–350
55. Bohm D (1980) Wholeness and the implicate order. Routledge & Kegan Paul, New York
56. Ord T (2006) The many forms of hypercomputation. Appl Math Comput 178:143–153
57. Hamkins JD, Lewis A (2000) Infinite time Turing machines. J Symb Log 65(2):567–604
58. Llyod S (2000) Ultimate physical limit to computation. Nature 406:1047–1054
59. Wang LJ, Kuzmich A, Dogariu A (2000) Gain-assisted superluminal light propagation. Nature 406:277–279
60. Musha T (2012) Possibility of hypercomputation by using superluminal elementary particles. Adv Comput Sci Eng 8(1):57–67 also in; Possibility of hypercomputation from the standpoint of superluminal particles. Theory Appl Math Comput Sci 3(2):120–128
61. Park M, Park Y (1996) On the foundation of the relativistic dynamics with the tachyon. Nuovo Cimento 111B(11):1333–1368

62. Kieu TD (2004) Hypercomputation with quantum adiabatic processes. Theor Comput Sci 317:93–104
63. vanMelkebeet D (2000) Randomness and completeness in computational complexity. Lecture notes in computer science. Springer, 2000 edition
64. Berkovich S (2010) Obtaining inexhaustible clean energy by parametric resonance under nonlocality clocking. www.chronos.msu.ru/RREPORT/berkovich_prime_energy.pdf (21 Sept. 2010)
65. Georgiev DD (2003) On the dynamic timescale of mind-brain interaction. In: Proceeding of quantum mind II: consciousness, quantum physics and the brain, Tucson, Arizona, 15 (March, 2003)
66. Dutheil R, Dutheil B (1992) L'hommesuperlumineux. Sand, Paris
67. Finkel O (2011) Some problems in automata theory which depend on the model of set theory. RAIRO-Theor Inf Appl 45(04):383–397

Chapter 6
Replicating a Learning Brain's Cortex in a Humanoid Bot: Pyramidal Neurons Govern Geometry of Hexagonal Close Packing of the Cortical Column Assemblies-II

Pushpendra Singh, Pathik Sahoo, B. Aswathy, Kanad Ray, Subrata Ghosh, Daisuke Fujita, and Anirban Bandyopadhyay

1 Introduction

Recently, an exciting surge of developing a neural brain simulator for an artificial brain model of humans provides features similar to biological, cognitive features [1]. The earlier studies of brain simulation discussed various approaches to consciousness and intelligence by constructing novel mathematical and experimental architectures of artificial neural networks [2, 3]. To simulate the behavior of the cortical columns, the performance of the neural network simulator is limited [4]. Holographic neural communications based on compatibility and dissolution have been used to develop artificial brain models [5]. However, International Business Machines (https://www.ibm.com/in-en) and Blue Brain Project (https://www.epfl.ch/research/domains/blu

P. Singh · P. Sahoo · D. Fujita · A. Bandyopadhyay (✉)
International Center for Materials and Nanoarchitectronics (MANA), Research Center for Advanced Measurement and Characterization (RCAMC), NIMS, 1-2-1 Sengen, Tsukuba, Ibaraki 3050047, Japan

P. Singh · K. Ray
Amity School of Applied Sciences, Amity University Rajasthan, Jaipur Delhi Highway, Kant Kalwar, Jaipur N11C, Rajasthan, India

B. Aswathy
Materials Science and Technology Division, CSIR-National Institute for Interdisciplinary Science and Technology (CSIR-NIIST), Thiruvanathapuram 695019, Kerala, India

S. Ghosh
Chemical Science and Technology Division, CSIR-North East Institute of Science and Technology, NEIST, Jorhat 785006, Assam, India

Academy of Scientific and Innovative Research (AcSIR), CSIR-NEIST Campus, Jorhat 785006, Assam, India

A. Bandyopadhyay and K. Ray (eds.), *Rhythmic Advantages in Big Data and Machine Learning* , Studies in Rhythm Engineering,
https://doi.org/10.1007/978-981-16-5723-8_6

ebrain/) are attempting to artificially create synapses and the brain's neural network using a supercomputer and link the artificially constructed neural network to the biological brain behavior. However, they did not find a proper relationship between the behavior and activity shown by the brain. The brain model constructed with 2.5 million neurons via a large-scale neural network can show various cognitive behaviors [6]. We have modeled the whole brain as dielectric and cavity resonators [7].

1.1 The Role of Cortical Columns in the Brain Models

A cortical column is the arrangement of neurons in a vertical form that shares the same input signal tuning features at different brain regions. In order to understand the organization of the cortical columns, there is a need to describe the branch circuits made of columns and their responses thoroughly [8]. Neurons in the upper layer of the human brain are organized into microscopic columns. The cortex layer is the basis of brain functions and is a major part of the brain. Recent studies have explained the role of a particular type of protein N-cadherin (Ncad), in developing the donut-shaped columnar structure in the larval medulla, the largest ganglion of the Drosophila melanogaster visual center. These structures sit on top of each other in a mature stage of the fly and turn into 3D columns. In 2004, symmetric and asymmetric sections of cortical columns and their migration along different stages at the cortex layer were discussed.

In the cortex layer of the brain, clusters of neurons are arranged in a specific order. The functional properties of cortical columns depend on the organization of neurons sharing similar signal transmissions. Most neurons make equal angles along the axis of the column to provide a better signal transmission. In the fully mature brain, it has been observed that neurons change their orientation after a sensory experience, and re-orientation edits the communication modes of neurons with each other. Some researchers had a negative realization about the cortical column that it could have no function. Initially, the column was characterized as a discrete structure.

The model for understanding the emergence of the intelligence of the cortical columns has been proposed [9]. Each cortical column integrates changing input over time to learn a complete model of objects under-recognition. The location signal for an object represents the relative location of the sensed object. Entire objects can be learned and identified by a single cortical column. The cortical column produces the location signals by integrating sensory and motor input over time [10]. Simulators with A.I. models are used to understand the function of cortical columns, but these are not sufficient to understand the functional mechanism of the human brain. The artificial brain, if made of integrated chips, would consume more power than the biological brain. Some issues related to the artificial brain, such as power consumption, its dimensions, and complexity, remain the subjects for discussion [11]. The neocortex has many types of neuron cells, layers with varied geometries, and complex connections. Such connections are present distinctly as Brodmann's regions in all cortex regions [12]. Creating biological cortical columns via organoids in a petri dish is more difficult and challenging because the basic features of neuron arrangement change while establishing connections with neighboring neurons.

1.2 Defects in Hexagonal Close Packing of Cortical Columns is Fundamental to Learning

A single cortical column consists of thousands of neurons perpendicular to the cortex layer and shows homogeneous stimulus properties all across the cortex domains. A column consists of neurons' connections in the seven layers. Some layers have excitatory lateral connections between neurons over a considerable distance. Cortical columns with excitatory lateral connections interfere rapidly with the surrounding columns. Reliable measurement of individual cortical columns with non-invasive functional brain imaging is difficult. A sub-millimeter range of spatial resolution permits us to barely detect the larger cortical columns.

Most cortical columns are arranged on the cortex layer as hexagonal close packing, and their patterns have also been seen in the somatosensory cortex area. The barrel cortex of the mouse [13] has been studied in terms of cortical column arrangement. However, the functional significance of the barrel cortex is not yet understood [8]. Cortical columns are particularly important in neurobiology because they help in understanding synaptic plasticity. The hexagonal symmetry of cortical columns has an important role in the decision-making process in the human brain. In the 1993s, there was a debate over the hexagonal symmetry of cortical columns, stating that a single column could function independently. However, the output would be concluded by the collective response of the cortical columns [14]. The neurons in the cortical column rearrange asymmetrically as they transmit signals through different phases. We aim to determine the correlation between the electromagnetic resonance peaks of a cortical column made of dielectric materials and its geometry. We are also interested to know how the distribution of electric and magnetic fields on the cortical column depends on the orientation angle of neurons.

1.3 A Map of 3D Clock Assemblies of a Cortical Column

In the 1970s, when the concept of time crystal first arrived, it had only one clock. It was used to explaining the life-like feature of viruses and neurons. All biological clocks in the brain-body system are controlled by the brain that continuously regenerates the living cell. Biological clocks or parodic oscillators typically run all cognitive features of the brain. Here, the cognitive functions of cortical columns are integrated into running clocks, as shown in Fig. 1 with many tied clocks in the 3D architecture. To test whether functional responses of cortical columns are not artifacts, we

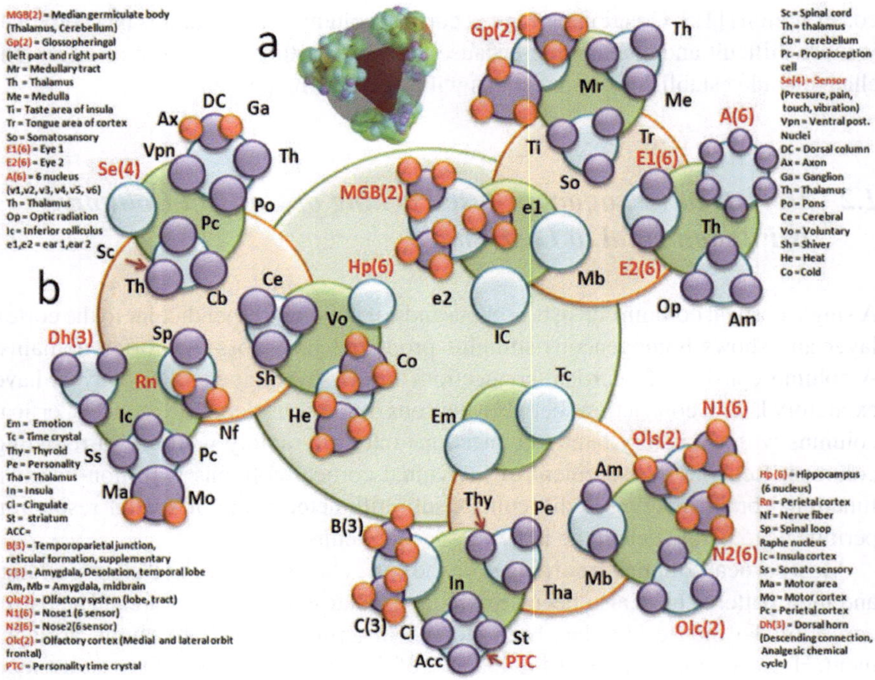

Fig. 1 3D and 2D clocking maps of cortical branches [7, 17] are shown in panel **a** and panel **b**, respectively. Each circle on the map shows a clock. Three circles on the circumference of a circle mean that three clocks run a loop. Following this protocol, a complete clocking map of a cortical column is developed and shown here. The clock assembly of elements in the connectome has geometric and self-assembled features

collected the number of cognitive responses available in columns by literature study [12, 15]. Communication pathways for the brain have been documented [16], and we have selected 20 very well-known conscious features of the human brain [17, 18]. Their communication pathways are presented in terms of the clock. Here, we presented the clocks-based architecture (Fig. 1) to show the cortical column as the operational machine. It is easy to build the clock-based circuit hardware where a set of clocks running soon as a composition of A.C. signal is triggered by a resonant antenna. Cortical column clocking map can sense many patterns following the phase prime matric [19] and geometrical musical language [20]. The clocks included in the clocking map of a cortical column are phase-related and may add a different type of dynamic feature. The map of the presented cortical column is already an outline of maximum clocks. During the brain's learning process, such architecture can be edited by adding some new clocks or removing some. Clocking architecture is a tool for mapping EEG and optical vortex patterns on the HBS brain.

1.4 Fractal Arrangement of Cavity and Dielectric Resonators Are Basic Features of the Brain Construction

There are two dominating types of brain research—accurate spatial mapping of biomaterials and detailed information about its neural wiring and identifying signal pathways running in a loop as clocks or rhythms. Neuron membrane forms a cavity inside. Suitable forms of energies get trapped within this cavity, and all the filaments inside it form an antenna like a dielectric resonator [21–24]. Each filament is also a cylinder hence a cavity resonator. The cavity surface is made of protein dimers; each dumbbell shape dimer is a dielectric resonator. The protein dimer is also a cavity resonator; inside, there is a secondary structure; each secondary structure is a dielectric resonator. The alfa helices are hollow cylinders, hence cavity resonators; however, they are helical dielectric resonators [7, 25]. In this way, the alternate combinations of dielectric and cavity resonators form the entire brain. A chain of resonance frequencies connects all components spanned over a wide time range that satisfies the conditions necessary for building a brain's clocking model. It is easy to construct a large resonator or clocks (<1 mm). Earlier, based on the clocking model of the brain, we have identified twenty types of cognitive features in the human brain. An integrated human brain model with 537 clocks holds cortex and cortical columns as a subset [7]. Most of the clocks used in the 3D clock assembly are taken from the standard literature [12, 17].

In order to find the natural resonant oscillations of the material in terms of dielectric resonance and cavity resonance, both the reflection coefficient (S11) and the transmission coefficient (S21) are simulated for various CST simulator built structures [25]. In a symmetric arrangement of a hexagonal close packing of cortical columns, a center column is surrounded by six cortical columns. In the cortex, in a unit of hexagonal close packing all seven cortical columns are surrounded by 12 columns. Thus the symmetric arrangement of columns in the cortex layer consists of a total of 19 cortical columns. As the human brain learns, the symmetry of the hexagonal arrangement deforms. From the literature, we theoretically created two types of deformed cortical column assemblies to measure learning and memory effects on those symmetric arrangements [26, 27]. Deformed means hexagonal close packing structures do not have an ideal geometric shape. Corner points of the packing are shifted. In order to match the transfer power between the connectome model and the cortical columns, the arrangement of the cortical column is changed so that it resonates in the same frequency domain as the connectome model (~MHz frequency range). In the current work, we have concentrated on the cortex layer alone. However, the cortex layer that holds approximately 200,000 cortical columns in the brain in a few millimeters thick layers is interconnected by connectome. The cortex layer and connectome are integral parts, and one could not be analyzed without the other. A theoretical and experimental discussion of the connectome model has been discussed in our previous work [18].

2 Material and Methods

2.1 Creation of Mimicked Geometry of Neurons and Cortical Columns

Interactions of electromagnetic waves with the neurons of cortical column assemblies are theoretically simulated using the CST (CST = Computer Simulation Technology) simulation software. To find the structural resonance peaks, we selected a wide frequency range from Hz to THz. We monitored the reflection and transmission spectra in MHz frequency range for the cortical column, and Hz–MHz frequency domains for neurons. We purchased pyramidal, horizontal, bipolar, and axon cells from the TurboSquid library (Louisiana, U.S.). Overall, 26 types of neurons were designed as shown in Fig. 2, then we have simulated the resonance properties of

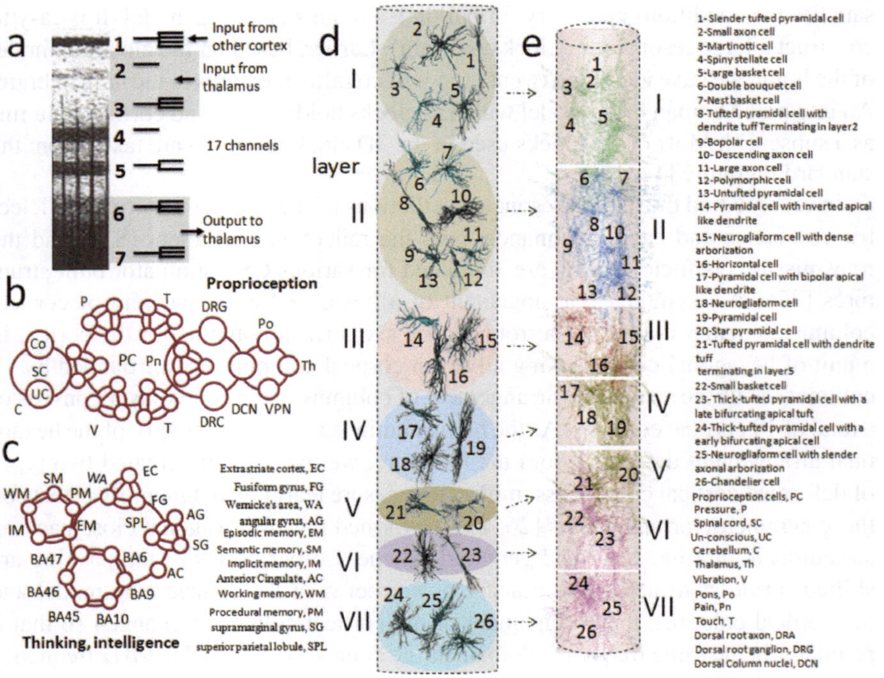

Fig. 2 **a** There are 26 classes of neurons in 7 layers of cortical columns. The column's layers behave as i/p or o/p channels to propagate the signal. Detailed structures of neurons are given on the right side of panel **e**. **b** and **c** Rhythms architectures of proprioception and thinking and intelligence. **d** A schematic diagram of the cortical column. Arranged neurons inside the column layers are given as- 5(1–5)_ layer I, 8(6–13)_ layer II, 3(14–16)_ layer III, 3(17–19)_ layer IV, 2(20–21)_ layer V, 2(22–23)_ layer VI, and 3(24–26) _ layer VII. **e** Photograph of a mimicked cortical column Details of purchased neuron: Neuron's name-pyramidal neuron: purchased site: https://www.turbosquid.com/; product id-1097217; polygons-1,041,365; vertices-1,060,927

neurons one by one (Fig. 3). Once we understand the fundamental properties of each neuron, we build columns in the simulator from real human brain cortical column neuron imaged data. We packed each neuron one by one inside each cortical column, with a particular orientation to replicate the cortex map developed by several brain-building projects. We do not cite particular literature for this purpose. While placing the neuron, we took utmost care that the energy transmission within and in the neighborhood of neurons in the cortical column is favored. Since we deviated from the reported designs eventually, the cortical column became our design. By rearranging neurons, we created different neural assemblies within a single cortical column. Figure 2 shows ten types of neurons created using dielectric material made of C, H, and O; dielectric constants are chosen based on the original materials structural simulation. Finally, we calculated EM field distribution on the structure at the

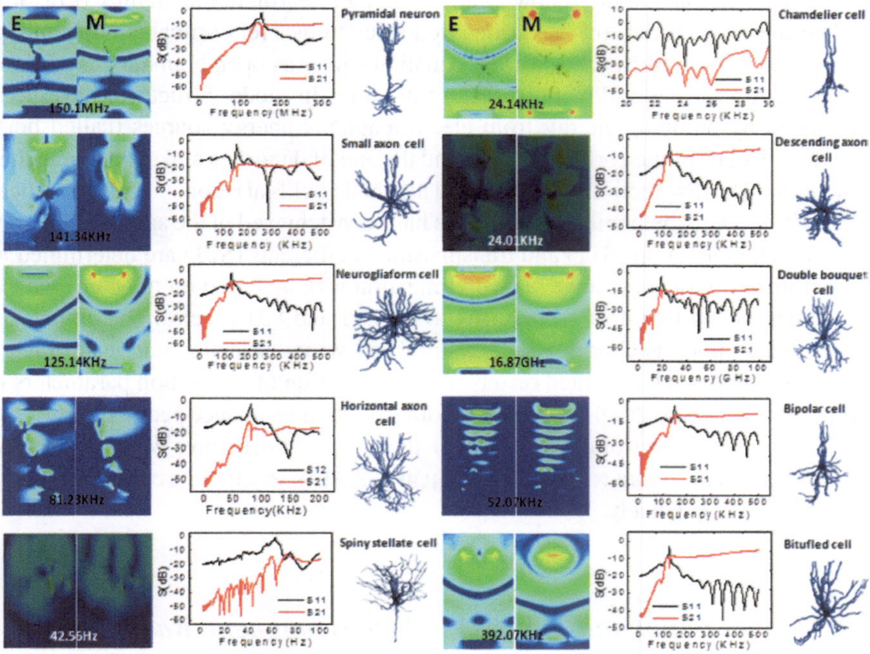

Fig. 3 Here simulated profiles of 10 individual neurons; pyramidal neuron, small axon cell, neuroglia, horizontal axon cell, spiny stellate cell, chandelier cell, descending axon cell, double bouquet cell, bipolar cell, and bitufled cell, are detected in term of resonance spectrum and EM field distribution. All neurons have been stimulated by placing the energy source (port 1 and 2) on their axon and dendrite branches. The EM field distributions are observed due to port 1 at the resonance frequency of neurons. Resonance frequencies of all 10 neurons are given as; 150.1 MHz-pyramidal neuron, 41.34 MHz-small axon cell, 125.14 MHz-neuroglia, 81.23 MHz-horizontal axon cell, 42.56 MHz-spiny stellate cell, 125.14 MHz-chandelier cell, 24.01 MHz-descending axon cell, 460.87 MHz-double bouquet cell, 52.07 MHz-bipolar cell, and 392.07 MHz-bitufled cell. Electric and magnetic fields are noticed as E and M, respectively **Simulation details**: Used solver-Maxwell equation solver; selected mode-time domain; boundary condition-open space; waveguide port dimension-2.3 µm × 2.3 µm; frequency domain-Hz to MHz range

resonance frequencies. A schematic overview of the cortical column is shown in Fig. 2. In Fig. 2a, the bottom two layers 6–7 have a different function from the upper three layers 1–3. Bottom layers send and receive signals to other cortical columns and distant cortex regions. In contrast, upper layers 1–3 receive input signals from bottom layers 6–7 and transmit the signal to the thalamus and other sensory regions instead of bottom layers. Figure 2d shows a single cortical column schematic with 26 classes of neurons interconnected to each other and organized into 7 layers.

The electromagnetic field distribution of each neuron at resonance (Fig. 3) and the reflection and transmission coefficients suggest the proper orientation of each neuron that we need to follow to engage a pair of neurons inside the cortical column. The orientation of each neuron has a specific direction from the horizontal axis of the cortical column (see the Fig. 4 and Table 1). Figure 4b suggests that by changing direction, one could tune resonance frequency emitted by the neuron, apart from the angular orientation that maps how to target another neuron in the vicinity (Fig. 4c).

The energy transmissions at various frequency ranges for cortical column and its assemblies are solved by the Maxwell equation solver as open boundary conditions (open space-x, y, and z-direction) set in time-domain mode. Structures have been stimulated by pumping energy from electromagnetic energy sources (called port) kept at both ends of the cortical column and the end of dendrites and axonal branches of neurons. The distribution of electric and magnetic fields at resonance in the cortical column suggests whether the convergence has been achieved in the simulation or not (Fig. 5). The reflection (S11) and transmission coefficients (S21) are determined in the optimized frequency range. The optimization is done based on the convergence of simulation and finding a proper peak in the S11 and S21 spectra. The dimension of waveguide ports is chosen very carefully, and wrong selection can lead to large simulation times and unwanted results. Proper selection of simulation parameters is required to detect the sharp or low bandwidth structural resonance peaks and periodic oscillations of the electromagnetic field on the dielectric structure. We have selected port dimensions of 4 mm × 4 mm and 2.3 μm × 2.3 μm to simulate cortical columns and neurons, respectively.

2.2 *Longitudinal and Transverse Electromagnetic Radiation at Resonance from the Handmade Giant Cortical Column*

Biomaterials are very different from the conventional materials we use to build electronic components. Yet, we created them in the CST simulator to understand if the ratios of resonance frequencies would be the same though resonance frequencies would differ by orders of magnitude. We could create biological samples like neuron and cortical columns in the simulator by connecting bits and pieces of dielectric slabs. At the same time, we could create 100 times larger structures using similar materials, and by varying size we could understand effective changes in the resonance band (S11, S21, S11, S22). Finally, cortical columns and neurons that are

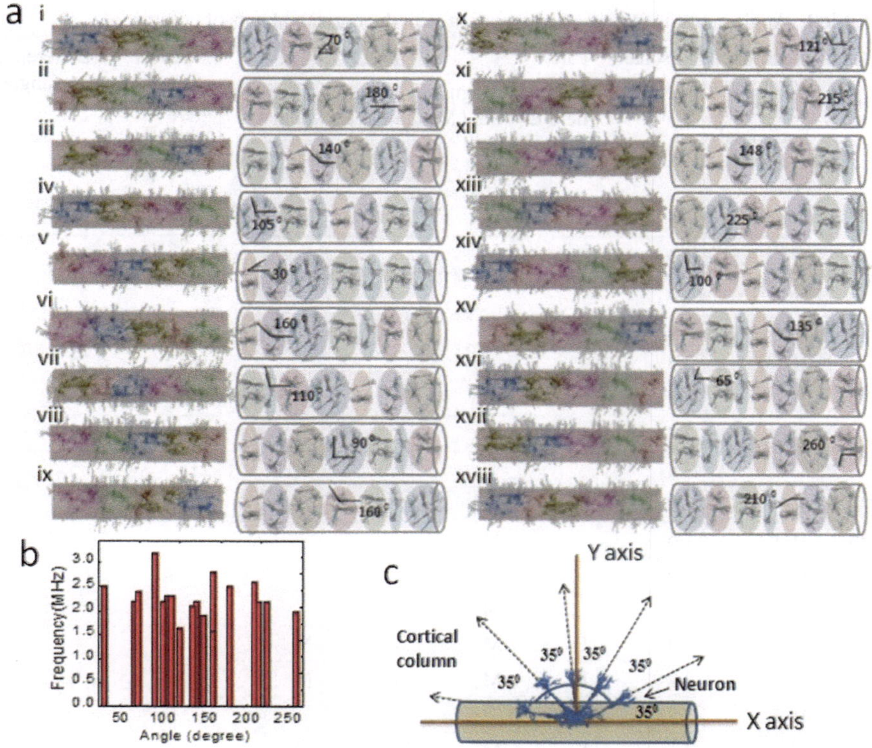

Fig. 4 Orientation angle of 26 types of neurons along the axis of 19 types of cortical columns **a** Orientation angle of the neuron relative to the horizontal (length) axis of the column, such angles have been considered as the dominating angles made by neurons with columns' axis. All 26 types of neurons have been mapped to the real human brain, and four purchased neurons make these neurons; whose names are as follows—pyramidal, horizontal, bipolar, and axon cells. All neurons are purchased from https://www.turbosquid.com/; product id-1097217; polygons-1,041,365; Vortices-1,060,927). **b** The graph between the resonant frequency of the column and the angle created by the neuron from the column axis. **c** Angle analysis of the neuron inside the cortical column. The resultant angle (35°) is obtained from the angle difference where the EM field dominates (see Table 1)

100 times larger than their actual size are built to verify theoretical results. We used conducting and semiconducting cables with suitable rigidity so that complicated neural branches could be replicated. We need to rearrange neural branches to form an analog of synaptic branches; we could do that easily. By following the biological cortical column design available in the literature, all cable-made neurons are properly 3D oriented inside the artificial cortical column. During the construction of cortical columns at a reduced ratio of 1: 1 million fibers, we created neuron branches and soma using flexible insulated copper wire. Since neurons were made by a heavy wire cable, we used foam and folded papers to keep them properly oriented inside the cylindrical cortical column. In the eventual cortical column, all neurons

Table 1 Angle analysis of neurons orientation with the horizontal axis of column

S.no	Neurons orientation angle from horizontal axis of cortical column	Resonant frequency (f) (MHz)	Part 1		Part 2		Neurons orientation angle at which E and M field get concentrate	Difference b/w angle
			Electric field E	Magnetic field M	Electric field E	Magnetic field M		
i	70°	2.4					30°	35
	180°	2.5					65°	35
	140°	2.2					100°	35
	105°	2.3					135°	25
ii	30°	2.5				M	148°	13
	160°	2.8					160°	12
	110°	2.3					215°	55
	90°	3.2					225°	10
iii	160°	1.6		M	E			
	121°	1.6						
	215°	2.2		M	E			
	148°	1.9	E	M	E	M		
iv	225°	2.2	E	M	E	M		
	100°	2.2			E			
	135°	2.1	E	M	E	M		
	65°	2.2	E	M	E	M		
v	260°	2						
	210°	2.2						
	170°	2.6						

were partially short-circuited, ends of their dendrites were soldered, and beyond the threshold, smooth electromagnetic signal transmissions were confirmed by a multimeter. A long helical wire antenna and function generator have been used to work as a port of energy sources used during theoretical simulation. Thus we could replicate the stimulation of dielectric resonance-based transmissions in such devices.

Essential changes in the hardware were made simply by changing the geometry of neuron arrangement in the cortical column. The electromagnetic resonance response of a handmade cortical column is observed experimentally in the MHz frequency range. A photograph of the homebuilt cortical column is depicted in Fig. 7a while the background image of the actual experiment setup is shown in Fig. 8. The cortical column is stimulated by pumping the GHz frequency signals through helical antennas connected at both ends of the cortical column. The signal is pumped by a function generator (R.F. 10 MHz—26 GHz) for the entire frequency band; amplifiers were used if necessary. The output responses of cortical columns are measured by an

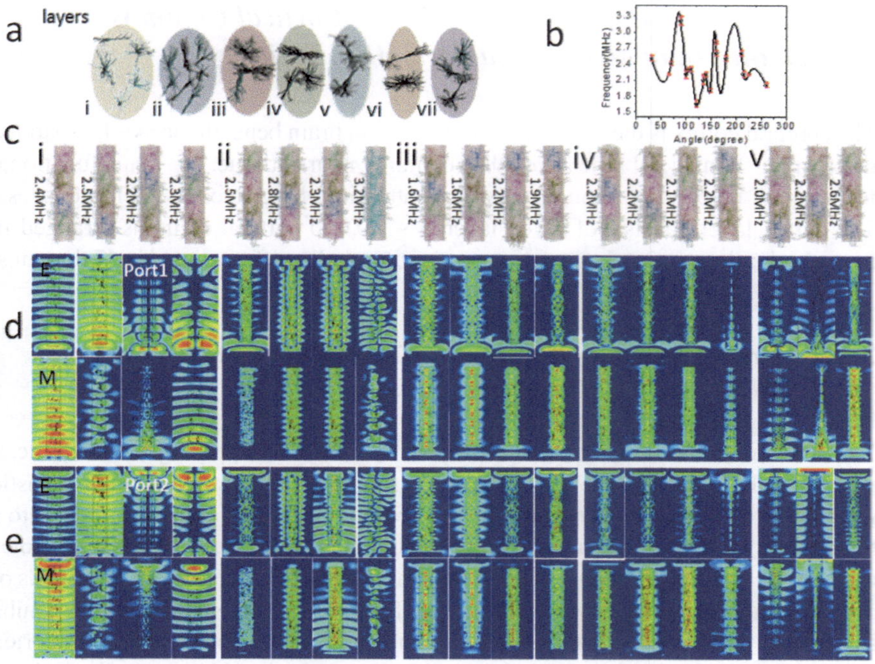

Fig. 5 **a** 7 cortical layers of different neuron orientations. **b** 5 sets of 19 different cortical columns. Resonance peaks of 5 sets are given as: (i) 2.4, 2.5, 2.2, and 2.3 MHz; (ii) 2.5, 2.8, 2.3, and 3.2 MHz; (iii) 1.6, 1.6, 2.2, 1.9 MHz; (iv) 2.2, 2.2, 2.1, and 2.2 MHz; (v) 2.0, 2.2, and 2.6 MHz. **c** The curve between neuron orientation angle on the horizontal axis of the column and the resonance peaks of the cortical column. **d** and **e** Electric and magnetic field distribution at their resonance peaks; set (i) 2.4, 2.5, 2.2, and 2.3 MHz; set (ii) 2.5, 2.8, 2.3, and 3.2 MHz; set (iii) 1.6, 1.6, 2.2, 1.9 MHz; set (iv) 2.2, 2.2, 2.1, and 2.2 MHz; set (v) 2.0, 2.2, and 2.6 MHz of applied ports 1 and 2 at column both ends. Electric and magnetic fields are noted as E and M, respectively **Simulation details**: Used solver-Maxwell equation solver; selected mode-time domain; boundary condition-open space; waveguide port dimension-4 mm × 4 mm; frequency domain-MHz range

R.F. explorer (300 MHz–1.2 GHz) as a 3D resonance spectrum from 600 MHz to 1.2 GHz. The third dimension was time, along with the frequency and amplitude of EM transmission. Resonance spectra are measured with distance from the column to understand the emitted signal strength. The signal is tested at every 90 cm distance interval for 5 m, repeatedly in radial, longitudinal, and transverse directions. Once we learned longitudinal variations of signal emission, the signal is measured radially all around the cortical column at 30° phase difference for a complete phase cycle (0°–360°).

2.3 *Creation of Capillary Tubes-Based Cortical Columns on the Humanoid Bot Subject, HBS Brain*

The cerebral cortex is the outer layer of the human brain beneath the skull, separated by a triplet thin membrane layer that contains neural tissue. Its main functional features are perception, thought, memory, attention, language, and consciousness. The cortex layer consists of seven-layered ~200,000 cortical columns arranged in hexagonal order within a large surface area, structurally differentiated in 47 domains.

2.3.1 Wireless Connection Between the Connectome and the Cortex Layer

We have chosen thermoplastic with proper flexibility. By applying moderate heat, we can easily fold it to make any geometry to suit biocomponents. We cut this plastic resembling the biological structure surface of the brain, and by transforming it into a tube shape, we created the sulcus and gyrus part of the brain. We made the small holes at a gap of 1 cm in a hexagonal lattice shape in the entire plastic to insert thousands of cortical columns prepared by capillary tubes. The entire arrangement of capillary tube assembly is divided into 47 isolated regions known as Brodmann regions on the cortex surface. To mimic the left and right lobes of the human brain with precise dimensions, we took the 3D printed model of the biological brain, wrapped thermoplastic on it, and heated it to the shape of folds on distinct lobes. The sulcus and gyrus energy transmission folds follow the concept of cavity networks described in our previous work [7, 18, 25]. Though we put a replica of connectome neural branches, we do not make any physical contact between connectome neural branches with the cortical columns of the cortex layer. These cortical columns are wirelessly triggered through the Yagi antenna at the ends of connectome fibers that extend in the folds vertically, directing upwards to the cortex layer. The electromagnetic interaction of connectome fibers is measured in terms of optical and magnetic vortices. The vortices form when we wirelessly pump the sensors of the humanoid bot HBS using a pulse stream through ten sources of function generators controlled by the Arduino board. A fixed bandwidth signal from the 35 kHz–4.4 GHz frequency range is passed through the Peripheral Nervous system (PNS) and distributed throughout the brain-body system of the HBS.

2.3.2 The Emission of Electrical Vortices from the Cortical Columns for Non-invasive Study: Shining Different Monochromatic Laser Lights for Acquiring Spatial Data of Different Dimensions

When monochromatic laser light (633.5 nm, He Ne laser) was shined on the capillary glass tube-based cortical columns, the optical vortex-like rings of fields generates. The capillary tubes are filled with helical carbon nanotube solutions, and these nanowires interact with the monochromatic light. Here, we have two kinds of interactions with the helical nanowire (helical CNT ~200 nm dissolved primarily into Hexane). First sensory signals from all over the brain-body circuitry of HBS are wirelessly pumped into the CNT system as electromagnetic wave exposure. To understand the mechanism of vortex modulation in the H-CNTs, we used 12 laser sources of different wavelengths directed at 45° angles along the cortex layer. These laser sources were attached all around the head of HBS. The signal from function generators and sensors embedded in the humanoid bot is used to write the electromagnetic field distribution analogous to the input information in the CNT solution. From sensors to the connectome, HBS is wired, but there is a physical gap from the connectome to the cortex. That gap is bridged by wireless antenna-like communication from the connectome edge to the capillary tubes filled with helical CNT. The electromagnetic field distributes in a 3D pattern in the glass cavity. It resonates with CNTs differently, and electromagnetic fields created the 3D pattern of fields is read by CNTs. On the one hand, the input information creates a unique 3D electromagnetic field pattern, and we let CNTs get changed in the presence of the electromagnetic field.

Input optical vortices get modified by CNTs. We read output hologram as an assembly of vortices as information content in the semiconductor camera. We read output information in two ways. First, establishing a one-to-one correspondence between the output vortex rings and the input electromagnetic signals. Second, estimating the possible structural changes by taking out the solidified CNTs and taking SEM images. Finally, connecting the input sensory signal with the EEG pattern of the HBS brain. Therefore, with training by basic sensory signals, we could correlate external information with the structural changes of the material and the output. Cortical columns do not generate singular but wireless communication between the cortical columns. Finally, the projected optical vortices and electromagnetic field radiate to the upper surface and produce energy distribution read by EEG.

3 Results and Discussion

Here we considered the dielectric and cavity resonator model of the human brain-body system and explained the role of geometry in the 'cortical columns and its assembly' in terms of phase and frequency shifts between the input and output signal. We analyzed the angular orientation of twenty-six filled neurons inside the cortical column. Our study reveals the distribution of electric and magnetic fields over the cortical columns and their assembly and the constituent neurons (see Fig. 1). Considering the same geometrical ratios as a biological cortical column, we have created cortical columns using metal and semiconductor wires and have experimentally detected EM interactions with them. Inside the column cavity, each neural network is also a dielectric resonator. The cavity of the cortical column vibrates as a whole, and the constituents vibrate as an individual unit. Both resonances occur in different frequency domains. Different types of resonances can be understood by localized and global trapping of the transmitted signals inside the cavities. The absorbed, reflected, and the transmitted signals can be in the form of electrons, ions, etc. These carriers vibrate with maximum amplitude at a particular frequency inside the cavity. Following rapid oscillations, the encoded cavity system vibrates with a lower amplitude that particular frequency is the resonant frequency. In the same way, the dielectric material vibrates when it absorbs the signal with a matching frequency.

A rhythm map of 'proprioception' and 'thinking and intelligence' is shown in Fig. 3b and c, respectively. A schematic model of the cortical column containing 26 type neurons (right of Fig. 3e) in 7 layers is shown in Fig. 3d. Figure 3e shows a single simulated map of the cortical column. Figures 5c and 6c, d show the 19 cortical columns, each column with vertically stacked seven layers of neuron assemblies following a hexagonal symmetric arrangement.

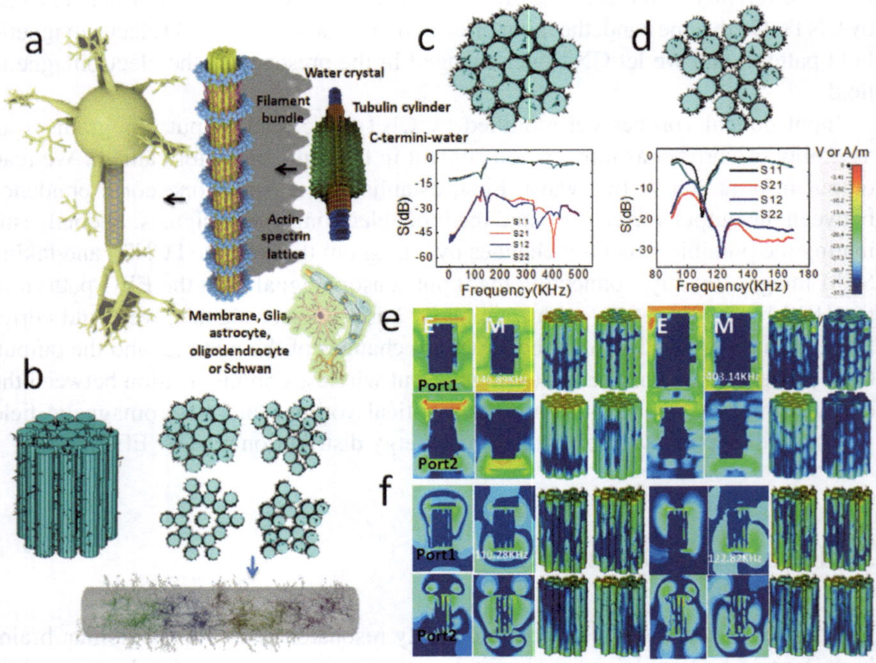

Fig. 6 Panel **a** and **b**. A step-by-step representation of components formed to build cortical column assembly. A single assembly of columns is a bunch of 19 classes of individual cortical columns. Each neuron has an axon which is formed by actin-spectrin lattice and microtubule bundles. Microtubule contains water crystals and alpha–beta tubulin dimmers. 4 classes of column assemblies and their resonance spectrums are shown in panel **b** and panel **c** and **d**. In contrast, the corresponding 2D and 3D EM field profiles are shown in panels (**e**) and (**f**). The EM field is observed at detected resonance peaks; 146.89 kHz, 403.14 kHz-panel (**c**), and 110.28 kHz, 122.82 kHz (**d**) for applied energy ports at both ends. The field intensity scale is shown on the right end of the panel (**d**). E and M symbolize electric and magnetic fields **Simulation details**: Used solver-Maxwell equation solver; selected mode-time domain; boundary condition-open space; waveguide port dimension-25 mm × 25 mm; frequency domain-kHz range

Thalamus receives the information from cortex and other sub-cortical area and sends to the layer 4–5 which divides the column into upper and bottom layers. Mid-layers 4–5 create the feedback circuit with upper layers 1–3 by passing signals through the thalamus and returning to layers 4–5. The thalamus integrates additional information. Layer 1–3 and layer 6–7 have a longer width of neuron's connection that shows the signal communication with other cortical columns in terms of input and output channels [9]. The body's orientation, location, position, and movement are free, while thinking and intelligence show one's ability to understand and interact with others [12].

We purchased pyramidal, horizontal, bipolar, and axon cells from the TurboSquid library (Louisiana, US), and created all essential neurons and 19 different cortical columns (Fig. 5c; i, iii, iv) by following their biological details. The combination of different layers distinguishes created cortical columns. By computer simulation, resonance peaks of all 19 cortical columns were found at the different frequencies in the MHz domain. Figure 5b shows the curve between columns resonance frequencies and angles formed by neurons that dominate more toward the length of columns. Each red point (Fig. 5b) represents the resonance peak of an individual cortical column against the orientation angle of the neuron. The curve has successive peaks and troughs, or it has reflected changes in the structural symmetries of the materials used in building the cortical columns. The changes are periodic, seen as an oscillatory wave as phase angles change 0°–250°.

The cortical column generates the magnetic field by stimulating the human brain. Such emission is detected by magnetic field detectors [28]. In that case, the electrode was placed on the human skull, recorded the output response as a potential difference between the electrode pairs, and measured the magnetic field perpendicular to the skull surface [29, 30]. The concept of information processing through the cortical column is based on the properties of synaptic connectivity sufficient to understand the fundamental properties of the cortical network. Vertical synaptic connectivity is stronger than horizontal synaptic connectivity in horizontal cases. The probability of synaptic connections between pairs of neurons sharply decreases with the distance [31]. We have detected the electric and magnetic field distribution inside the 19 types of cortical columns. Cortical columns are stimulated from both ends by applying waveguide sources, port1 and port 2, and EM field distributions are seen at their resonance peaks.

In Fig. 5e, field simulation profiles show that the neuron's orientation angle is the integral multiple of 35° angles ($n \times 35°$, where $n = 1, 2, 3, \ldots$) (Fig. 4c) along the axis of the column (Table 1). The electric and magnetic fields are periodically concentrated along with the column layers. One could notice alternate high and low-intensity domains. Everywhere inside the brain, electric and magnetic fields have the periodical transmission nature over the brain components such as connectome, blood vessel, neurons, microtubule, ventricle, etc. [25]. However, at some orientation angles, either electric or magnetic field concentrate due to port1 and port2. For example, from Table 1, at 30°, 160°, 215°, 100° orientation angle: M; M, E; M, E; E (M-magnetic field; E-electric field) concentrate for port 2; port 1, 2; port 1, 2;

port 1 while at 65°, 135°, 148°, 215° orientation angle: electric and magnetic fields concentrate for both ports.

The cortical column self assembles and forms the 2D sheet in a hexagonal form that twists in several ways and creates geometrical shapes, for example, nucleus, hippocampus, limbic system, etc. Hippocampus looks like a special structure, but hexagonal close packing of cortical columns on the cortex layer folds uniquely into the hippocampus geometry. However, we have created neuronal assemblies from the cortical column, which are listed in Fig. 6b whereas in Fig. 6b, two column assemblies are stimulated by applying the EM energy sources at both ends. The reflection and transmission coefficients (S11, S21, S12, and S22) of both structures (Fig. 6c, d) were found at frequencies 146.89 kHz, 403.14 kHz, and 110.28 kHz, 122.82 kHz, respectively (bottom panel of Fig. 6c, d). Resonance occurs in a unique way inside the brain; for example, in the case of protein; vibration occurs in the secondary structure of a protein, in the case of microtubule; vibration is caused by change occurs in length, diameter, and pitch of the microtubule, in the case of cortical columns; resonance means neurons symmetry in the organization. The EM field distributions on the column structures (listed in Fig. 6c, d)at their resonance peaks (146.89 kHz, 403.14 kHz, and 110.28 kHz, 122.82 kHz, respectively) are shown in 2D cross-sectional plane and in a 3D form as shown in Fig. 6e, f, respectively. Field distributions follow the resonance of multiple carries in multiple time domains inside the cavity of brain components such as cortical column assemblies, blood vessels, etc. In the symmetrical shape or the hexagonal shape of columns (Fig. 6c), energy is not transmitted inside the column assembly or in the 2D cross-section plane. However, signal propagation appears on the surface of columns assembly or in 3D plane for both waveguide ports1, 2 (Fig. 6e). In contrast, in the asymmetric organization of the cortical columns, the signal passes in and out of the assembly surface in pulse form.

Twenty-six classes of neuron geometries deliver distinct resonance peaks, as shown in Fig. 7. The resonant oscillations of neurons with multiple dynamic modes provide neuronal information processing modes [32, 33]. Neurons reduce the resonant channel noise to organize the clocking structure as it transits from random to an ordered time architecture [34,35]. The transition from the neuron's spike model to the clock model has been reported [36]. A recent study on neurons revealed the time crystal-like nested clock architecture of neurons [21, 37]. The time crystal map of neurons has wide ranges of clocks, some of them are fastest, and some are the slowest clocks that set the limits for the temporal sensitive response. In other words, the neuron behaves as the phase editor [38]. Neuron firing is associated with phase modulation. Such modulations are associated with various kinds of rhythms, e.g., cardinal rhythms [39, 40]. The concept of integrated clock assembly, as shown in Fig. 1 adds phase coupling feature to the basic concept of neuron firing. According to the conventional neuron model, the ionic channel regulates the membrane potential that reflects neuron firing. All the filaments inside remain silent [24].

We have included additional electromagnetic oscillations of the secondary and tertiary firing control parameters such as microtubule bundles. Most importantly, act in and beta spectrin inside the neuron core inspired by STORM and cryo-TEM studies of the internal structure of neurons [41, 42] which regulate the time map of neurons

[43]. Microtubules with different lengths, diameters, and pitches generate specific resonance peaks. Thus, in the human brain, a rhythm-generating circuit assembles different compositions of microtubules, produces the complex signal transmission behavior across the membrane, and regulates the membrane potential for the neuron firing [44]. The geometries and functions of neurons usually control the order and disorder of the human brain. To understand the role of neuron oscillation inside a cortical column, we have replicated the geometry of the neuron considering the dielectric material and observed the relationship between structural oscillation as reflection and transmission coefficients (S11, S21), and also detected the field distribution at their peaks. In Fig. 7, oscillation peaks of neurons have been observed from Hz to GHz frequency domains. Electric and magnetic fields are concentrated along with the soma, axonal, and dendritic branches. The examples of a few cells we studied small axon cell (41.34 kHz), horizontal axon cell (81.23 kHz), spiny stellate cell (42.56 Hz), and descending axon cell (24.01 kHz), while for other neurons; pyramidal neuron (150.1 MHz), neuroglia form cell (125.14 kHz), chandelier cell (24.14 kHz), double bouquet cell (460.87 GHz), bipolar cell (52.07 kHz), and bitufled cell (392.07 kHz). The input energy propagates along with the structure in a rippled form, or these neurons do not resonate and show only the noisy profile.

Figure 7 shows the experimental setup and measurement protocol. Direct experimental testing of the function of the column inside the biological system is challenging. To overcome this limitation, we made fundamental changes in the testing cortical column, compromising the biological properties significantly. Neither there is any ion flow nor mechanical vibration of neural components (Fig. 6a). Eventually, the electric and magnetic field distributions at the resonance frequencies in the cortical column show 3D communication channels without wiring. Intertwined field lines connect different parts of the neighboring and far distant neurons. This is the most interesting discovery that we have made in all our studies. We do admit that our study is not replicating the true biological environment. However, the resonance of the cortical column along longitudinal, radial, transverse, and circular directions suggests a new form of communication that is scarcely discussed in biophysics exploration. We have found that the radiating or emitted field strength is limited to a certain distance in all three directions; in longitudinal direction-6.7 m, redial direction-3.8 m, and transverse direction-2.2 m. However, structural resonance profiles (x-axis-time (μs), y-axis-frequency (MHz), z-axis-intensity amplitude (dBm)) are measured at intervals of 90 cm distance in each direction. For all 3 cases, the electromagnetic resonance profiles of a cortical column are nearly identical in looking and appear to be of equal magnitude. A common resonance peak is detected at 744.95 GHz frequency (white arrow). The circular loop has two resonance peaks of 744.95 GHz (red arrow) and 957.33 GHz (blue arrow) frequency (top of panel Fig. 7a). At the same time, the field intensity appears and disappears (dBm) at resonance frequencies (744.95 GHz-red curve and 957.33 GHz-blue curve) or shows the periodical nature of the field during the variation in the circular angle (Fig. 7b). However, in longitudinal (black), redial (red), and transverse (blue) direction, field strength curves have a non-periodically relationship with the distance at the 744.95 GHz resonance peak.

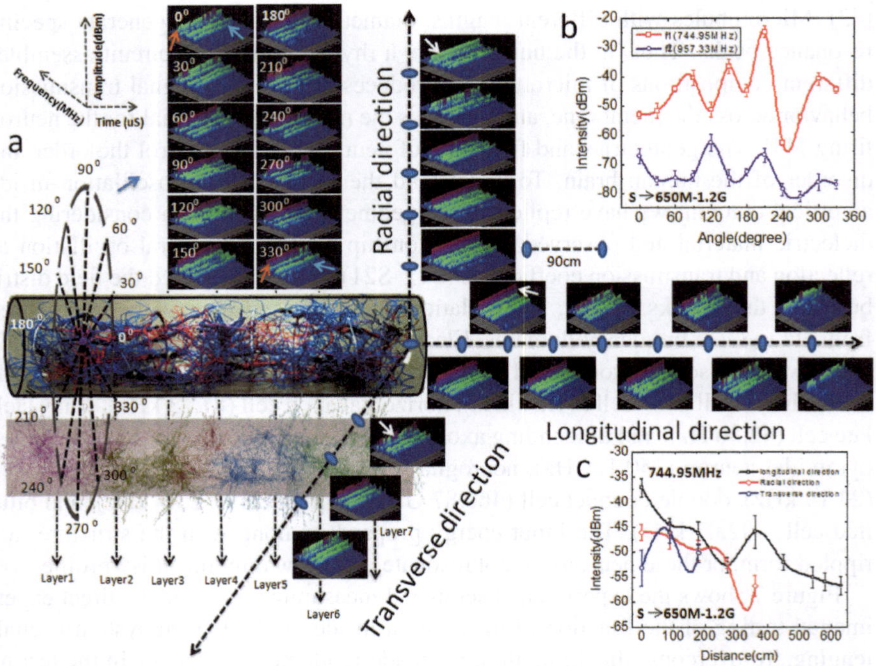

Fig. 7 **a** A photograph of a handmade cortical column. A short-circuited map of 26 classes of neurons is packed inside the cortical column using a single wire cable at a ratio of 1:10^9. 3D resonance spectrums are mapped in cyclic (0°–330° angle), radial, longitudinal, and transverse directions. Radial, longitudinal, and transverse direction emitted radiation distances are 3.8 m, 6.7 m, and 2.2 m, respectively. The dominant resonance peaks are observed at 744.95 GHz (red arrow) and 957.33 GHz (blue arrow) frequency and a common resonance peak (white color) at 744.95 GHz frequency in all three directions. **b** Variation in radiation intensity (dBm) at both resonance peaks 744.95 GHz and 957.33 GHzwith the phase angle 0°–330°. **c** Intensity variation (dBm) plot with the distance of the radiation emitted in radial, longitudinal and transverse direction at 744.95 GHz resonance peak

The cortical column is a unit for signal processing in a complex manner through input signals via interconnected neurons and produces the output.

3.1 H-CNTs-Based Cortical Column

To verify the electromagnetic resonance band, we conducted an experimental study using an analog of the biological cortical column, formed by a capillary tube and containing seven different layers of liquids toluene, Hexane, ethanol, benzene, methanol, chloroform, and heptane, which dissolve in CNTs. When light passes through such layers, due to solvents, an electromagnetic field modulates. Helix-filled H-CNT solutions of various lengths, pitches, and diameters [17] are self-assembled

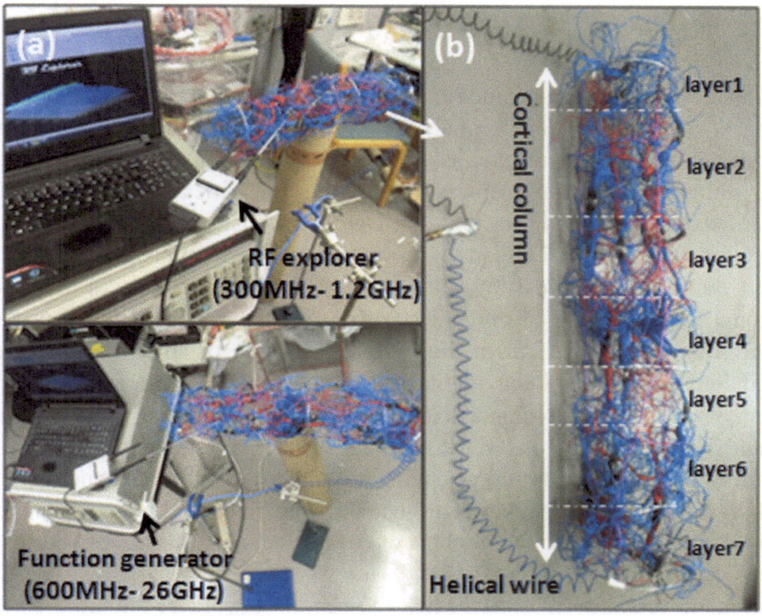

Fig. 8 **a** A picture of the actual experimental setup. We pumped the signal of the 600 MHz–26 GHz frequency range to both ends of the cortical column through a function generator. Single cable helical wires were used to feed the signal into the cortical column. Output responses were detected by R.F. Explorer (300 MHz–1.2 GHz) on a computer screen. We need to select the frequency range in which we want to see the responses of the cortical column, so we tested well in the 300 MHz–1.2 GHz frequency range. **b** The handmade cortical column is made up of semiconductor wire. In this, the signal passes through 4 layers composed of 26 types of neurons. Neurons orient differently. We tried to place several neurons (Soma) at 35° angles along the horizontal axis of the column, inspired by the theoretical case. At this orientation angle, it shows maximum efficiency for emitting a high-distance signal

through connectome fibers by sensory input signals. Information sent to these tubes edits to the self-assembled H-CNTs. Such columns are sensors that transmit and edit sensory information through the overall body as human cognitive features. The end terminals of the connectome fibers are shaped like the Yagi antenna that feeds the signal into the capillary tube.

We have tried to explain the ratio of resonance frequencies in cases of theoretical-based, classical-based, and H-CNT-based cortical columns. We have obtained the resonance peaks for a theoretically constructed cortical column by solving the Maxwell equations in the electromagnetic simulation software CST (computer simulation technology). For H-CNT-based cortical columns, we detected resonance peaks using frequency R.F. explore (see details in Fig. 8). In contrast, we determined the frequency at which a high-intensity optical and magnetic vortex is produced for the cable-based homebuilt cortical column. The electromagnetic resonance response for all cases is the same, at least in terms of the frequency ratio.

3.2 Non-linear Properties of Cortex

Most of the studies in the geometric correlation of signal transmission in a cortical column layer describe spatial connections of neurons limiting the possibilities of communications within the domains of linear processes [45, 46]. Various researchers have explored Non-linearity. However, these non-linearities are limited to 1D variations of sequential signal transmission. Here, we consider the entire system made of the alternate cavity and dielectric resonators, packed one inside another. Consequently, the signal transmissions are vertical, one inside another. In the deformed hexagonal close packing of cortical columns, we have looked into neurons with cortical columns in the term of the orientation angle of structural projection of emitted fields and effect of that holographic or antenna-like projection on the change of symmetry of cortical column assemblies [47–49]. We made an effort to match the resonant signal transmissions in the artificial structures in terms of optical vortices with the EEG patterns captured when the artificial brain of the humanoid bot learns new things. Linking the outputs as projected by the humanoid bot to the uppermost cortex layer of the brain to the rhythms of the cortical columns has not been studied before.

Brodmann found in 1900–1920 that the neural network of the brain cortex has distinct geometries. The neuron arrangement in particular domains of the 2D sheet of the cortex is different. He initially suggested that the cortex layer be divided into 53 regions based on the typical geometric features. Later, with high-resolution mapping of neuron configurations, the number of distinct functional regions was decreased from 53 to 47. Here, we tried to understand the effect of electromagnetic field distribution on 47 Brodmann areas or thin cortex layer [50, 51] when we apply MHz signals from a channel of 10 R.F. sources on the assemblies of cortical columns. We could not take many neurons for a particular area, but using 7 neurons in a single unit could understand functional variations. The finding for us in the CST simulation for the Brodman area was significant as far as the symmetry of resonant oscillation frequencies is concerned. It appears that the thin cortex layer adopted 47 distinct symmetries to cover a wide range of integration of resonant oscillation frequencies. In a particular area, to find out the role of the cortex column, we recorded the EEG pattern on the artificial cortex surface that we built, and simultaneously we monitored the modulation of optical vortices transmitted by the cortex layer to estimate the structural changes in the cortical column structures.

The solitary cortical column operates independently, but the output responses emerge as a collective response of a selected region of connectome from the cortex. In the theoretical studies, we found that high-intensity electric and magnetic fields dominate in the cortical columns, with most neurons making 35° angles along the longitudinal axis of the cortical column. In the experimental studies, we have placed wire-made neurons at 35° angles along the axis of the cortical column device. It emits signals up to a distance of 6.7m in the longitudinal direction. In the biological experiment, we poured in an assembly of helical wires of different densities in the capillary tube. The emitted signals generate high-intensity vortices as a function of input electromagnetic signal frequencies.

Figure 6 accounts for our effort to understand the effect of deformation in geometry on the electromagnetic field distribution. We read the field distribution from point to point on the cortex layer just above the humanoid bot head made of thermoplastic by placing electromagnetic detector sensors. Brain learning leads to changes in cortical column assembly. Here in our bot, the lattice structure does not change, its internal materials conformation changes significantly. Our humanoid bot used to gather information from outside using its artificial eyes, nose, ear, skin cells, etc., as outlined in Fig. 9a. Therefore, our software-free bot could sense external world and affect the cortical column assembly.

Two types of resonances are present inside the biomaterials. The cavity resonance and dielectric resonances (see Sect. 1.1) periodically oscillate energy in a localized spatial domain. These spatiotemporal clocks are often intertwined in the hardware of the material. These loops are phase-locked, and therefore equivalent clocks are used to find an assembly of 3D clock architecture or a clocking model of biomaterials. We have shown earlier that thermal noise drives these clocks [7, 17]. Electrical noise is often a combination of electromagnetic signals of different frequencies. During transmission of signals into a periodic loop, the noises are filtered by a biomaterial, which turns into quantized energy. Here we described the 3D clock assembly of all primary components or clocking model of the cortical columns as shown in Fig. 2. The signals of appropriate frequencies along the cylindrical cavity

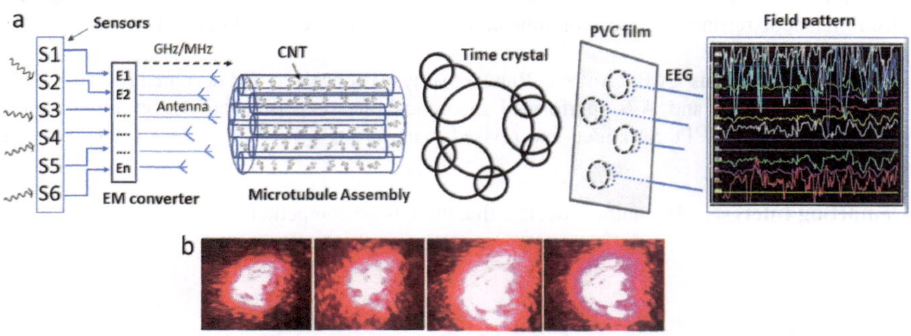

Fig. 9 The output of the hardware-humanoid bot. **a** Signals coming from the surrounding space and function generators are sensed by sensors mounted on the body of the HBS. These signals pass through the entire nervous system of the HBS body and then into capillary tubes filled with H-CNTs, arranged in a hexagonally form. Such signals activate the helical nanowires present in the capillary tubes. We can edit the information of helical nanowires by triggering the solution in the MHz or GHz range via Yagi antennas. The reactions of the cortical column assemblies are read by placing a 16 probe EEG machine above the HBS brain, made of PVC plastic. The output of EEG was found in the α, β, γ, δ waveform, as shown at the most right. **b** Photographs of optical vortices are shown here. The vortices are generated by illuminating the cortical column assemblies using light of different wavelengths through a 12 laser sources arranged in a circular form around the bot head. To read the information from the vortices, we have created a vortex analyzer simulator which converts these vortices into time crystals. We read the information in form of time crystals [17]

cause its resonant oscillations in various nested loops. Using the existing literature on the energy transmissions and correlating the periodic oscillations with the 3D clock architecture, we have identified the possible sub-clock structures for energy interaction. Readers could compare Fig. 1 with Fig. 9b optical vortices.

4 Conclusion

Here, we validated theoretical simulations of resonance bands by dissolving various liquids such as toluene, Hexane, ethanol, benzene, methanol, chloroform, heptane into H-CNT and place them in the capillary tube. We found more accurate electromagnetic responses in emitted vortices by the cortex assembly layers than the wire-made cortical column. Here, we do not expect to replicate all biological features but only one, which has exchanged the geometric feature as a ratio of frequencies. The theoretical and experimental results derived are consistent with each other. The spontaneous and simultaneous emission of a signal from the cortical columns has remained a mystery for a long time. Developing a map of clock assemblies enables the representation of phased, de-phased, coherent, conditional coupled clocks in the form of neural code as time. The cortical map in a functional Brodmann region circuit is also an integrated module of clocks.

Acknowledgements Authors acknowledge the Asian office of Aerospace R&D (AOARD) a part of United States Air Force (USAF) for the grant no FA2386-16-1-0003 (2016–2019) on electromagnetic resonance-based communication and intelligence of biomaterials.

Author Contributions A.B. (Anirban Bandyopadhyay) conceived and designed research; Pu.Si. (Pushpendra Singh) and A.B. performed theoretical and experimental study; Pu.Si. drafted manuscript; A.B. and Pu.Si. edited and revised manuscript, K.R. AsB and P.Sa. S.G. reviewed the work.

Competing Interests The authors declare that there is no competing interest.

References

1. Fildes J (2009) Artificial brain 10 years away. http://news.bbc.co.uk/2/hi/technology/8164060.stm
2. Goertzel B (2012) Special Issue on mind uploading. Int J Mach Conscious 4(1):1–3
3. Bamford S (2012) A framework for approaches to transfer of a mind's substrate. Inter J Mach Conscious 4(01):23–34
4. Bergman D et al (2019) JANOS: an integrated predictive and prescriptive modeling framework. https://arxiv.org/pdf/1911.09461.pdf
5. Manger R (1991) Holographic neuronal networks. Mathematical Communications
6. Eliasmith et al (2012) A large-scale model of the functioning brain. Science 338(6111):1202–1205

7. Singh P et al (2020) A self-operating time crystal model of the human brain: can we replace entire brain hardware with a 3D fractal architecture of clocks alone? Information 11(5):238. https://doi.org/10.3390/info11050238
8. Horton JC, Adams DL (2005) The cortical column: a structure without a function. Philos Trans R Soc Lond B Biol Sci 360(1456):837–862
9. Hawkins J, Ahmad S, Cui Y (2017) Why does the neocortex have layers and columns, a theory of learning the 3D structure of the world. bioRxiv 162263. https://doi.org/10.1101/162263
10. Lewis M, Purdy S, Ahmad S, Hawkins J (2019) Locations in the neocortex: a theory of sensorimotor object recognition using cortical grid cells. Front Neural Circuits 13:22
11. Makin S (2019) The four biggest challenges in brain simulation. Nature 571(7766):S9–S9. https://doi.org/10.1038/d41586-019-02209-z
12. Carter R (2014) The human brain book: an illustrated guide to its structure, function, and disorders. D.K.; United Kingdom, London
13. Woolsey TA, der Loos HV (1970) The structural organization of layer IV in the somatosensory region (SI) of mouse cerebral cortex. The description of a cortical field composed of discrete cytoarchitectonic units. Brain Res 17(2):205–242
14. Gawne TJ, Richmond BJ (1993) How independent are the message s carried by adjacent inferior temporal cortical neurons? J Neurosci 13:2758–2771
15. Zeki S, Shipp S (1988) The functional logic of cortical connections. Nature 335(6188):311–317. https://doi.org/10.1038/335311a0
16. Achard S, Salvador R, Whitcher B, Suckling J, Bullmore E (2006) A resilient, low frequency, small-world human brain functional network with highly connected association cortical hubs. J Neurosci 26:63–72
17. Bandyopadhyay A (2020) The making of an artificial brain from a time crystal. Taylor & Francis Inc. Imprint CRC Press Inc., Publication City/Country Bosa Roca, United States, 372
18. Singh P, Sahoo P, Ray K, Ghosh S, Bandyopadhyay A (2020) Building a non-ionic, non-electronic, non-algorithmic artificial brain: cortex and connectome interaction in a humanoid bot subject (HBS). Advanced intelligence systems and computing. Springer Nature
19. Reddy S et al (2018) A brain-like computer made of time crystal: could a metric of prime alone replace a user and alleviate programming forever? Soft Comput Appl (Springer, Singapore) (2018)
20. Agrawal L et al (2016) Fractal information theory (FIT) derived geometric musical language (GML) for brain inspired hypercomputing. Advances in intelligent systems and soft computing (AISC), vol 2. Springer, pp 37–61
21. Ghosh S et al (2016) Inventing a co-axial atomic resolution patch clamp to study a single resonating protein complex and ultra-low power communication deep inside a living neuron cell. J Integr Neurosci 15(4):403–433
22. Sexena et al (2020) Fractal, scale free electromagnetic resonance of a single brain extracted microtubule nanowire, a single tubulin protein and a single neuron. Fractal Fract 4(2):11. https://doi.org/10.3390/fractalfract4020011
23. Singh et al (2020) Reducing the dimension of a patch-clamp to the smallest physical limit using a coaxial atom probe. Progr Electromagn Res B 89:29–44
24. Singh et al (2021) Electrophysiology using coaxial atom probe array: live imaging reveals hidden circuits of a hippocampal neural network. J Neurophysiol. https://doi.org/10.1152/jn.00478.2020
25. Singh P, Ray K, Fujita D, Bandyopadhyay A (2018) Complete dielectric resonator model of human brain from MRI data: a journey from connectome neural branching to single protein. Lecture notes in electrical engineering, pp 717–733. https://doi.org/10.1007/978-981-13-1642-5_63
26. Chklovskii DB, Mel BW, Svoboda K (2007) Cortical rewiring and information storage. Nature 431:782–788
27. Chklovskii DB, Schikorski T, Stevens F (2002) Wiring optimization in cortical circuits. Neuron 34:341–347

28. Arlinger S, Elberling C, Bak C, Kofoed B, Lebech J, Saermark K (1982) Cortical magnetic fields evoked by frequency glides of a continuous tone. Electroencephalogr Clin Neurophysiol 54(6):642–653. https://doi.org/10.1016/0013-4694(82)90118-3
29. Elberling C, Bak C, Kofoed B, Lebech J, Saermark K (1981) Auditory magnetic fields from the human cortex. Influence of stimulus intensity. Seand Audiol 10:203–207
30. Hari R, Aittoniemi K, Jirvinen M-L, Katila T, Varpula T (1980) Auditory evoked transient and sustained magnetic fields of the human brain. Localization of neural generators. Exp Brain Res 40:237–240
31. Boucsein C, Nawrot MP, Schnepel P, Aertsen A (2011) Beyond the cortical column: abundance and physiology of horizontal connections imply a strong role for inputs from the surround. Front Neurosci 5:32
32. Llinás R (1988) The intrinsic properties of mammalian neurons: insights into central nervous system function. Science 242:1654–1664
33. Wang XJ (1994) Multiple dynamical modes of thalamic relay neurons: rhythmic bursting and intermittent phase-locking. Neuroscience 59:21–31
34. White JA, Rubinstein JT, Kay AR (2000) Channel noise in neurons. Trends Neurosci 23:131–137
35. Enright JT (1980) Temporal precision in circadian systems: a reliable neuronal clock from unreliable components? Science 209:1542–1545
36. Aviel Y, Gerstner W (2006) From spiking neurons to rate models: a cascade model as an approximation to spiking neuron models with refractoriness. Phys Rev E 73:051908
37. Agrawal L (2016) Inventing atomic resolution scanning dielectric microscopy to see a single protein complex operation live at resonance in a neuron without touching or adulterating the cell. J Integr Neurosci 15(4):435–462
38. Lundstrom BN et al (2009) Sensitivity of firing rate to input fluctuations depends on time scale separation between fast and slow variables in single neurons. J Comput Neurosci 27:277–290
39. Andersen P, Eccles JC (1962) Inhibitory phasing of neuronal discharge. Nature 196:645–647
40. Benson JA, Jacklet JW (1977) Circadian rhythm of output from neurons in the eye of Aplysia. J Exp Biol 70:151–211
41. Xu K, Zhong G, Zhuang X (2013) Actin, spectrin and associated proteins from a periodic cytoskeleton structure in axons. Science 339:452–456
42. Hirokawa N, Sobue K, Kanda K, Harada A, Yorifuji H (1989) The cytoskeletal architecture of the presynaptic terminal and molecular structure of synapsin 1. J Cell Biol 108:111–126
43. Arvanitaki A, Chalazonitis N (1968) Electrical properties and temporal organization in oscillatory neurons. In: Salanki J (ed) Neurobiology of invertebrates. Plenum, New York, pp 169–174
44. Liebovitch LS et al (1987) Fractal model of ion-channel kinetics. Biochim Biophys Acta 896:173–180
45. Striegel DA, Hurdal MK (2009) Chemically based mathematical model for development of cerebral cortical folding patterns. PLoS Comput Biol 5(9): e 1000524
46. Reimann MW et al (2015) An algorithm to predict the connectome of neural microcircuits. Front Comput Neurosci. https://doi.org/10.3389/fncom.2015.00120
47. Braitenberg V, Braitenberg C (1979) Geometry of orientation columns in the visual cortex. Biol Cybern 33:179–186
48. Klyachko VA, Stevens CF (2003) Connectivity optimization and the positioning of cortical areas. Proc Natl Acad Sci 100:7937–7941
49. Mountcastle VB (1997) The columnar organization of the neocortex. Brain 120:701–722
50. Brodmann K (1909) VergleichendeLokalisationslehre der Grosshirnrinde. Johann Ambrosius Barth, Leipzig
51. Garey LJ (2006) Brodmann's: localisation in the cerebral cortex. Springer, New York

Chapter 7
How the Biofield is Created by DNA Resonance

Ivan Viktorovich Savelev, Richard Alan Miller, and Max Myakishev-Rempel

1 Introduction

1.1 Biofield

Although there is plenty of evidence for the existence of the biofield, the evidence that DNA interacts with it is very limited. Therefore, the idea that the biofield is created by the mass of DNA in the organism remains a hypothesis. We will first briefly summarize the existing evidence and then briefly review our studies, in which we used computational genomics to reveal the traces of resonance signaling in the genome and provided statistical evidence for this resonance signaling.

The idea of the biofield is ancient. Indian yoga teaches about prana as life energy and chakras as centers of resonant vibration. Traditional Chinese medicine teaches about the meridians, which work as conduits of information, and herbal medicine for balancing the energies. Yoga and Chi-Gong teach the art of mastering the energies and redirecting them. Chi-Gong practitioners use body movements to send and direct the biofield energy through the air.

I. V. Savelev · M. Myakishev-Rempel (✉)
DNA Resonance Research Foundation, San Diego, CA, USA
e-mail: max@dnaresonance.org

Localized Therapeutics, San Diego, CA, USA

R. A. Miller
OAK, Inc., Grants Pass, OR, USA

A. Bandyopadhyay and K. Ray (eds.), *Rhythmic Advantages in Big Data and Machine Learning* , Studies in Rhythm Engineering,
https://doi.org/10.1007/978-981-16-5723-8_7

1.2 Brief History of Biofield Research

Scientific biological experimentation with biofields ran in parallel with the development of electromagnetic science. Franz Mesmer called it "animal magnetism". In his experiments, water and plants were charged by him with magnetism, and hypnotized (mesmerized) people were able to detect the presence of this energy [1, 2]. The chemist Jöns Jakob Berzelius proposed in the early nineteenth century that a regulative force exists within and maintains the functions of living matter [3]. A widely used physiology textbook of the nineteenth century by Johannes Peter Müller [4] taught that the presence of a soul makes each organism an indivisible whole and that the behavior of light and sound waves showed that living organisms possessed a life-energy unexplainable by physical laws. The concept of biofield was introduced by the experimental biologist and metaphysicist Hans Driesch. In 1892, Driesch proposed the field properties of organisms as a conclusion of his experiments with fertilized sea urchin eggs [5]. In the early 1920s, Alexander Gurwitsch, Hans Spemann, and Paul Weiss independently postulated the idea of "morphogenetic fields" that guide the organism's development [5]. This was experimentally proven by Gurwitsch and others [6–8]. Modern experiments demonstrated biological information can travel through the air [9–12]. Burlakov demonstrated that biological systems can strongly affect embryo development and that by manipulating the biofield it is possible to produce developmental abnormalities in fish embryos, causing them to develop extra heads and tails [13, 14]. The effects of the biofield on embryos on each other were reproduced technologically—and the fields were produced by a device that caused similar developmental abnormalities in fish embryos [13, 14].

1.3 DNA as a Source of the Biofield

We (RAM) proposed the idea that the mass of the DNA of the organism is a source of the biofield 46 years ago [17, 18]. This idea makes a lot of sense. All of the genetic information about an organism is located in DNA. The genome is a complex instruction, 3.2 billion bases long, located in every cell of the body except red blood cells. DNA is one of the most harmoniously structured biomolecules; it comprises a substantial amount of the body. An average adult body contains about 300 g of DNA or 0.3% of the body weight. Since we already know that the biofield exists, it makes much sense that it should be directly defined by the genome of each cell and for the genomes of all cells to communicate with each other through it. This would also be in line with the hypothesis of Gurwitsch, suggesting that the genome creates the biofield and uses it to guide the development and homeostasis of the organism.

As we mentioned, the evidence for the involvement of DNA in the creation of the biofield is very limited. Although experiments that need to be done are quite feasible and inexpensive, they have been traditionally stigmatized by mainstream science in part because biofield therapy undermines the value of pharmaceuticals. Experiments that would prove the involvement of DNA and the biofield are pretty basic. They would involve manipulation of the DNA sequence, spectroscopic studies, and cell culture studies in model organisms. The experiments of Gurwitsch and Burlakov about the effects of the biofield on embryo development should be reproduced and the role of DNA in these effects should be tested. Specifically, we need to demonstrate that biofields affect the chromatin structure of DNA and, inversely, that modifications of DNA produce changes in biofields. Toward that goal, we and others have previously demonstrated that specific frequencies of weak nonthermal radiations of electromagnetic waves change gene expression [15]. We have also found that electromagnetic radiation with specific wavelengths protects DNA from damage in cell culture (not yet published). Yet, much more remains to be done to demonstrate the role of DNA in the creation of the biofield.

The main challenge is that the resonance language of DNA is still not deciphered. We find it very unlikely that frequencies by themselves carry information. We believe that, as it is in technological communications, certain frequencies serve the role of carrying waves, and the information in the biofield is transmitted through the modulation, shape, and interference patterns of electromagnetic waves. In addition, as we discuss below, we believe it is possible that DNA transmits information through other types of waves and structures.

1.4 Our Computational Studies of DNA Resonance

We will now review our results on the computational tracing of resonance signaling in the genome. Our research starts from a series of conjectures, but then it comes to testable hypotheses, the testing of which produces substantial evidence. The human genome was sequenced 20 years ago. By now, the complete genomes of 5,300 animal and 1,500 plant genomes have been sequenced [16] and much of the functional genomic data have been accumulated. By conditions of academic grants, the researchers are obligated to upload their genomic data into public databases. So, the amount of useful genomic information is exceptional. A large number of large genetic studies have been accomplished in the last 25 years, including studies for psychiatric, neurological, metabolic, and other genetic disorders. With all this achievement, much is still unknown and this causes disappointment among funding organizations, the researchers, and the public because even with all this powerful data, many questions remain unanswered and many therapies are still not possible. For example, there is no cure for cancer, AIDS, MRSA, and many other infections, arthritis, autoimmune and psychiatric disorders. Existing drugs often don't use the vast genomic information and are still selected by chemical trial and error. We view this to be a result of the limitation of mechano-chemical materialistic scientific dogma. Mainstream scientists refuse to accept already proven results on biofield and stigmatize research in biofield and bioresonance therapies. Although the ideas of DNA resonance introduced by us 46 years ago [17, 18] are embraced by some of the public, they remain prohibited and largely unknown to mainstream scientists. One of the main mistakes of mainstream science is the failure to adequately research the function of so-called junk DNA.

1.5 Noncoding 99% of the Genome

The initial publication of the human genome 20 years ago demonstrated that protein-coding sequences comprise only 1% of the genome but there is an additional 2% of the genome that are noncoding but highly conserved (as is the coding 1%), suggesting that these sequences are functionally significant. Yet, mainstream science almost exclusively focused on the coding part of the genome, ignoring the vast 99% of the noncoding part of the genome. The ENCODE project initially was able to publish the evidence for the importance of noncoding parts of the genome [19], but then this work was criticized by the more conservative side of the mainstream science [20] and thus the funding for the research of noncoding parts of the genome was slowed down [21]. We reason that these functional noncoding sequences may function as resonators, antennas, and electronic circuits in natural DNA resonance signaling. This is in agreement with the experimental results published by us and others that light affects gene expression [15, 22]. Also, our yet unpublished experiments demonstrate that pretreatment of cultured mouse fibroblasts with certain wavelengths of the low power millimeter waves protects DNA from damage by subsequent treatment with UV light.

2 The Model

Our model of the genomic biofield is based on thorough literature research, discussions with biophysics researchers, laboratory experiments, and on our own abilities to perceive biofields. Each author of this chapter is trained in the Eastern healing energy arts. RAM and MMR taught the subject and have published books on the topic. Here we describe the current version of our genomic biofield model. Currently, it is largely hypothetical. We found some supporting evidence for it, but more research is needed to verify and prove it.

2.1 Electron and Proton Chains

Every cell in our body except red blood cells contains two copies of the genome. The genome sequence consists of 3.2 billion basepairs A-T and G-C. Each basepair oscillates between its tautomeric states [23] with an estimated frequency in the megahertz-terahertz range. GC basepairs oscillate between three tautomeric forms and AT basepairs oscillate between two tautomeric forms. Each purine (A or G) of the basepair contains an aromatic ring. In the tautomeric oscillations, the aromatic ring oscillates between the aromatic and non-aromatic states. The GC basepairs exist in a predominantly nonaromatic state and switch to an aromatic state in short pulses. AT basepairs, on the other hand, exist predominantly in an aromatic state and switch to nonaromatic states in short pulses. Stretches of purines (like for example, AAAAAAA or AGAGAGAG) are frequent in the genome. Aromatic rings in purine stretches are united in one system and their electrons delocalize along the entire stretch. Oscillations of electrons in purine stretches are coordinated and thus formed aromatic electron chains are among the main oscillators in the genome. While stacks of aromatic electron chains of purine stretches are negatively charged, there are also oscillations of proton chains that are positively charged. These proton chains are longer on average than electron chains made of stacked aromatic rings and are formed by longitudinal hydrogen bonds.

In contrast to classical transverse hydrogen bonds, which have been well-known since the discovery of the double helix, longitudinal hydrogen bones have been described, but are not well-known [24]. We have developed an algorithm that predicts the structure of longitudinal hydrogen bonds in DNA and demonstrated that repetitive elements (called HIDERs) containing stretches of longitudinal hydrogen bonds are enriched by evolution and conserved in evolution [25]. Therefore, the chains of protons are also defined by the DNA sequence and are frequent in a genome. The hydrogen bond protons are partly delocalized along these chains. We believe that these proton chains are a second major type of oscillators in a genome alongside the aromatic electron chains. The oscillations of electron chains and proton chains in DNA are coordinated between each other. This coordination is must be complex since protons are 800 times heavier than electrons. These oscillations are likely fueled in part by thermal motion. Importantly, the oscillations of charges within the base stack are insulated from surrounding water by the hydrophobicity of the base stack and highly charged sugar-phosphate backbone. Due to this insulation, the oscillations of electric charges in DNA don't necessarily lead to the mechanical oscillations in DNA and this may be protected from being dumped by the viscosity of the nucleoplasm.

Therefore, according to our model, the genome contains a large number of electron and proton chain oscillators. These oscillators are interconnected into one computer-like system by the conductivity of DNA. The conductivity of DNA was experimentally demonstrated and quite well-studied in the early 2000s [26, 27]. It has been demonstrated experimentally that DNA is highly conductive, has little or no resistance to electric current, and the base stack of DNA works as an insulated wire, ensuring that charges do not escape from the base stack.

2.2 DNA Resonances in Chromatin

Based on our molecular models of charge oscillations, we observed that DNA methylation, the epigenetic markup language in eukaryotes, doesn't block the resonances but likely affects their frequencies. The DNA in our cells is packed in a complex chromatin structure. In each cell nucleus, the double-stranded DNA is coiled into 30 million nucleosomes made of histone proteins. Each nucleosome is shaped like a short spindle of hockey puck proportions. In nucleosomes, the negative charge of the DNA is balanced by the positive charge of the proteins of the core nucleosome particle. These nucleosome DNA protein complexes are highly organized and likely serve as units of oscillations as well. Nucleosomes are very mobile, they roll along the DNA strand. We believe that nucleosomes might read the resonance code of DNA while rolling along the DNA strand. To illustrate the idea, consider videotape players which were popular around the 1980s. The videotape was positioned at an angle and coiled around a spinning hockey puck-shaped reading head. The head rolled through the tape, inducing electrical oscillations in the tape, and retrieving the information stored in the tape. This is pretty much how we envision reading of resonance patterns by nucleosomes: DNA is wrapped around nucleosomes in a very similar fashion to the reading head and the videotape player (except that videotape makes 0.5 turns around the reading head, while DNA makes 1.7 turns around nucleosome). As the nucleosome rolls through DNA, it reads the resonance information from it and converts it to acoustic signals that it broadcasts to the nucleoplasm. Since there is a large number of nucleosomes in the nucleus (30 million), they collectively create a moving sound pattern, which directs the work of the cell and controls the movement and function of proteins. In other words, nucleosomes play a symphony using DNA as sheet music, and proteins dance to this music. The principle of using sound to direct biological processes is known as cymatics [28]. Sonotherapy is known from ancient times and recently was revived [29]. We believe that sound oscillations created by multiple cells direct the process of embryogenesis.

2.3 DNA Resonances in the Body

In addition to sound signals that are produced by the cells, there are also electrical signals, electrical oscillations, voltage oscillations, and electromagnetic waves that are exchanged between the cells. The tissues of the body are connected into a unified fiberoptic network provided by fascia tissue. This network is known from ancient times as traditional Chinese meridians. The extracellular matrix contains protein filaments that work as fiberoptic guides. They collect information from the outside of the cell membrane and direct it into the unified fiberoptic network of the fascia tissue.

This fiberoptic network is directly connected to our brain and is part of our mind. DNA of the whole body is directly involved in the thinking process. This way, a well-known thinking process that goes through the electrical neuronal firings is supplemented by a faster and more complex fiberoptic network thinking process. The electrical neuron firing component is responsible primarily for the control of movement, sensation, and communication. The fiberoptic component is responsible for the thoughts, intuition, and nonlocal communication of the brain.

2.4 Resonances in Noncoding DNA

Only 3% of our genome is coding for the protein. The rest is termed "junk" DNA. It's puzzling for materialist scientists why the genome contains so much noncoding DNA. Especially puzzling for them is the fact that a large portion of noncoding DNA is conserved in evolution indicating that it is functional. We believe that noncoding DNA is responsible for resonance signaling and it is a biocomputer. It performs computational and coordination functions responsible for the control of development and metabolism of the organism.

2.5 Resonances of Repetitive Elements

Over 50% of our genome is represented by highly repetitive sequences. Some of the repetitive sequences are tandem repeats. In tandem repeats, short sequence units are repeated periodically as wagons in a train. Other repetitive elements are interspersed repeats, which are scattered around the genome without any perceivable order. The Alu element is the most abundant and most interesting of the repetitive elements. It belongs to the category of interspersed repeats. There are over 1.1 million copies of the Alu element in the genome. Alu sequence is unique to primates. Other species have Alu-like elements but they differ substantially in sequence. The DNA string of the Alu element coils around almost two nucleosomes: one complete nucleosome and a part of another nucleosome. The Alu element is also bound by a special protein called an Alu-binding factor. Alu elements are enriched in gene promoters, and minor variations in Alu element sequences are responsible for variations in gene activity. We believe that Alu elements are the "control elements" predicted by Barbara McClintock 70 years ago. We believe that Alu elements receive the wave information and convert it into biological information by controlling gene expression. Also, since Alu elements are unique to primates, it's likely that it is the main DNA sequence that makes us human and that they are responsible for the abilities of our mind. We also think that Alu elements may function as memory units of our mind. Although 1.1 million Alu elements per genome (2.2 million per cell) are similar in sequence, patterns of their methylation vary. Since methylation patterns are stable for years, Alu elements could serve as long-term memory units of our mind.

As we showed previously [23], each basepair is working as a logical unit switching between intuitive and choice-making states. In the process, it turns on and off the aromaticity of the purine aromatic ring. Multiple basepairs share protons and electrons in an organized discrete fashion, thus coordinating and making interdependent tautomeric oscillations in different basepairs. An Alu element contains 300 basepairs and this makes it a minicomputer. Since the Alu element occurs 2.2 million times per cell and occupies about 10% of the genome, it's likely that it is responsible for a large portion of the cell's computation.

Several other interspersed repetitive elements likely have unique vibrational properties and functions. The seemingly random localization of the repetitive elements in the genome is likely nonrandom. The repetitive elements are likely the words in a highly organized text of the genome, and the combinations of the repetitive elements make sentences. Local groups of repetitive elements likely serve as vibrational units performing a specific function and having a specific logic.

Genes in a genome are written sequentially. The gene has a beginning and is read from the beginning to the end. However, the vibrational information doesn't have to be read from the beginning to the end. The whole genome could be vibrating at once and be read as a holographic record [18].

2.6 Coordination of Chromatin Dynamics

As mentioned above, chromatin is highly mobile in living cells. When a gene is expressed, a piece of DNA containing the gene is unwrapped, and when the expression is over, it is compacted back to its original location. The genome is stored in chromosomes and each chromosome is a separate macromolecule. We believe that each chromosome has its unique set of frequencies.

Each chromosome is located in a specific chromosomal territory occupying a part of the nucleus [30]. Movements of the parts of the chromosome are highly coordinated. We believe that DNA moves by itself by using nucleosomes as acoustic propellers. The nucleosome, which is a complex of DNA and protein, creates vibrations which propel the nucleosome in a specific direction within the chromosomal territory. The nucleosomes or groups of nucleosomes might be sensing their position in the nucleus via vibrations and coordinate their movements using this vibrational information. The nucleosome might serve as an interface between the vibration in the DNA and the biochemistry of the cell. We believe that the nucleosome converts the DNA sequence information into the vibrations which are sent outside into the nucleoplasm and vice versa: the nucleosome may receive the vibrations from the nucleoplasm and direct it into the DNA, thus modifying the pattern of its oscillations.

We believe that collectively, the genome within the nucleus directs the work of the cell, it creates a field that coordinates the locations and the movements of the macromolecular complexes and similarly, the totality of the DNA (300 g) of an organism creates the field which coordinates the growth and the health of the organism.

2.7 *Experiments Needed to Test the DNA Resonance Hypothesis*

The above model is based on a series of conjectures. It is a hypothesis that requires extensive verification and all of its components should be experimentally verified. A typical experiment would require discovering and recording the fields and frequencies produced by the genome, reproducing them, and demonstrating that these frequencies and fields are responsible for the biologic effect. In order to prove that these fields are produced by the genome, manipulation of the DNA sequence is required. Also, this manipulation is needed to decipher and read the vibrational language (code) of the genome.

Genetics and genome manipulations already firmly established that that genomic sequence defines the shape, physiology, and behavioral qualities of the individual. Yet, the current inability of science to decipher the 99%-noncoding part of the genome doesn't yet allow scientists to predict the shape, physiological, and behavioral qualities of the individual. Nonetheless, even without understanding the vibrational code of the genome, the emerging field of personalized medicine is now developing algorithms for this purpose based on the empirical associations of genomic variations (SNPs) with health and behavior. This has now become possible due to the availability of high throughput genotyping and vast electronic medical records. Understanding the vibrational code of the DNA would greatly facilitate this development. Once the vibrational code of the genome is understood, the sequence of the genome of a person would allow recreating the photograph, health, and behavioral profile of the person, or more precisely, to the extent in which these qualities which are defined genetically, and this genetic component for many qualities is over 50%.

The initial experiments for deciphering the vibrational code of DNA may start from spectroscopic studies of natural DNA in physiological conditions. By manipulating the sequence of the DNA, it should be possible to demonstrate that its vibrational properties change when its sequence is changed. This set of basic experiments will be the key to deciphering the resonance code of the genome. Spectroscopic studies of reconstituted nucleosomes are likely to demonstrate their functions as vibrational units. These are basic experiments that need to be done for the verification of the DNA resonance model.

Although the experimental verification of the model remains to be done, we were able to confirm some of its basic statements using computational genomics. Computational genomics is cheaper and faster than experimental research.

2.8 *Our Computational Results on DNA Resonance*

Many academic funding sources require that scientists upload their experimental genomics data into public databases. So, the deciphering of the resonance code of the genome can be started and did start from using the public data.

Based on the resonance model, we approached the genomic sequence as a program that contains resonating elements. We assumed that the main resonating elements of the genome are repetitive sequences, but we also assumed that there must be some resonating elements among the nonrepetitive unique parts of the genome. Here, we utilized the principle of redundancy. The principle of redundancy is known from the genetic code, which is partly redundant: the same aminoacid is often coded by different triplets. For example, Tryptophan aminoacid is coded by AGA, ACC, ACA, and ACG triplets. We assumed that in the resonance code, it is also possible that multiple nucleotide sequences could code for the same resonating structure. For example, purines A and G are strikingly similar to each other. Similarly, pyrimidines C and T are also similar to each other. Since purines A and G contain the aromatic ring in the same position, we assumed that the basic resonating structure made out of purine rings depends only on the purine/pyrimidine structure and would be independent of the specific nucleotides that compose this structure. For example, the sequence GTGT would resonate with the sequence ATAT because A and G are purines with a similar chemical structure and have similar positioning of the aromatic ring. Therefore, we formulated the concept of HIDERs (Homologous If Decoded Elements Repetitive). A HIDER is a special kind of repetitive elements. Sequence fragments are called "purine HIDERs" if they have a different primary sequence (comprised of A, G, C, and T letters) but have an identical purine sequence (comprised of R and Y letters). In physical terms, HIDERs are resonating sequences that vary in outside appearance but have similar forms of their resonators (antennas). To illustrate this, consider smartphones. They vary in their appearance but their antennas (located inside) are tuned to the same carrying frequency allowing them to connect to the same network.

Next, we explored the possibility that our genome might be enriched with HIDERs, especially in functionally important regions. We did a computational search for purine HIDERs in the genomes of various organisms and found that purine HIDERs are enriched by evolution with high statistical significance and also are located in conserved areas of the genome. This gave support to our resonance model and pointed to the functional significance of HIDERs since evolutionary conservation is a sign of function.

In addition to delocalized electron chains, we investigated the delocalized proton chains in DNA. Based on molecular structure, predicted the existence of longitudinal hydrogen bonds in the DNA and the existence of delocalized proton chains created by hydrogen bonds [25]. We formalized the algorithm by which the DNA sequence can be converted into the sequence of the hydrogen bonds and formulated the idea of proton HIDERs. Proton HIDERs are defined as two DNA sequences that differ in the primary DNA code and have a common sequence of hydrogen bonds. We have screened genomes of multiple species using this algorithm and demonstrated that proton HIDERs are strongly enriched by evolution and are located in the conserved areas of the genome, suggesting their functional importance.

Typically, sequence elements located in transcription start sites control gene expression by modifying the chromatin structure near the transcription start. We observed that proton HIDERs are colocalized with gene transcription start sites with high statistical significance, suggesting that they are involved in gene regulation. Most likely, HIDERs located near the transcription start receive vibrational resonance signals at a distance and convert this vibrational information into biochemical information by controlling gene expression. The reverse information flow should be also considered. The resonance elements located in transcription starts will receive biochemical information from the gene expression and convert it into vibrational information and transmit this vibrationally at over a distance. This way, the computational genomics approach provided support for our model of DNA resonance signaling.

3 Discussion

3.1 Approaches to the Research of DNA Resonance

Traditionally, the research of biofields is done either through experimental or theoretical approaches. In the above research, we went the third way, which was somewhat hybrid. We used 3D molecular models and computational genomics. In the 3D molecular modeling approach, we utilized public high-precision crystallographic experimental data, and in the computational genomics approach, we utilized the experimentally identified public genome sequences of multiple organisms. We also utilized public data about conserved genomic regions and experimentally identified locations of the transcription starts. Currently, we are utilizing experimentally identified public data on mutations in the genome and their functions.

We now also utilize the extensive published experimental data on the biological effects of electromagnetic fields, on the physics of electromagnetic and sound waves, quantum mechanics, cellular biology, and many other pieces of information that were obtained experimentally by others. While experimental data in biology is often imprecise and requires many replication studies, the data on genomic sequences and 3D molecular structures is digital and highly reproducible. It is rare to find reproducible digital data in biology, but the genomic sequence is an exception. It's highly reproducible and is composed of very specific digital sequence information.

3.2 Morphogenetic Function of Biofield

The biofield has unique properties at distances comparable with the size of the organism and yet its effects are not mediated by chemical signaling or neuron firing. The idea of DNA resonance signaling implies that specific sequences exchange information at a distance without using chemical messengers. This is a radical and timely departure from the conservative materialistic worldview. While the details of chemical messaging, morphogene gradients, and neuronal firing are well-established, we find these mechanisms to be insufficient to explain the complexity, precision, and speed with which organisms develop their structure and function. Similarly, we find neuronal firing and biochemical regulation in neurons to be not sufficient to explain the work of the mind.

The theories and experiments of the biofield pioneered by Gurwitsch and others [6–8] demonstrated the existence of field-based morphogenetic mechanisms and their importance for organism development. Some 45 years ago we have proposed that it is the mass of the organism's DNA that produces this biofield. Now, with the emergence of experimental and computational genome research methods, it is necessary to investigate how does the 3.2 billion nucleotide-long genomic sequence produces the biofield and interacts with it. The key to that will be deciphering the resonance language (same as vibrational language or resonance code) of the genome.

3.3 Opportunities for Research of Resonances in Noncoding DNA

As with any book, the genome is far from being uniform. Genome sequence recorded as a computer file occupies about 0.8 Gigabytes. When the first draft of the human genome was completed, the participants of conferences were given CDs with the complete genome sequence on them. Our genome could be printed in about 6000 books of average size.

There's a treasure trove of public genomics databases and functional genomics data that can be used to start deciphering the resonance code of the genome. Among this data, there is much data that can not be explained under the purely chemical paradigm. It is waiting for being reinterpreted using the DNA resonance signaling paradigm. As mentioned above, the importance of the non-coding DNA, also referred to as junk DNA, has been highlighted by the first publication of the human genome and by the publications of the ENCODE project [19].

As we previously mentioned, repetitive elements are the best candidates for being the universal resonating sequences in the genome. Alu elements are present in large numbers (1 million copies, 10% of our genome), colocalized with gene promoters, and are involved in the regulation of gene expression. We believe that the large numbers of repetitive elements of the genomes of complex organisms have several explanations.

1. These repetitive elements are used for DNA resonance signaling.
2. They define the program of the organism's development. More complex organisms have more repetitive elements.
3. We believe that repetitive elements also participate in the work of the mind, and this agrees with the observation that organisms with more complex behaviors have more repetitive elements.
4. A large number of repetitive elements also helps the problem of dissipation of the signal. For example, if only two identical sequences (resonators) are located

in distant areas of the nucleus, the waves coming from the first resonator would dissipate before reaching the second resonator. This would make the communication impossible. On the other hand, the 1 million Alu sequences occupy the whole nucleus, are connected to each other by chromosomal DNA, which serves as an electric conductor of waves, and by the nucleoplasm which we believe also serves as a conductor of waves; the nuclear membrane works as a reflector, thus containing the waves produced by the mass of the Alu sequences. This prevents the waves from dissipating and makes them self-sustainable, allows them to shape the liquid crystalline structure of the nucleoplasm, and provides the main vibrational pattern for our biofield.

Ideally, experimentation is required to isolate waves produced by the Alu element. It's quite possible that for the appropriate vibration, Alu DNA would need to be in its natural environment, bound to nucleosomes and to Alu-binding protein, but it is also possible that just a purified Alu DNA could produce a signal in spectroscopic studies. Proper controls should be done to make sure that the signal is specific to Alu.

Another approach to assemble Alu with nucleosomes in vitro thus creating reconstituted chromatin. By modifying the DNA sequence of Alu, it should be possible to isolate the vibrations of the natural Alu sequence. Once the vibration is isolated and identified, then it should be reproduced and its biological effects should be studied. It is quite possible that the field produced by the Alu elements would have therapeutic and psychological effects.

3.4 Discussion of Our Computational Results

On the other hand, even without the experimentation, it is possible to see the traces of resonance signaling in the genomic data. To identify the resonance elements, we utilized the redundancy principle similar to that known from the aminoacid genetic code.

We started with 3D DNA modeling and formulated an algorithm for the conversion of the primary genomics sequence into the sequence and shape of proton and electron chains. We identified repeated proton and electron chains (called HIDERs) and found that with high statistical significance these are HIDERs enriched by the evolution of the genomes of multiple species, colocalized with conserved areas indicating that they are functional and colocalized with transcription start sites of genes indicating that they are involved in the regulation of gene expression.

It is not possible to explain the obtained results via chemical signaling because HIDERs have different chemical structures on the outside and can not be recognized by transcription factors. The only quality that connects them is that they harbor proton and electron chains of similar shapes. So, the observed statistically significant enrichment of HIDERs in evolution and in conserved regions supports the idea of that HIDERs function as resonators in DNA resonance signaling.

Also, the evolutionary conservation of HIDERs is another indication of their functional importance. Conservation of sequence elements between species is a reliable indication that the element is important and has not changed during the million years of evolution. Thus, the fact that HIDERs are located in conserved areas suggests their functional importance and that they also have been conserved during the million years of evolution.

We also observed that HIDERs colocalize with transcription starts with high statistical significance. This suggests that HIDERs receive the field signal, and convert it into biochemical information by controlling the gene expression.

These computational genomic results are supportive of the idea of DNA resonance signaling, but more work is needed to obtain more evidence. Wave signals from HIDERs need to be identified. The creation of HIDERs using DNA synthesis and modification of them using CRISPR should be used to demonstrate their function. For example, it should be shown that those mutations of HIDERs that don't change the electron and proton chains, don't affect the function, but those modifications of HIDERs that do affect electron and proton chains do change the function. Signals produced by HIDERs should be identified and technologically reproduced to prove their function. It is likely that HIDERs are the important words in the resonance language of the genome. Identification of them was the first step in deciphering this language. Once the resonance code of the genome is interpreted, it would be possible to predict the structure and function of the organism based on its genome. Based on the genomic sequence of the individual, it would be possible to predict the genetically determined components of their appearance, health, and character.

3.5 On the Physical Nature of DNA Resonances

Once we realized that the mass of DNA of an organism creates a biofield, this led to a number of questions and predictions [25, 31–33]. The questions were: What is the frequency of oscillation? How is it that the oscillations are not dumped by the viscosity of nucleoplasm? How are these frequencies protected from being absorbed nonspecifically by other biomolecules? How is biological information included in the waves? What are the mechanisms of converting biochemical information into wave information and conversely—from the wave information to biochemical? In other words, how could the genome sense the biochemical information and how could the genome control and direct the biochemical processes, including morphogenesis?

Let's now consider the role of sound in physiology and the organism's development. One of the reasons we turned our attention to sound, is because the electromagnetic resonances in DNA don't seem to be sufficient to explain the ability of the biofield to control the movements and activity of proteins. For example, we described a vast network of resonators in the genome and the system of transmission of the information through electromagnetic waves from one part of the genome to another and from one group of cells to the whole organism, yet this information has to be expressed into material action. For the biofield to regulate physiology and development, a mechanism is required that would allow the biofield to enforce its informational decisions onto the proteins. In other words, the question is how can the DNA control the proteins within the nucleus and within the cytoplasm beyond the nuclear wall? Can the electromagnetic waves alone be responsible for driving the protein molecules to their location and controlling their activity?

We find that unlikely. Although there is a phenomenon of laser tweezers where laser light is used to hold tiny drops of water containing biomolecules. We found it hard to extrapolate the idea of laser tweezers to the control of proteins by biofield. The power density of the electromagnetic fields in an organism doesn't seem to be strong enough to explain the control of the proteins. On the other hand, electroacoustic and optoacoustic explanations seem more plausible to be responsible for the control of proteins by biofield.

3.6 Electroacoustic Model

Cymatics is a technique that creates lifelike patterns using interference patterns of sound. There are many cymatics experiments on YouTube where the sound is used to create interference patterns that develop in real-time which strikingly resemble chromosomal rearrangements in mitosis, embryo development, the morphogenesis of plants, and the formation of the skeleton in vertebrates. In such experiments, platforms of the square, round, or other geometric forms are vibrated by a sound speaker attached to the bottom of the platform, the sound frequency is varied, and rearrangements of sand or magnetic particles on a vibrated platform is recorded on video. Alternatively, the water wave arrangements vibrated by ultrasound are recorded using a microscope. The platforms of about 80 cm size use audible frequencies. The microscopic experiments use ultrasound frequencies, and therefore, if the sound is used by nature to drive and direct the motion of the proteins in the cell, the frequencies of this sound must be extremely high.

We believe that DNA creates sound interference patterns, using optoacoustic and electroacoustic principles. These interference patterns can be called "vibronic field" implying that the sound oscillations are coupled with the oscillation of charges and electromagnetic oscillations. This idea is in line with the recent paper on the ability of DNA to engage in vibronic oscillations (Farhi and Dogariu 2021). The genome consists of billions of electromagnetic oscillators, which are basepairs, chains of protons, and electrons. These oscillators are vibrationally linked into one complex oscillator. These oscillators are insulated from each other electrically but are coupled to each other electromagnetically. Together, they create complex oscillating currents in the DNA. There possibly are some insulated domains within the chromosome that represent complex vibrational units and are vibrationally insulated from each other by so-called genomic insulators. We believe that the unit that converts electrical oscillations in the DNA into sound using the electroacoustic principle is the nucleosome. The nucleosome looks like an induction coil and the DNA, which is a conductor, is wrapped around it 1.6 turns. The nucleosome is surprisingly symmetrical in structure. It consists of a highly negatively charged sugar-phosphate backbone of the DNA and a highly positively charged nucleosome core particle made out of histone proteins. Charge oscillations in DNA cause charge oscillations in the nucleosome. In the nucleosome, these charge oscillations result in mechanical oscillations, which create and radiate the sound beams in specific directions from the nucleosome. The nucleosome has a number of cavities, which may serve as resonant cavities and work as speakers shooting the sound beams into the nucleoplasm.

The genome contains about 300 million nucleosomes, and we believe that together they create a three-dimensional cymatic interference pattern. This pattern is responsible for the movement and regulation of the proteins and loops of DNA within the nucleus, as well as the compaction and decompaction of chromatin. This vibration pattern is structured and integrated by the network of protein fibers called the nuclear matrix, and these vibrations are directed to the cytoplasm and control the protein movements and function in the cytoplasm. It is likely that the nuclear membrane serves as an integrator of the vibrational information and creates sound interference patterns in the cytoplasm using the holographic principle. The nuclear membrane is known to harbor lipid rafts organized in complex patterns. These lipid rafts on the nuclear membrane are organized by the sound and serve as a moving holographic record, which creates the holographic patterns in the cytoplasm.

3.7 Electroacoustic Frequencies

Consider a model of electroacoustic conversion: When the electromagnetic wave of a certain frequency creates a resonance charge oscillation in a proton or electron chain, the oscillation frequency remains the same as in the incident wave, and when this charge oscillation creates a sound wave in the nucleoplasm, the frequency also remains the same. MAKE FIGURE OPTOACOUSTIC CONVERSION. Therefore in the electroacoustic conversion, the frequency remains the same. It's only the wavelength that changes. The conversion of the electromagnetic wavelengths to sound wavelengths and the reverse is quite simple. The wavelength is equal to the speed of the wave divided by its frequency. To calculate the optoacoustic conversion we used 1500 m/s sound speed in saltwater and $224 * 10^6$ m/c speed of light in saltwater.

Figure 1 demonstrates the conversion of electromagnetic oscillations to sound and back via the oscillation of a charged particle. Shown are wavelength conversions for the popular therapeutic frequencies and essential biological objects of various sizes.

3.8 A Rough Estimate of DNA Frequencies

Although the existence of the biofield is well-established in experiments [34, 35] and the energy field is used every day by energy healing practitioners, its physical nature

light wavelength	frequency	sound wavelength	
1.5 mm	150 GHz	10 nm	nucleosome diameter
3.6 mm	63 GHz	24 nm	tetranucleosome width
3.7 mm	60 GHz	25 nm	millimeer wave therapy frequency
4.2 mm	53 GHz	28 nm	mm wave therapy, chromatin fiber diam.
5.3 mm	42 GHz	36 nm	millimeer wave therapy frequency
7.5 mm	30 GHz	50 nm	millimeer wave therapy frequency
1.5 cm	15 GHz	100 nm	Alu repetitive element, 300 bp
3.74 cm	6 GHz	250 nm	millimeer wave therapy frequency
15 cm	1.5 GHz	1.0 um	mitochondrion length
30 cm	753 MHz	2.0 um	Line repetitive element, 6 kbp
1.0 m	214 MHz	7 um	mammalian nucleus
3.0 m	75 MHz	20 um	mamallian cell
75 m	3 MHz	500 um	ultrasound imaging frequency
224 m	1 MHz	1.5 mm	ultrasound imaging frequency
2.2 km	100 kHz	1.5 cm	many biological objects
22 km	10 kHz	15 cm	chromosome length
224 km	1 kHz	1.5 m	human height
299 km	750 Hz	2 m	human height and genome length
12742 km	18 Hz	85 m	Earth diameter
28637 km	7.83 Hz	192 m	Schumann resonance frequency

Fig. 1 Electroacoustic conversion of wavelengths. (Currently, the shortest described wavelength for ultrasound is 250 nm, so the shorter wavelengths are shown in grey as tentative)

is largely unclear. And since we believe that the biofield is generated by DNA, the physical nature of DNA resonance is also far from being clear. As we previously discussed, we have demonstrated that traces of DNA resonance are imprinted on the genomic sequence. In other words, the patterns of genomic sequence evolved in such a way to support DNA resonances and generate the biofield. For our predictions of proton and electron chains as resonance oscillators in DNA, we use molecular modeling, but our molecular models are as yet very vague about the frequency of such oscillators. More precise modeling and experimentation are needed to find the frequencies of the DNA resonances. Nonetheless, since our models rely on the oscillations of charge chains of known shapes and dimensions, with proper modeling, it should be possible to estimate the oscillation frequencies. Clearly, these are charge oscillations and as such they must produce electromagnetic waves. The frequency of these waves must depend on the size of the proton and electron chains. Since, as we believe, proton and electron chains are vibrationally linked together in part through

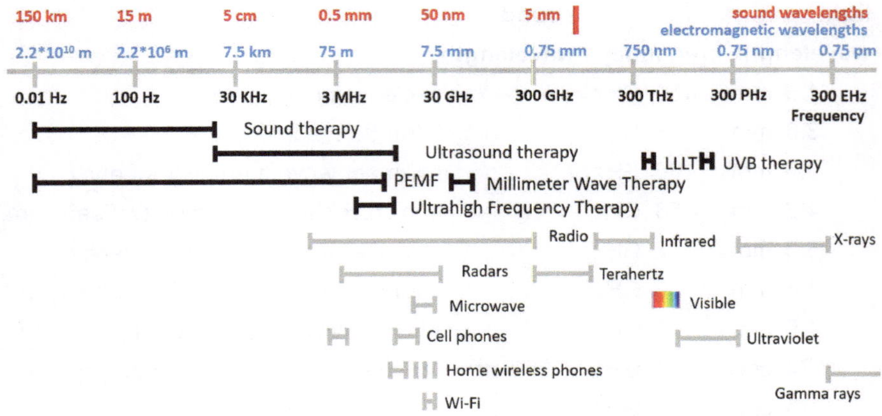

Fig. 2 Therapeutic electromagnetic frequencies and optoacoustic conversion. The sound wavelength scale stops at around 1 nm, the theoretical limit for the optoacoustic conversion. The middle of the chart is magnified to show commonly used frequency ranges

electrical DNA conductivity, these oscillations must have complex patterns. In Fig. 2, we estimate approximate oscillation frequencies of various genomic elements and compare them to known therapeutic electromagnetic frequencies.

For therapeutic electromagnetic frequencies, it is helpful to distinguish high power therapies such as heat, light, and radio waves, which act nonspecifically, and low power therapies such as UVB, low-level light, infrared, millimeter-wave, and ultrahigh-frequency, which act specifically. With low-level energy therapies, a low dose amount of electromagnetic field causes a substantial health improvement, suggesting that this frequency is of signaling in nature and that it taps into the existing signaling language of the DNA. We and others have published experiments suggesting that a low-level light therapy at wavelengths of 660 nm and 850 nm affects gene expression. We have also discovered experimentally that pretreatment with low-level millimeter waves of cell culture protects it from DNA damage. Yet many more experiments are required to decode the vibrational DNA language.

3.9 Frequencies by Themselves Are Not Therapeutic

We think that it is a misconception in the physical therapy field that certain frequencies are therapeutic, which we think is an oversimplification. We believe that in most cases it's not the frequencies that convey information to the body. The waves of the biofield generated by the DNA must have complex patterns. For example, we rarely use single continuous tones to convey the information in our speech. Our speech is highly modulated and contains variable tones, pulses, clicks, and hissing sounds. Even similar simple tones are often colored by harmonics and complex shape waves. Similarly, in music, many notes are combined and producing a balance of harmony and disharmony. Each instrument colors the tone, adding harmonics and modulating the shape of the wave. The patterns of electromagnetic waves can also be complex. Multiple frequencies can be combined. Waves can be polarized and circularly polarized. The phases of various frequencies and frequency patterns can be shifted relative to each other. The directionality of the waves is important too. Waves from multiple sources can create interference patterns. Accordingly, we believe that biofield patterns are also of complex interference nature and find it very unlikely that the natural language of DNA can be represented by simple frequencies. Most likely, simple therapeutic frequencies open the communication channels with the genome of the body, but the direction of the therapy must be dependent on the intentions of the therapist and of the patient. This must be one of the reasons why many wave therapy experiments fail double-blind tests. We believe that simple frequencies don't convey therapeutic information. They only open the channels for the intentions of the healer and the patient. Keeping in mind these limitations, electromagnetic wave therapy provides us with starting points toward understanding the physics of DNA resonance signaling and its language.

3.10 Local Nonelectromagnetic Component of the Biofield

For practical research purposes, it is very important to differentiate between local and nonlocal effects of the biofield. We don't know whether they have similar or different nature, but it is clear that local biofields can be studies using traditional experimental approaches and nonlocal effects are very hard to study using traditional approaches and so, they would require very different approaches. The existence of non-local information transfer and intentional influences is well established experimentally [36–38], but the information on local nonelectromagnetic components of the biofield is scarce. When we speak about "local nonelectromagnetic components of the biofield" we imply that there are likely fields that are attached to the body like an aura, or an energy field, but which can not be measured by common electromagnetic measurement devices and which penetrate shields such as metal and water barriers which common electromagnetic waves do not. Nonlocal effects of consciousness certainly penetrate electromagnetic shields and often are independent of distance, but these are hard to study experimentally. The local nonelectromagnetic fields of the body even though may not be registered by electromagnetic meters, still could be detected using energy-sensitive people and biological model systems such as DNA solutions, cell culture dishes, and plants. The key property distinguishing the local biofield from nonlocal is its attachment to the body. It should move with the body and fade with distance.

One of the authors of this paper, MMR, is a Reiki healer; as such, he works with the biofield frequently and is attuned to feel it. In Reiki healing, many practitioners, about half of them are able to feel the energy of a patient by moving their hands at about 2–10 inches from the body. For those who are able to feel it, it is normal to feel the energy of the patient about 80% of the time. When energy-sensitive Reiki practitioners are being treated by other healers, they feel the energy from the hands of the healer. Reiki energy is not unique to Reiki. MMR experienced energy healing from many other forms of energy healing including Chi-Gong, Johrey, "healing touch", Pranic Healing, and traditional Ukrainian energy healing—and the healing energy felt very similar in all these modalities. From the physics perspective, the nature of this healing energy is puzzling since the healing sessions are done through clothes and often through blankets. About half of these healing treatments are done from a distance between 1 inch and 6 feet. Many of the electromagnetic frequencies are absorbed by clothing and blankets: UV, visible light, near-infrared, infrared, and millimeter waves. Moreover, if it was a classical electromagnetic field, the healing treatment would be more effective at shorter distances, but for example, Johrey practitioners prefer to do the treatments at about 6 ft distance.

It is very normal for Reiki healers to direct Reiki energy from one hand to another as a focused biofield beam as a warmup exercise. During this exercise, the receiving hand feels the incoming energy beam as a gentle physical sensation resembling a gentle blow off the air. To verify whether this sensation is real, MMR tested it with other people and with Reiki patients. Very often, beams of energy can be detected coming out of Reiki patients in certain locations near the body. In group training and healing sessions, multiple Reiki practitioners independently scan the energies of the patients and report their observations. Using such approaches, MMR was able to verify for himself that the sensation was objective and verified by the other Reiki practitioners.

Next, to see whether the Reiki energy beam was of physical nature, MMR tested whether it penetrates the water. For that MMR used repeated simplistic experiments in a swimming pool, a bath, and a freshwater lake. MMR was sending a beam of Reiki energy from the fingers of one hand to the open palm of another and he would move one or another hand to make sure the sensation of the beam is moving across the open palm. While doing this, MMR was immersing himself and hands in the water. Surprisingly the perceived position of the beam on the open palm didn't change from the immersion of both hands in water. This experiment was repeated many times and produced the same result. It suggests that the Reiki energy might be local and yet, not of classical electromagnetic nature. An infrared, millimeter wave, microwave, and gigahertz waves would be absorbed by water. A near-infrared beam would be somewhat deflected as does visible light when passed through water. Only extremely low frequency (ELF) 3 Hz–30 kHz radio waves can penetrate water, but since the diameters of Reiki energy beams are often small (5–10 cm in diameter), it is very unlikely that they can be made of ELF radio waves since the wavelengths of these waves are extremely large (10^4–10^5 km). We can in a preliminary way conclude that there is a type of field which is specially organized, but capable of penetrating a layer of freshwater unhindered and this suggests that this beam has unusual properties different from usual electromagnetic waves. Since (as we believe) biofield and Reiki energy, in particular, are generated by DNA, these simplistic experiment sheds a bit of light on the nonphysical nature of at DNA resonances, or at least one of the components of the fields produced by DNA.

In respect to DNA resonance and the biofield, it's important to differentiate:

(1) electrical currents that go through a conducting medium,
(2) sound (that is transmitted better through the tissue than the air),
(3) electromagnetic waves (that go through the air unhindered and are often scattered by the tissues)
(4) local nonelectromagnetic component of the biofield (that is goes through the air, water, and clothing unhindered), and
(5) nonlocal biofield and psychic phenomena (that pass through large distances unhindered).

In summary, to properly understand the nature of DNA resonance, it would be important to differentiate traditional physicochemical signals, such as chemical signaling, electric currents, electromagnetic waves, and sound, from nontraditional biofield waves and signals that pass through shields such as water. Among the nontraditional signals that pass through shields, it is important to differentiate the local ones and nonlocal ones. We believe that there are parts of the biofield that penetrate through shields but still are localized. They are generated by the body, are located in and around the body, and move with the body.

3.11 *The Nonlocal Nature of the Locality*

The question of locality is essential for the understanding of biofields. At the atomic scale, in living tissues, nonlocality is a norm, and localization of the wave function for each atomic-scale object occurs only occasionally due to decoherence which is caused by outside influence such as infrared radiation [39]. On the other hand, at the macro scale of the human body, the locality is a norm, in agreement with the quantum math, since the collapse of the wave function for large objects happens instantaneously [39]. The biofield seems to display the nonlocal properties since it penetrates clothing and water, as well as local properties since it is perceived to be attached to the body. In other words, it seems to have hybrid borderline properties between locality and nonlocality and is able to collapse from nonlocal into a local state under the influence of the intention of an observer.

3.12 *Parallels with the Subtle Bodies of the Ancient Eastern Teachings*

The idea that the health of the physical body is supported by the subtle bodies is ancient and is already mentioned in Upanishads dated before the fifth century BCE. The physical body is taught to be accompanied by 4 subtle bodies: vital energy, mind and memory, consciousness, and divine bliss.

Once a yogi traveled by air. When the airplane landed, the yogi stepped out, sat on the asphalt, and started to meditate. People asked, "Why don't you go?" "I'm waiting for my soul to catch up", said the yogi.

Speaking of the soul, we suggest that DNA may serve as the primary interface between the physical body and the soul. The soul is vast and multidimensional, and it interfaces with the physical body through DNA resonances at different frequencies corresponding to the subtle bodies.

Those who communicated to Alzheimer's sufferers might have noticed that much of the time, they are only partly conscious. We suggest that this is because a big part of the soul is disconnected from the body. Yet, once in a while life becomes interesting to them. Their eyes light up and they come to their normal senses. We suggest that this is when the DNA resonance has kicked in and the soul reconnects to the body.

Here we attempt to unite metaphysical and physical sciences, restoring the true unified science. The divide between the natural sciences and the science of spirituality was never scientifically justified. Much of the divide needs to be healed and much would need to be learned. Psychology and psychiatry would have to return to the study of the soul (the psyche), which gave the origin to their names, and understanding of DNA resonance would likely produce practical improvements in psychology and psychiatry.

3.13 *Brain–Computer Interface and Synthetic Telepathy*

Moreover, deciphering the DNA resonance code will result in deciphering the language of the brain, thus enabling another practical application: the brain–computer interface and synthetic telepathy. Brain–computer interface is a field of technology that is well-funded by agencies and industry and is making quick progress. Synthetic telepathy is the connection of minds of people via a technological link and circumventing the conventional modalities of communication: light, sound, and touch [40]. Without understanding the language of the brain, or the brain code, technological communication with the brain will be limited by the speed of perception of light and sound by the brain, which is very slow. Currently, scientists have only very crude ways of deciphering NMR (nuclear magnetic resonance) and EEG (electroencephalographic) readouts. Currently, the noisy spikes on the EEG plots are interpreted as noise. Deciphering DNA resonance code, we believe, will lead to deciphering the brain code and this way will radically improve the speed of technological communication with the brain. Once brain–computer interface and synthetic telepathy are healthy is of sufficiently high speed, this will radically improve the efficiency of communication between people. We believe that telepathy will be the most important leap that will change the future of humanity for the better.

3.14 *Nature of Local Nonelectromagnetic Waves*

The local and nonlocal biofield phenomena were scientifically studied with a few scientific approaches. The nature of nonchemical signaling was studied by several methods: (1) it was shown that one biological sample, such as a fish embryo, a plant, or a cell culture dish can transfer biological signals to another biological sample via air or a quartz glass [7, 9, 10, 41], (2) electromagnetic waves were recorded coming from a perturbed biological sample or a developing embryo, and the properties of the waves were found to be indicative of the state of the sample [9, 42, 43], or (3) it was demonstrated that electromagnetic waves of specific properties can produce specific effects in biological samples [5, 6, 9, 41]. The nonlocal information transfer between people, sending thoughts and messages at large distances, influencing people at a distance, and distant healing effects were demonstrated with high statistical significance [34–36, 38, 44, 45]. Remote viewing was found to be statistically significant [46, 47]. Local energy healing with hands at short distances was found effective with statistical significance [45, 48]. We were not able to find publications that specifically investigated the effects of barriers such as water and metal sheets on the penetration of Reiki-like energy fields, but there are publications describing the significant effect of distance on the perception of healing energies [45, 49].

3.15 *Naming*

So, far, there is no consensus for the naming of the special biofield waves that are local and nonelectromagnetic in the classical sense. Konstantin Meyl described local waves that form the biofield as "scalar waves", a special form of helical electromagnetic waves, where the magnetic and electric components of the wave are phase-shifted relative to each other and where frequency and wavelength change along the path of the wave. Other suitable names for local nonelectromagnetic biofield would be etheric or pranic field—after the etheric body of Theosophists which is the pranic body of Hinduism responsible for life energy, one of the five subtle bodies.

3.16 *DNA as a Dimensional Portal*

It is likely that DNA works as a dimensional portal, meaning that it mediates the information transfer and serves as an interface between the physical body and nonphysical biofield. This would explain why it is so difficult to detect the biofield with traditional physical devices. Since physical devices are located in this physical dimension, they typically are not able to register the signals transmitted in a parallel "subtle" dimension, the "subspace" and this would also explain why a biological entity is able to detect the biofield of another biological entity. Say one biological entity is disturbed; its DNA is specifically excited and sends the information into the subspace. If another biological entity is tuned to the proper subspace frequencies, its DNA will receive the proper frequencies, become vibrationally activated, and convert this specific vibrational signal into a specific structural change or a biochemical signal such as gene expression. Some of the subspace communications will be local and fade with distance, as do physical waves, which normally fade proportionally to squared distance. Other biofield components will be nonlocal and will not fade with distance. We believe that DNA resonances are multimodal; that is, capable of communication with sound, electric currents, electromagnetic waves, and nonphysical waves, local and nonlocal.

3.17 Scientific Method

There is widespread confusion between the scientific method and materialism. The overwhelming majority of scientists believe that the scientific method is applicable only to matter and material objects, and that anything immaterial such as the biofield and soul cannot be studied scientifically and are beyond science. This is an oversimplification. Quantum physics, for example, is a subject where the scientific method forces researchers to question the reality of the matter. From quantum physics, we learn that the experience of the material world is an illusion and can only be obtained at macroscales comparable to the size of our bodies. At microscales, such as the scale of DNA, things become less real. Positions and states of molecules become uncertain. Quantum nonlocality, entanglement, quantum coherence, and quantum delocalization have become a norm. This illustrates that the scientific method applied consistently does not lead to materialism. With our commitment to the scientific method, the consistent use of it inevitably leads us to DNA resonance and genomic biofield.

3.18 Structural Oscillations in Liquid Crystals

Until now, we considered three types of waves: electromagnetic waves, sound waves, and nonelectromagnetic (in the classical sense) local biofield waves. Let's now consider whether DNA resonance may happen through the waves which don't oscillate in time. Although it sounds like a misnomer, such waves exist. Consider waves in the sand. These clearly are waves and yet, they don't move. Consider also periodic patterns in crystals; these have many qualities of waves, but they are static.

The nucleoplasm (the thick inner solution of the cell nucleus) is a concentrated solution of DNA (about 1% [50]), histones, and many other proteins and molecules [51]. Unfortunately, there is no information on the biochemical content of the nucleoplasm, although, with modern methods of laser capture it is possible to measure it. Although it seems that no one has yet measured the biochemical content of the nucleoplasm, molecular biologists were able to make many enzymes work in vitro, such as polymerases, ligases, phosphatases, kinases, and other enzymes used in gene engineering. Since these enzymes are practically useful, much effort was placed into optimizing their work in vitro. So, the buffers used for the enzymes can give us a glimpse into what could be the natural content of the nucleoplasm. These buffers often contain sodium, potassium, magnesium and chloride ions, ATP, and a large amount of albumin which is considered inert. The pH is approximately neutral. One thing is clear: it's not fresh water. It's buffered salty water saturated with DNA, histones, and other proteins in such a way that DNA and histones undergo a constant process of crystallization and dissolution. It is very likely that nucleoplasm is a highly structured liquid crystal and that DNA plays a key role in structuring it. This way, it is possible that the mechanism for DNA resonance involves changing the structure of the nucleoplasm and sensing the changes in the structure. For example, one piece of DNA would send a signal by creating a special structure which would then propagate through the nucleoplasm and then be sensed by another piece of DNA in the same nucleus. This way, self-propagating periodic patterns would be created, but these patterns would not be wiggling as light and sound waves do. The periodicity would be semi-static. A structural change in one place of DNA would result in the propagation of the new structure through the nucleoplasm and lead to changes in another piece of DNA.

Although it seems a bit unconventional, there is a field of science that describes such events. It is the theory of chaos and self-organization [52] popularized by fractal pictures of Mandelbrot. Such self-organization would require that nucleoplasm is pumped with energy and eager to accept and propagate new structures. An example of this would be the propagation of new memes through social networks. Another example is fashion. Fashion starts in one place and then propagates through the system. So, many components copy it. People who live in a cold climate are familiar with the growth of crystals of ice in cold weather on the inner surfaces of windows. When the outside temperature is colder than the inside temperature, window frost condenses the water from the room and then crystals of ice start growing in this water, and this pattern propagates at long distances. Furthermore, as with daily cycles of temperature, this pattern creation repeats every day in a different way. Interestingly, videos of developing model embryos show that during the development embryos undergo oscillations. It is likely that the cell allows the patterns to form, propagate and then the cell dissolves them, and then, it allows new patterns to be created and propagated. And likely, the DNA is the primary source of these patterns.

This idea seems to be novel and testing it would require new experimental approaches. Possibly, it would require recreation of the nucleoplasm in vitro, stimulating the pattern creation, propagation, and dissolution via oscillations of temperature in the nucleoplasm (not unlike it is done in PCR) and using DNA as a template to prime the pattern formation in the nucleoplasm. Much in this respect can be learned from the science of liquid crystals, which are used for computer and smartphone displays.

3.19 Ideas for Experimental Testing of DNA Resonance Theory

The biological sciences have perfect tools for studying DNA resonances. For example, it would be interesting to compare the information transfer between various biological objects while varying their similarity on the genetic level. If our model is correct, organisms with identical DNA, such as twins and clones will be able to exchange vibrational signals with more ease, than genetically different organisms. The amount of genetic similarity could be easily varied in this order: identical clones and twins, siblings, distant relatives, same species, different species located close on the genetic tree, species located far apart on the genetic tree.

Genome manipulation using CRISPR technologies and artificial chromosomes would allow us to identify those sequences that are responsible for DNA resonances and signal transmission. It would be also interesting to see if the effects of signal transmission can be reproduced in cell culture or in reconstituted chromatin in vitro, in synthetic DNA in solution, or in nanoelectrodes with synthetic DNA strands anchored on the surface. It's quite possible that for proper vibrations, DNA would require the natural protein-rich broth of the nucleoplasm and even some of the electromagnetic and sound oscillations that are likely present within the nucleus. DNA might behave like a laser and would need to be pumped up with energy to properly resonate. The structure of DNA is also of importance. The level of supercoiling of the double helix is essential for its ability to form electron and proton chains. Therefore, purified DNA and short synthetic oligonucleotides might not be sufficient to produce DNA resonance. Longer DNA pieces in physiological conditions of the nucleoplasm would likely be needed. Pumping them with coherent electromagnetic and sound waves might be required. Yet, all these experiments are quite possible and would be trivial for a laboratory specializing in chromatin structure. We estimate that there are about 20,000 laboratories in the world having the expertise and experience to do such experiments.

The key to studying DNA resonance is understanding its language. That's where genetically modified biological systems such as cell cultures will be useful. It would be interesting to insert big fragments containing human Alu DNA into mouse cells and see if they gain the ability to communicate with human cells. This would be a properly controlled experiment, allowing us to isolate the exact sequences that are required for nonchemical communication between the cells.

To isolate the specific waves and wave patterns produced by DNA resonances, spectroscopic studies will be useful of various DNA sequences in solutions mimicking the protein content of the nucleoplasm, or in reconstituted chromatin and living cells to isolate the waves which are produced by the DNA resonances. Varying sequences of DNA it would be possible to identify a specific wave frequency and pattern for each sequence pattern. For example, telomeric and centromeric sequences contain large numbers of copies of repeats: telomeric TTAGGG and centromeric 169 and 171 bp alphoid repeats.

Currently, the researchers are puzzled by the fact that the number of single nucleotide polymorphisms NSPs in our genome (about 12 million), is somewhat too small to explain the large extent of variation among individuals. Unsurprisingly, novel methods of long-range mapping allowed the researchers to observe that telomeric and centromeric regions contain a large number of variations often involving regions in the range of 20 Mb size. Telomeric and centromeric regions are poorly studied because they are noncoding. Yet, due to their highly periodic nature, they make a perfect candidate to be carrying resonance information, and variation in their periodic patterns must be responsible for individual differences in biofield patterns. This opens a vast research opportunity for geneticists to discover the genomic variations explaining behavior, metabolic and physical variations between individuals.

3.20 Practical Applications of DNA Resonance Theory

The acceptance of the DNA resonance theory might take a long time since it shatters many misconceptions in biology and medicine, but its practical applications can be embraced by the market much sooner. That's what happened to the small RNAs and RNA silencing around 2010. The mainstream stigmatized the theories about the importance of small RNAs until a practical application of small interfering RNAs for gene silencing demonstrated their exceptional practical value. MMR's collaborator, the pioneer of small RNA research, Irina Konstantinova, died prematurely when her project, which published great works on small RNAs, was shot down. That was in the year when the RNA silencing was finally recognized by mainstream science [53]. We hope to see DNA resonance theory recognized in our lifetime, in particular, because of its practical potential.

The potential areas of application of DNA resonance include medicine and biomedical research. The immediate useful applications can be developed in computational genomics. The reason for this is that computational genomic development is much faster and computational applications often don't require regulatory approvals.

3.21 GWAS Application of DNA Resonance Theory

The first application is for GWAS (genome-wide association study) data analysis. GWAS became popular when genotyping of single nucleotide polymorphisms (SNPs) was automated. In the past two decades, several thousand large GWAS studies have been completed [54]. GWAS is a powerful tool for locating genes responsible for genetic diseases and traits, such as psychiatric disorders, autoimmune disorders, and metabolic disorders. GWAS studies point to genes responsible for addictions, obesity, aging, and other important traits. Many large GWAS studies were funded and continue to be funded with the hope to find drug targets for many important traits. Yet there is a problem with the GWAS approach, which was obvious from the very beginning: only a tiny fraction of found causative mutations result in mutations that affect proteins or the level of their expression. The vast number of causative mutations—over 99% of them—don't cause any changes in proteins, and this is where mainstream science and molecular biology exhausted their ability to explain their own results. Numerous bench-to-bedside conferences and programs failed to deliver practical results because the underlying molecular theory avoids considering DNA resonance and the biofield. We believe that the majority of causative mutations affect biofields via DNA resonance.

In our computational analysis of public GWAS data, we observe that causative mutations affect in one way or another the chains of protons and electrons in DNA (MMR, unpublished). We suggest that large numbers of noncoding SNPs serve signaling purposes and are part of a sophisticated quantum computer, which is the genome. Our computational algorithms allow us to prioritize the SNPs based on their resonance properties and predict which ones are functional and which variations are detrimental to health. This is the immediate practical result that can help GWAS researchers to prioritize their SNPs and select some for functional wet lab CRISPR testing. A typical GWAS study starts with over 10 million SNPs in the human genome, proceeds to genotype many of them in thousands of people, and, based on the obtained genotypes, identifies the so-called association peak. A typical association peak is between 0.2 and 5 MB and contains between 100 and 10,000 SNPs. These SNPs need to be prioritized. Our algorithms allow us to prioritize these SNPs and pinpoint the ones that are predicted to disrupt DNA resonance. This application can be expanded into genome data mining. A large number of GWAS results are publically available, and this data can be re-analyzed using our DNA resonance algorithms, providing candidate SNPs for functional CRISPR wet lab analysis.

3.22 Gene Expression Analysis Application

The next practical application is the analysis of gene expression data. High throughput gene expression analysis has been available for over 25 years, and by now, it has become much more affordable and precise. This analysis takes a small amount of biological material or even a single cell and measures the amount of RNA of every gene. This is a powerful tool that allows the correlation of gene expression with biological function and disease. There are a large number of software services and packages that analyze gene expression, yet molecular biologists are unable to predict gene expression changes based on current molecular models. This we believe is again because they take into account only molecular mechanisms for analysis and to account for DNA resonance signaling. We believe have developed computational DNA resonance models that should help to explain and predict gene regulation.

3.23 Personalized Medicine

The price of identifying the majority of mutations (about 11 million) in a genome of an individual (the nonrepetitive part of it), which is nearly equivalent to sequencing the individual's genome, is now about $10 K and is expected to go down to $1 K within a few years. It is already possible in still rudimentary ways, based on the genomic sequence to predict the metabolic profile of the individual, predisposition to diseases, sensitivity to medications, and some of the psychological and character traits, to prescribe individualized diet and healthy lifestyle. Currently, the algorithms for such predictions are imperfect. One reason for that is the lack of high-quality data that can be used for training of the algorithms. Another reason is the delay in governmental regulations that would properly enable full-genome diagnostics and recommendations. The other problem with current analysis algorithms is that they are developed by the scientists unaware of the DNA resonance layer of regulation in the genome and are not likely to discover it by chance. The incorporation of DNA resonance algorithms in the genome analysis will allow reading the genome "between the lines", once the DNA resonance language of the genome is known, it will become much clearer how the mutations disrupt the healthy resonance patterns. In our current analyses, we already see the evidence that the genome is patterned by evolution to enable resonances and how the mutations disrupt the resonance patterns.

3.24 Other Biomedical Applications

The next set of applications is biomedical. It is slower to develop but is even more promising. We believe that one of the primary purposes of the genomic field is to control an organism's development that is morphogenesis. The experiments of Burlakov with the manipulation of the biofield [41, 42], which manipulated biofield precisely enough to modify fish embryo development and control the number of heads and tails developed by the embryo, lead us to believe that therapeutic manipulation of the biofield would allow us to introduce previously unthinkable biomedical applications to control the shape of the body, including obesity, as well as to grow and repair organs including teeth, bones, joints, heart, kidneys, liver, spleen, and eye retina. This is in line with some of the scattered reports on millimeter-wave therapy results related to morphogenesis. Yet, in these experiments and therapies, the researchers so far have been using empirical approaches by modifying the waves without understanding the basic mechanisms of their origination. Our models of DNA resonators, which include proton and electron chains, provide the immediate key to the identity of the source oscillators, that is DNA sequences, and in particular Alu, centromeric and telomeric repeats. Therefore, the experiments can be devised where millimeter waves are used to treat synthetic DNA samples of varied sequences containing candidate source resonators. This way, the wave treatments can be optimized to achieve the desired outcomes using a spectroscopic or nanoelectronic readout.

3.25 Therapeutic Devices

There are already many electromagnetic and light therapy devices on the market. Light and megahertz therapy devices are already approved by FDA. Above, proposed experiments to identify wave patterns of the genomic field for various sequence elements. Once the proper parameters for the wave shape and wave modulation are identified, these can be embedded into the therapeutic devices. Similar approaches can find application in veterinary medicine, agriculture, food production, and biotechnological production. Since biotechnological production does not require sophisticated approvals, this application might get to the market first, followed by agriculture, food production, veterinary care, and human therapy.

3.26 Brain–Computer Interface

Brain–computer interface is another promising application for DNA resonance technologies. This R&D field is well-funded by public and private money, and it is aimed at the direct connection of electronic devices to the mind, circumventing the use of vision, hearing, voice, and touch. The intention is for people to be directly plugged into the computer and internet. The current speed of communication is limited by the speed of speech, which is about 5 bytes per second. The goal of the brain–computer interface is to speed it up manyfold. Although we see the dangers of abuse of this emerging technology, there is one potential benefit that we believe will outweigh the dangers: synthetic telepathy, see our review of synthetic telepathy [40]. Synthetic telepathy is the communication of one mind to another via the technological interface, which circumvents natural ways of communication including voice, hearing, vision, and touch. We believe that once people are able to communicate with each other at higher speed and precision than is currently possible, this will allow humanity to develop a hive-like mind and shift to a new evolutionary level. The benefit of the evolutionary shift outweighs the potential dangers of the misuse of the technology.

The problem with the brain–computer interface technologies currently developed by mainstream science is that scientists don't understand brain signals. As 99% of the genome looks like junk to molecular biologists, and similarly, the brain waves look like noise to neurologists. We believe that the genome is a biocomputer based on DNA resonance and the work of our mind is based primarily on the resonance of DNA sequences in the genome. So, the key to the brain–computer interface should be to understand the resonance language of the genome which we believe is the resonance language of the brain.

The first step in that direction is already done. We have already identified proton and electron chains, which are the key resonators [25, 31], and we already identified the logic circuits of the genome, which are tautomeric conformational changes of basepairs [23]. Therefore, the deciphering of the DNA resonance language will immediately lead to the understanding of brainwaves. Once the language of brainwaves is understood, the brain–computer interface can move from vocal language to the language of the brain, which will be manifold faster and more precise.

3.27 Therapeutic Device for Psychiatric and Psychological Disorders

Similarly, we believe that the major mechanisms of psychiatric and neurological disorders are based on DNA resonance. The psyche is the soul. Psychology and psychiatry are the sciences that should study the soul and DNA resonance is a mechanism with which the soul is attached to the body. Treatments of psychological and psychiatric disorders with red and near-infrared light and millimeter-wave therapy already produce substantial results [55, 56], but these results so far are obtained by empirical approaches. Understanding the DNA resonance-based mechanisms of disorders will allow us to treat them much more efficiently.

Transitional Chinese medicine practitioners and Reiki healers are taught to identify blockages in energy flow in the body, clear these blockages, and rebalance the energy flow. Introducing the devices that would be able to measure biofield imbalances and rebalance them using the biofeedback principle is a promising approach to psychological and psychiatric disorders. Some of these therapies have been developed using empirical approaches. The addition of DNA resonance codes to these modalities will improve the diagnostics and therapy.

3.28 Ecological Applications

Energy practitioners (intuitives and sensitives) can already clearly perceive the healthy energies of nature and the unhealthy energies of cities. We believe that electromagnetic pollution is nowadays the major cause of sickness. Similarly, the biofields of healthy foods are healthy and coherent, many processed foods are unhealthy due to discordant vibrations. These effects were measured by Fritz Albert Popp using biophotonic amplifiers [57]. We suggest that understanding the role of DNA resonance in the creation of the biofield will make it possible to measure and characterize the fields of environments of water and food, thus giving the public an objective measurement of biofields in food, and ecological fields in water, homes, and devices. One of the concerns is that 5G bands for smartphone communication are located near the therapeutic bands for millimeter-wave therapy, which we believe directly affects DNA. We cannot say that 5G smartphones produce necessarily negative effects, but based on millimeter-wave research, they should certainly produce biological effects positive or negative. Most likely, 5G radiations will affect the mood and behavior of people. This effect should be properly studied, especially keeping in mind their influence on DNA and chromatin structure.

3.29 Current Status

Here we will summarize the current status of DNA resonance theory. With all the progress in DNA resonance research summarized above, the question remains whether it is real. We don't have any doubt that the biofield is real. Well documented are the existence of the biofield [34, 35, 58] and of nonchemical cell-to-cell communication [9, 10]. But the experimental evidence biofield is produced by DNA is still lacking. At the moment, our theory is based on a series of conjectures aimed to explain and can solve a number of mysteries or unsolved problems that are typically swept under the carpet. The resulting model of DNA resonance provides a preliminary explanation for a number of mysteries outlined below.

3.30 Morphogenesis

Current mainstream biology provides only one explanation for the mechanism of morphogenesis, which is the chemical gradients of signaling proteins, called morphogens. Although these mechanisms are well established, they are clearly insufficient. The morphogenetic effects of the morphogen gradients are clearly real, but it's also clear that they cannot define the high precision and complexity of large complex organisms. The gradients are produced by the synthesis of the morphogenetic protein in one location and the gradual diffusion of it through the tissue. This can only work locally. It's irrational to believe that morphogen gradients can work in such sizes as the size of a human, a tree, an elephant, or a whale. Also, gradients are very imprecise and depend on the movement of the fluids and the temperature. There is clearly a need for a mechanism that will guide and sustain the complex structures. The morphogenetic field makes much more sense. Genetics provides solid evidence that the genome defines the shape of the organism. Developmental abnormalities are caused by many identified genomic mutations. It was experimentally demonstrated that the positions of organs in drosophila were swapped when the positions of genes were swapped. Therefore, it was logical to conclude that the genome defines and produces the morphogenetic field [17, 18, 59].

3.31 Genomic Field

So, our next question was: what is the mechanism for that? How could the genome define the structure of the morphogenetic field? If the field is created by oscillators, then we should look for ideal oscillators among the biomolecules. The most periodic and harmoniously structured biomolecules are DNA, microtubules, and centrioles, which are also made from microtubules. These are nearly perfect helixes with highly periodic structures within them. While microtubules are made out of identical monomers, DNA's monomers are organized in a sophisticated sequence. This makes DNA the most likely source of the biofield because the genetic program is organized into structures that should be capable of oscillation and the oscillations in those structures will be dependent on the genomic sequence.

3.32 Oscillators Within DNA

Next, we started to look for the identity of possible oscillators. The key requirement for these was that the oscillations should depend on the DNA sequence and should not be dumped by the viscosity of the nucleoplasm. This brought our attention to the chains of protons and aromatic electrons in the base stack. An additional advantage of these oscillators is that they are charged, so their oscillations would immediately result in the production of electromagnetic waves. The oscillations would require energy and we considered two major sources for this energy. One was thermal motion and another one was ATP, the universal fuel of the cell. Next was the problem of the possible dissipation of the signal within the cell. This brought our attention to the repetitive elements. It is very rational to suggest that repetitive elements are the key resonators in the genome. For example, the Alu element comprises 10% of our genome, making it the main resonator. When the number of resonators per cell is low, and they're diluted by other resonators, the signal will be lost due to scattering and nonspecific absorption. But when resonators are present in large numbers and in high concentration, they become linked in a unified resonating system and less energy is required to sustain the signal.

There is much evidence (unnoticed by mainstream science) that Alu elements may play the key regulatory role in the genome and may be those control elements predicted 65 years ago by the Nobel prize winner, Barbara McClintock [60]. We thoroughly reviewed this evidence in Ref. [61].

It would be interesting to use spectroscopy to identify the specific waves produced by Alu elements and use cell culture experiments to measure the effects of those waves. Next, we realized that we can use the redundancy of the resonance code to prove its existence using computational genomics. We devised the algorithms that predict which different sequences (HIDERs) will resonate with each other based on the similarity of their proton and electron chains. We demonstrated that those sequences are strongly enriched by evolution and located in conserved regions, suggesting their functional importance. We also demonstrated that part of them is colocalized with the transcription start sites, suggesting that they take part in regulating gene expression.

This is the best supportive evidence for DNA resonance we have so far. Direct experimental studies with manipulation of DNA sequences are required to prove the existence and function of HIDERs. Next, we looked at the detailed structure of proton and electron relocations in their chains, and based on molecular structure, we discovered that each basepair (among 6.4 billion basepairs per cell) works as a miniature logic circuit. The sequence of basepairs defines how these logic circuits are connected. Therefore, we discovered the principle of how the DNA is organized in a quantum biocomputer [23]. In our models, we see the initial glimpses of how protons and electrons are moving along the base stack and how these movements are interdependent. For example, the relocation of protons turns on and off the aromaticity of the aromatic electron rings in purines. We found substantial evidence that these oscillations of aromaticity have key importance for the function of DNA and the work of the mind. Therefore, even without direct experimental evidence, the main principles of the design of the quantum biocomputer within the genome begin to become clarified without experiments, just based on the existing three-dimensional molecular structures of the DNA. More experimental and computational work is required to further detail those structures.

Moreover, these studies begin to reveal the resonance language of the DNA. This solves the next mystery: the vast amount of noncoding DNA in the genome. Only 1% of DNA is coding for the proteins. The function of the remaining 99% is a mystery to biochemists. On the other hand, our models of DNA resonance already laid the foundation for deciphering the resonance language of DNA by identifying HIDERs and Alus as elements in this language.

Another mystery of biochemistry is how macromolecules find their way and propel themselves within the cell. The complexity of the cell is so immense that it is very irrational to think that macromolecules find their locations by random diffusion. We believe that the real mechanism is very different. The cell and the nucleus resemble a three-dimensional city where each macromolecule is a car minding its own business and navigating to its own location using an optimized trajectory. We believe that the biofield within the cell directs macromolecules to their locations and regulates the speed of the work of the enzymes. When more components are needed, the biofield senses the lack of the components, increases their production, and redirects them to their destinations. This model is supported by experiments in cymatics where sound waves are used to arrange and move patterns of particles. We suggested that DNA produces three-dimensional patterns of ultrasound. We further suggested that these patterns are moving and they move the macromolecules to their location according to the cymatic models. We proposed that nucleosomes work as sound generators by converting DNA sequence into patterns of sound.

3.33 *Mind and Memory*

The next mystery is the mystery of mind and memory. Current neurobiology teaches that (1) the mind works through electric impulses and action potential propagation in the brain, (2) that learning occurs through neuroplasticity or in other words, because a network of neurons is changing its topology, and (3) that memories are recorded in a topology of neurons and strength synaptic connections. Although we accept the functional importance of the above mechanisms, we find them largely incomplete. It is unlikely that the topology of neurons and the strengths of synaptic connections are retained for a lifetime. Neuron topology and synaptic connections are pretty malleable while memory lasts for a lifetime. Proteins are also very short-lived whereas DNA is very stable and remains unchanged for a lifetime. So, it's very likely that memory is recorded and methylation marks in DNA and that Alu elements serve as universal memory units in the genome and store methylation marks for long periods of time. As we mentioned above, we believe that DNA is a biocomputer and is directly involved in the work of the mind and that Alu elements, which are specific to primates, provide a universal resonator circuit for our mind, working as universal memory and logic units. An additional advantage of this model is that it moves the mind from an amorphous continuous model into a digital one. The genome is digital while the neuronal network is not. The advantage of a digital mechanism of the mind is that it can be programmed and reprogrammed that the information can be copied and may be interpreted as a language. The genome has addresses, which are defined by unique sequences, and linear, providing a linear structure to the code.

3.34 *Gene Regulation*

The next mystery is gene regulation. Although genomics produces excellent data on gene expression, the current paradigms based on the binding of transcription factors to DNA are insufficient for modern biochemistry to be able to explain and predict gene expression results. We believe that the DNA resonance model will provide the missing mechanism and will improve the prediction ability of the gene regulation models. We believe that genes and different pieces of DNA communicate with each other through DNA resonances and that similar sequences resonate with each other, synchronize and exchange information. We believe that the incorporation of DNA resonance-based algorithms in the gene expression analysis will produce much better results and allow us to explain the existing data and improve the predictive ability of gene regulation models.

3.35 *Energy Healing*

Energy healing involves the use of hands to send energy to the patients and provide healing for those disorders that do not well respond to allopathic medicine. We believe that DNA resonance plays a key role in energy healing. The healer senses the energy patterns of the patients and modifies these patterns by creating new energy patterns. There is resonance communication between the healer's DNA and the patient's DNA, and this is the key to the ability of the healer to change the patterns in the energy of the patient.

4 Conclusions

In summary, the DNA resonance theory illuminates the mechanisms for a number of puzzling subjects: the nature of biofield, morphogenesis, coordination of the work of proteins in the cell, gene regulation, mind, memory, and energy healing. While these explanations are hypothetical, they illustrate how DNA resonance theory can improve our understanding of a number of scientific problems once it is proven. We hope that this promise will motivate further research of DNA resonance and its funding.

Acknowledgements We thank Alexey Tovmash, Vadim Guschin, and Lev Shishkin for brainstorming the subject. We thank Yelena Zhigmitova for the literature work. We thank Lev Shishkin, Anton Klimov, and Lilya Yulmetova for their assistance in computational work. We acknowledge with gratitude the moral support of our research by Peter Gariayev who recently departed. We thank Stanley Krippner, Burt Webb, Dean Radin, Rupert Sheldrake, Daniel Winter, and Glen Rein for their advice and moral support of our work. We thank Pablo Andueza for the feedback on the manuscript that leads to its improvements. The work was funded solely by MMR.

Authors Contributions MMR developed the DNA resonance theory to its current state and wrote the chapter. IVS, as a research assistant, helped with the development of the theory, literature research, and computational work. RAM contributed to the discussion of the nonlocality of the biofield.

References

1. Mesmer FA (1781) Mémoire sur la découverte du magnétisme animal. Chez Michel Maklot. https://play.google.com/store/books/details?id=t6E9AAAAYAAJ
2. Best M, Neuhauser D, Slavin L (2003) Evaluating Mesmerism, Paris, 1784: the controversy over the blinded placebo controlled trials has not stopped. Qual Saf Health Care 12:232–233. https://doi.org/10.1136/qhc.12.3.232
3. Ede A (2007) The rise and decline of colloid science in North America, 1900–1935: the neglected dimension. Ashgate Publishing, Ltd. https://play.google.com/store/books/details?id=Xk0OeY8vVLwC

4. Müller J (1840) Handbuch der Physiologie des Menschen: für Vorlesungen. Bd. 2. J. Hölscher. https://play.google.com/store/books/details?id=ij4UAAAAQAAJ
5. Bischof M (1994) Vitalistic and mechanistic concepts in the history of bioelectromagnetics. In: Biophotonic: non-equilibrium and coherent systems in biophysics, biology, and biotechnology. Proceedings of international Alexander Gurwitch conference, pp 3–14. http://www.marcobischof.com/media/art/art_3d1c571e8aec4/VIT_MECH_MOSCOW.doc
6. Gurwitsch A (1922) Über den Begriff des Embryonalen feldes. Wilhelm Roux Arch Entwickl Mech Org 51:383–415. https://doi.org/10.1007/BF02554452
7. Gurwitsch AA (1988) A historical review of the problem of mitogenetic radiation. Experientia 44:545–550. https://doi.org/10.1007/BF01953301
8. Volodyaev I, Beloussov LV (2015) Revisiting the mitogenetic effect of ultra-weak photon emission. Front Physiol frontiersin.org. 6:241. http://dx.doi.org/10.3389/fphys.2015.00241
9. Cifra M, Fields JZ, Farhadi A (2011) Electromagnetic cellular interactions. Prog Biophys Mol Biol (Elsevier) 105:223–246. https://doi.org/10.1016/j.pbiomolbio.2010.07.003
10. Scholkmann F, Fels D, Cifra M (2013) Non-chemical and non-contact cell-to-cell communication: a short review. Am J Transl Res ncbi.nlm.nih.gov. 5:586–593. https://www.ncbi.nlm.nih.gov/pubmed/24093056
11. Trushin MV (2004) Distant non-chemical communication in various biological systems. Riv Biol researchgate.net. 97:409–442. https://www.ncbi.nlm.nih.gov/pubmed/15754593
12. Xu J, Yang F, Han D, Xu S (2017) Wireless communication in biosystems. arXiv [physics.bio-ph]. http://arxiv.org/abs/1708.02467
13. Burkov VD, Burlakov AB, Perminov SV, Kapranov YS, Kufal GE (2008) Correction of long range interaction between biological objects using corner-cube reflectors. Biomed Radioelectron elibrary.ru. 41–48. https://elibrary.ru/item.asp?id=11678936
14. Burlakov AB, Kapranov YS, Kufal GE, Perminov SV (2012) About possible influence on biological object electromagnetic fields. In: Weak and ultraweak fields and radiation in biology and medicine-proceedings of IV international congress, pp 111–112. http://www.biophys.ru/archive/congress2012/proc-p112-d.pdf
15. Myakishev-Rempel M, Stadler I, Polesskaya O, Motiwala AS, Nardia FB, Mintz B, Baranova A, Zavislan J, Lanzafame RJ (2015) Red light modulates ultraviolet-induced gene expression in the epidermis of hairless mice. Photomed Laser Surg 33:498–503
16. Genome List - Genome - NCBI. https://www.ncbi.nlm.nih.gov/genome/browse/#!/eukaryotes/
17. Miller RA, Webb B, Dickson D (1975) A holographic concept of reality. Psychoenerg Syst (Gordon and Breach Science Pub Ltd.) 1:55–62. http://www.nwbotanicals.org/events/real.pdf
18. Miller RA, Webb B (1972, 2002) Embronic holography: an application of the holographic concept of reality. DNA Decipher J 2. http://www.dnadecipher.com/index.php/ddj/article/view/26
19. ENCODE Project Consortium (2012) An integrated encyclopedia of DNA elements in the human genome. Nature 489:57–74. https://doi.org/10.1038/nature11247
20. Germain P-L, Ratti E, Boem F (2014) Junk or functional DNA? ENCODE and the function controversy. Biol Philos 29:807–831. https://doi.org/10.1007/s10539-014-9441-3
21. Shreeve J, Gardner G, Audio RH. The genome war: how craig venter tried to capture the code of life and save the world. https://www.amazon.com/The-Genome-War-James-Shreeve-audiobook/dp/B0001I1KTQ/ref=sr_1_1?dchild=1&keywords=genome+war&qid=1612942768&sr=8-1
22. Szezerbaty SKF, de Oliveira RF, Pires-Oliveira DAA, Soares CP, Sartori D, Poli-Frederico RC (2018) The effect of low-level laser therapy (660 nm) on the gene expression involved in tissue repair. Lasers Med Sci 33:315–321. https://doi.org/10.1007/s10103-017-2375-7
23. Myakishev-Rempel M, Savelyev I (2020) How Schrödinger's mice weave consciousness. https://www.researchgate.net/publication/346012232_How_Schrodingers_mice_weave_consciousness
24. Šponer J, Mokdad A, Šponer JE, Špačková N, Leszczynski J, Leontis NB (2003) Unique tertiary and neighbor interactions determine conservation patterns of Cis Watson–Crick A/G base-pairs. J Mol Biol 330:967–978. http://www.sciencedirect.com/science/article/pii/S0022283603006673

25. Savelev I, Myakishev-Rempel M (2020) Evidence for DNA resonance signaling via longitudinal hydrogen bonds. Prog Biophys Mol Biol. https://doi.org/10.1016/j.pbiomolbio.2020.07.005
26. Holmlin RE, Dandliker PJ, Barton JK (1997) Charge transfer through the DNA base stack. Angew Chem Int Ed Engl 36:2714–2730. http://doi.wiley.com/10.1002/anie.199727141
27. Schuster GB (2000) Long-range charge transfer in DNA: transient structural distortions control the distance dependence. Acc Chem Res (ACS Publications) 33:253–260. https://doi.org/10.1021/ar980059z
28. Meijer DKF, Geesink HJH (2016) Phonon guided biology: architecture of life and conscious perception are mediated by toroidal coupling of phonon, photon and electron information fluxes at discrete eigenfrequencies. Neuroquantology 14. https://www.academia.edu/download/52162519/Phonon_guided_biology_NQ.pdf
29. Christianto V, Susilo K, Smarandache F (2020) Roles of cymatics and sound therapy in spirituality and consciousness. Infinite Study. https://play.google.com/store/books/details?id=2tX7DwAAQBAJ
30. Dorier J, Stasiak A (2009) Topological origins of chromosomal territories. Nucleic Acids Res 37:6316–6322. https://doi.org/10.1093/nar/gkp702
31. Savelev I, Myakishev-Rempel M (2020) Possible traces of resonance signaling in the genome. Prog Biophys Mol Biol 151:23–31. https://doi.org/10.1016/j.pbiomolbio.2019.11.010
32. Savelyev IV, Zyryanova NV, Polesskaya OO, Myakishev-Rempel M (2019) On the existence of the DNA resonance code and its possible mechanistic connection to the neural code. NeuroQuantology 17. http://dx.doi.org/10.14704/nq.2019.17.2.1973
33. Polesskaya O, Guschin V, Kondratev N, Garanina I, Nazarenko O, Zyryanova N, Tovmash A, Mara A, Shapiro T, Erdyneeva E, Zhao Y, Kananykhina E, Myakishev-Rempel M (2018) On possible role of DNA electrodynamics in chromatin regulation. Prog Biophys Mol Biol 134:50–54. https://doi.org/10.1016/j.pbiomolbio.2017.12.006
34. Kafatos MC, Chevalier G, Chopra D, Hubacher J, Kak S, Theise ND (2015) Biofield science: current physics perspectives. Glob Adv Health Med 4:25–34. https://doi.org/10.7453/gahmj.2015.011.suppl
35. Hammerschlag R, Marx B, Yamamoto M, Aickin M (2012) P02.42. Non-touch biofield therapy: a systematic review of human randomized controlled trials reporting use of only non-physical contact treatment. BMC Complement Altern Med 12. http://dx.doi.org/10.1186/1472-6882-12-s1-p98
36. Radin D, Schlitz M, Baur C (2015) Distant healing intention therapies: an overview of the scientific evidence. Glob Adv Health Med 4:67–71. https://doi.org/10.7453/gahmj.2015.012.suppl
37. Storm L, Sherwood SJ, Roe CA, Tressoldi PE, Rock AJ, Di Risio L (2017) On the correspondence between dream content and target material under laboratory conditions: a meta-analysis of dream-ESP studies, 1966–2016. Int J Dream Res 10:120–140. https://psycnet.apa.org/fultext/2018-14483-003.pdf
38. Cardeña E (2018) The experimental evidence for parapsychological phenomena: a review. Am Psychol 73:663–677. https://doi.org/10.1037/amp0000236
39. Ball P (2018) Beyond weird: why everything you thought you knew about quantum physics is different. University of Chicago Press. https://play.google.com/store/books/details?id=Nk1vDwAAQBAJ
40. Miller RA (2001) Synthetic telepathy. Electro-Magick & Neuroshamanism, OAK, Consciousness Technologies II
41. Burlakov AB (2000) Distant physical interactions between the developing fish embryos. In: Biophotonics and coherent systems. Moscow, Univ. Press, pp 298–304
42. Belousov LV, Burlakov AB, Luchinskaia NN (2002) Statistical and frequency-amplitude characteristics of ultra weak emissions of the loach eggs and embryos under the normal conditions and during their optic interactions. I. Characteristics of ultra weak emission in normal development and the optic role of egg envelope. Ontogenez. europepmc.org. 33:213–221. https://europepmc.org/abstract/med/12068723

43. Beloussov LV, Burlakov AB, Louchinskaia NN (2003) Biophotonic patterns of optical interactions between fish eggs and embryos. Indian J Exp Biol nopr.niscair.res.in. 41:424–430. https:// www.ncbi.nlm.nih.gov/pubmed/15244263
44. Sheldrake R (2009) Morphic resonance: the nature of formative causation. Inner Traditions / Bear & Co. https://play.google.com/store/books/details?id=F75gxeTBgocC
45. Nelson LA, Schwartz GE (2005) Human biofield and intention detection: individual differences. J Altern Complement Med 11:93–101. https://doi.org/10.1089/acm.2005.11.93
46. Marks D, Kammann R (1978) Information transmission in remote viewing experiments. Nature 274:680–681. https://doi.org/10.1038/274680a0
47. Roe C, Cooper C, Hickinbotham L, Hodrien A, Kirkwood L, Martin H (2020) Performance at a precognitive remote viewing task, with and without ganzfeld stimulation: three experiments. J Parapsychol 84:38–65. http://nectar.northampton.ac.uk/13569/1/Roe_etal_JoP_2020_Performance_at_a_Precognitive_Remote_Viewing_Task_with_and_without_Ganzfeld_Stimulation_Three_Experiments.pdf
48. Dyer NL, Baldwin AL, Rand WL (2019) A large-scale effectiveness trial of Reiki for physical and psychological health. J Altern Complement Med 25:1156–1162. https://doi.org/10.1089/acm.2019.0022
49. Labuschagne DB (2003) BIOFIELDS: their detection, measurement and behaviour. Victoria University - ResearchGate. https://www.researchgate.net/profile/David_Labuschagne/publication/265966720_BIOFIELDS_Their_Detection_Measurement_Behaviour/links/542257650cf26120b7a02223.pdf
50. Milo R, Jorgensen P, Moran U, Weber G, Springer M (2010) BioNumbers—the database of key numbers in molecular and cell biology. Nucleic Acids Res 38:D750–D753. https://doi.org/10.1093/nar/gkp889
51. Milo R, Phillips R (2015) Cell biology by the numbers. Garland Science. https://play.google.com/store/books/details?id=9NPRCgAAQBAJ
52. Gleick J (1997) Chaos: making a new science. Vintage. https://play.google.com/store/books/details?id=FZmSJPfTQVwC
53. Myakishev M, Polesskaya O, Kulichkova V, Baranova A, Gause L, Konstantinova I (2008) PCR-based detection of Pol III-transcribed transposons and its application to the rodent model of ultraviolet response. Cell Stress Chaperones 13:111–116
54. Buniello A, MacArthur JAL, Cerezo M, Harris LW, Hayhurst J, Malangone C, McMahon A, Morales J, Mountjoy E, Sollis E, Others (2019) The NHGRI-EBI GWAS catalog of published genome-wide association studies, targeted arrays and summary statistics 2019. Nucleic Acids Res (Oxford University Press) 47:D1005–12. https://academic.oup.com/nar/article-abstract/47/D1/D1005/5184712
55. Chang J, Wang R, Li C, Wang Y, Chu XP (2018) Transcranial low-level laser therapy for depression and alzheimer's disease. Neuropsychiatry. jneuropsychiatry.org. https://www.jneuropsychiatry.org/abstract/transcranial-lowlevel-laser-therapy-for-depression-and-alzheimers-disease-12428.html
56. Lin F, Hu W, Li A (2012) Millimeter-wave technology for medical applications. In: 2012 IEEE MTT-S international microwave workshop series on millimeter wave wireless technology and applications, pp 1–1. http://dx.doi.org/10.1109/IMWS2.2012.6338252
57. Popp FA, Gu Q (1992) Recent advances in biophoton research and its applications. World Scientific. https://play.google.com/store/books/details?id=RdXO0o5f4iwC
58. Rubik B, Muehsam D, Hammerschlag R, Jain S (2015) Biofield science and healing: history, terminology, and concepts. Glob Adv Health Med. 4:8–14. https://doi.org/10.7453/gahmj.2015.038.suppl
59. Miller I, Miller RA, Webb B (2011) Quantum bioholography. DNA Decipher J 1. https://dnadecipher.com/index.php/ddj/article/view/10
60. McClintock B (1956) Controlling elements and the gene. Cold Spring Harb Symp Quant Biol symposium.cshlp.org. 21:197–216. http://symposium.cshlp.org/content/21/197.short
61. Polesskaya O, Kananykhina E, Roy-Engel AM, Nazarenko O, Kulemzina I, Baranova A, Vassetsky Y, Myakishev-Rempel M (2018) The role of Alu-derived RNAs in Alzheimer's and other neurodegenerative conditions. Med Hypotheses 115:29–34

Chapter 8
How Schrödinger's Mice Weave Consciousness

Max Myakishev-Rempel and Ivan Viktorovich Savelev

In the Introduction, we will review the previously published research that connects DNA resonances with the work of the mind. This area is not well known so it will be useful to revisit some of the published theories and evidence supporting them. Yet, the novelty of this paper is in the proposed molecular mechanism for the work of the mind, which will be presented in the Hypothesis section followed by the Discussion where we will link the proposed hypothesis with some of the theories of consciousness, quantum mechanics, and self-organization.

1 Introduction

DNA Resonance and the Mind

The role of DNA in the work of the mind is currently thought to be limited to protein-coding genes, which are dynamically regulated and which, in turn, regulate the levels of proteins that, in turn, regulate the work of the brain. This mechanism is indirect, slow, and seems insufficient to explain the complexity and speed of our thinking. Previously, we suggested that there is a much faster and more direct mechanism by which DNA is involved in the thinking process [1]. It involves charge oscillations in DNA and the exchange of electromagnetic signals between cell nuclei via a network of microtubules and other fibers. In this way, the old picture of slow and indirect involvement of DNA in the work of the mind is supplemented by a model of direct and fast signaling between the DNA of all nuclei of the body via electromagnetic

M. Myakishev-Rempel (✉) · I. V. Savelev
DNA Resonance Research Foundation, San Diego, CA, USA
e-mail: max@dnaresonance.org
URL: http://dnaresonance.org

Localized Therapeutics, San Diego, CA, USA

A. Bandyopadhyay and K. Ray (eds.), *Rhythmic Advantages in Big Data and Machine Learning* , Studies in Rhythm Engineering,
https://doi.org/10.1007/978-981-16-5723-8_8

waves. This conceptual transformation could be likened to the supplementation of snail mail with the instantaneous connection of billions of people into one constantly active internet network via electromagnetic waves.

Subcellular Thinking Structures

The exclusivity of the neuronal signaling mechanism for thinking is challenged by simple organisms that do not have neurons or have only a few neurons. The nematode Caenorhabditis elegans has only 302 neurons but displays several complex behaviors including predator escape and mating. Some single-cell organisms that have no neurons also demonstrate complex behaviors and the ability to learn. Moreover, free-living single-cell ciliates such as Stentor roeselii are capable of learning [2] as is Plasmodium, which is a single large cell with many nuclei [3]. Paramecium, a single-cell organism, can swim, learn, display complex behaviors, and sexually reproduce [4]. This demonstrates that there are subcellular structures capable of thinking and making decisions [4].

DNA Resonance Signaling

The early theories for resonance signaling in biology were developed long before the discovery of the role of DNA in coding genetic information. The role of electric fields in morphogenesis was developed by Mathews [5], Morgan and Dimon [6], Lund [7], and others over 100 years ago. The existence of the morphogenetic field was proposed nearly 100 years ago [8]. It was proposed that the morphogenetic field is produced by the union of the cells of the organism, and this field guides the development of the shape of the body and regulates the function of each part and organ. Consider a modern analogy: GPS navigators tell drivers their location and centralized systems such as Uber wirelessly guide the drivers. Similarly, it was proposed that the body generates a morphogenetic field, which tells every cell its location and guides its actions. The existence of the morphogenetic field was experimentally demonstrated by independent groups [9, 10]. In these experiments, perturbing one of the chemically separated biological samples leads to measurable effects in another [11–14]. The electromagnetic oscillations in the cells were proposed to be driven by the constant chemical energy flux and were estimated to be in the millimeter-wave region [15].

Miller and Web proposed that genomic DNA is the main source and receiver of the morphogenetic field, allowing the genomic program to direct the morphogenesis directly via a holographic electromagnetic field (Miller et al. 1975; Miller and Webb 1973). Moreover, it was proposed that through the same field, the genomic DNA of brain cells is directly involved in the work of the mind [16]. Hameroff proposed that microtubules in axons work as light guides and transmit information in neurons, thus explaining the high speed and bandwidth of the mind [17]. We combined and expanded the ideas of Miller, Webb, and Hameroff by suggesting electromagnetic information transfer between the DNA in the nucleus, the microtubules in the cytoplasm, and the fibers of the extracellular matrix in the fascia [1].

An important factor to consider for the theory of the genomic biofield is the dissipation and scattering of electromagnetic signals in the tissues. Morphogenetic field and biofield is usually perceived as an unstructured field generated by the tissues and going in all directions. Let us call it the "Field Model". While we accept that some of the signaling is likely to occur via the Field Model, this model has a problem of dissipation and scattering of the signals. Therefore, we proposed a "Fiberoptic model" [1], in which genome copies of all cell nuclei of the organism exchange electromagnetic signals via a network of protein fibers serving as optical guides.

For an analogy of the field model, consider smartphones that send signals in all directions and contact cell towers in any direction: this information exchange is less specific and much of the signal is wasted. For the analogy of the fiberoptic model, consider broadband modems connected to internet hubs via fiberoptic cables: this exchange is specific, there is little interference and data loss, and the information transfer rate is much higher than for the smartphones that send and receive signals in all directions. So far, experimental evidence is published only for the Field Model [11–14]. In these publications, the field and information transfer was demonstrated, yet the role of DNA in its generation and reception was not tested. We find it likely that both the field and the fiberoptic models coexist side by side, with some signals exchanged via microtubules and other signals exchanged via the field.

The fiberoptic model [1] has the advantage that it minimizes data loss and crosstalk: information can be exchanged between specific locations with high specificity. A substantial body of experimental evidence suggests the existence of a system that exchanges electromagnetic signals via fiberoptic-like tubular structures of fascia tissue [18, 19]. These tissues wrap and penetrate the entire body and regulate its growth and health. This system closely corresponds to the placement of meridians in traditional Chinese medicine [18, 19]. We proposed that genome copies of all cells of the body are vibrationally coupled with the signaling system of meridians in the fascia and thus are linked into a single fiberoptic network [1]. The frequencies of the waves in this network may be in the infrared and millimeter-wave range [1].

For genome copies to communicate via electromagnetic waves, DNA fragments should be able to resonate in a sequence-dependent manner. Although mechanical oscillations in DNA have been proposed [20, 21], we reasoned that the mechanical oscillations would be damped by the viscosity of the nucleoplasm. Instead, we proposed that there must be oscillations of delocalized charges in the nucleobase stack, which would be protected by the DNA backbone from oscillation dumping. In this model, DNA harbors vibrationally coupled oscillations of delocalized proton and electron clouds in the base stack. We modeled their approximate shapes and, based on multiple genome sequences, produced statistical evidence for evolutionary selection and conservation of DNA sequences predicted to harbor repetitive electron and proton cloud patterns [22, 23]. Thus, based on the genomic data from various species, we provided the initial evidence for the existence of resonance signaling in DNA.

Furthermore, in this model, the key oscillators serving as transmitting and receiving antennas are repetitive elements in DNA that comprise over 50% of our genome; the vibrational information is coded in positions of repetitive elements, variations within them, and in their flanking sequences; the repetitive elements work as radios by converting biomolecular information into electromagnetic wave messages and vice versa; repetitive elements create an interference pattern of waves that is united between all cells of the organism, guides its development, and is an integral part of the work of the mind; the wave signals that are received by the DNA resonance elements are guiding the expression of genes and chromatin dynamics. Much in this model remains to be proven. In our previous publications [22, 23] we only provided the initial computational-genomic evidence for the evolutionary selection for certain electron and proton cloud patterns, which suggests the existence of DNA resonance signaling. In this paper, we will further develop the aspects of this model that offer a more detailed mechanism for the link between DNA and consciousness.

2 Hypothesis

Our main focus here will be on the hydrogen bonds and electrons in tautomeric forms of the DNA basepairs. The main question we asked was: which charged particles are delocalized in the basepairs and how do the shapes of delocalized charge clouds depend on the DNA sequence? The ultimate goal here was to understand the sequence dependence of the delocalized charge oscillations that potentially mediate our thinking process and other signaling in the body.

The normal coexistence and interconversion of tautomeric forms of normal Watson–Crick basepairs are known from experiments in model systems [24], Fig. 1.

From the classical chemistry perspective, basepairs oscillate between their tautomeric forms with the frequency in GHz–THz range [24–26]. Yet, from the quantum chemistry perspective, the tautomeric forms coexist in the state of quantum superposition until they are forced to make a choice in which state they exist. The choice of a tautomeric form could be forced by the background infrared irradiation, which is high in living tissues, or by an interaction with small molecules, that are in Brownian motion, constantly bump the double helix, and sometimes reach the base stack.

Both the classical and the quantum chemistry perspectives are true at the same time and the description of the model depends on which perspective one looks at it. As one can see in Fig. 1, it takes two proton relocations (a double proton transfer) to switch the basepair tautomeric form from GC1 to GC2, from GC2 to GC3, and from GC3 to GC1. The transition between forms GC1, GC2, and GC3 is also accompanied by electron relocations. As one positively charged proton jumps one step, two or more electrons in a chain jump one step each toward this proton to rebalance the electrical charge and to keep the charge of each of the larger atoms neutral.

Importantly, although the two hexagonal and one pentagonal rings in the basepair look similar and have alternating double bonds, only the central ring is aromatic (see Supplement 2 for details). Its aromaticity is classical and characterized by the unification of 6 pi electrons of the ring into one delocalized double-ring-shaped cloud, Fig. 3. The other two rings do not have enough pi-electrons to create a delocalized ring and therefore are not aromatic.

The key observation here is that the central ring is only aromatic in GC1 and is not aromatic in GC2 and GC3. This happens because the relocation of protons causes

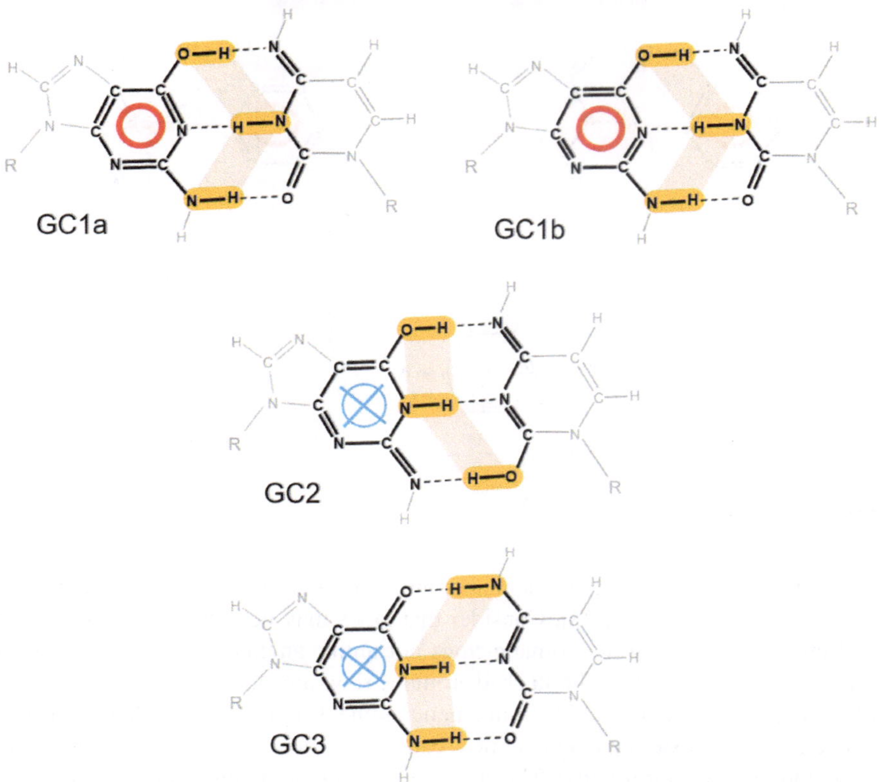

Fig. 1 Tautomeric forms of GC basepair. The hexagonal heterocycles of purines are called here "the central ring" and labeled with circles. The uncrossed circles signify the presence of aromaticity and the crossed circles signify the loss of aromaticity. The remaining heterocycles are not aromatic. The black lines signify the structures that undergo tautomerization and grey lines signify the structures that stay unchanged during tautomerization. Links to the backbone are shown as "R"

the relocation of electrons and the ring loses an electron to a proton that attaches to the purine. In GC1, the electrons of the central ring exist in a superposition of two configurations GC1a and GC1b.

A similar dependence of the aromaticity of the central ring from the relocations of protons is observed in the classical Watson–Crick's basepair AT, Fig. 2.

The main difference is that in AT there are only two tautomeric forms: AT1 and AT2. Tautomer AT1 is aromatic and AT2 is not.

Once we noticed the oscillation of aromaticity in basepairs, we realized that it may be involved in the work of the mind. Previously, we developed a model in which resonances in DNA are united over the entire organism by a network of optical fibers and thus participate in the work of the mind. Here, the process of enabling and collapsing aromaticity in basepairs reminded us of spinning and stopping a roulette wheel or throwing dice. It is well known that in the aromatic state the 6 pi electrons are delocalized, forming a stable ring that freely spins. The spinning of

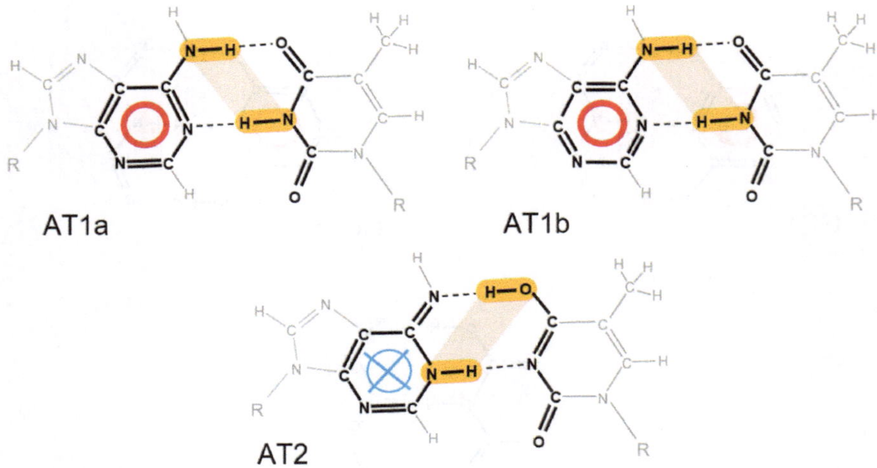

Fig. 2 Tautomeric forms of basepair AT

the ring creates a magnetic moment and, inversely, applying a magnetic field to the electron ring spins it [27, 28]. Consider that DNA may perform logical operations and thinks by spinning its aromatic rings not unlike an electronic machine, as we previously proposed [29]. The stacked aromatic rings are attracted via a known effect of aromatic ring stacking and their magnetic moments unite, causing them to behave collectively. The external magnetic field can switch the polarity of the stretch of the rings and reverse their rotation. The attraction of rings spinning in the same direction and magnetized in the same direction would likely straighten and shorten the double helix, while repulsion of the rings spinning in the opposite directions and magnetized in the opposite directions would likely bend or expand the double helix. Therefore, supercoiling of DNA may be under the control of high-frequency waves. This is relevant to gene expression since it requires uncoiling of the regions harboring the genes to be expressed. Typically, in this process, the loops containing multiple genes to be transcribed are uncoiled by gyrases, the genes are transcribed, and then the loops are coiled back. Similarly, uncoiling of DNA happens during the process of replication. Thus, regulation of coiling is an essential process in gene and chromatin regulation. We suggest that possibly, in addition to gyrases, the cell uses voltage oscillations to control coiling of DNA and it is likely that this mechanism is used by DNA to control its own coiling.

Wave Function Collapse

We suggest that proton jumping coupled to the oscillation of aromaticity of purines is a natural mechanism for the ability of DNA to process information and think. As we illustrated above, jumping of protons causes the collapse and expansion of the wave function of the aromatic electrons of the central ring. Or in other words, the proton jumps force these aromatic electrons to localize and delocalize.

Such collapses of the wave function were proposed to be the mechanism underlying the work of the mind and consciousness [30, 31]. On the molecular level, this idea was further developed by Hameroff [32] for microtubules. Here, we propose that DNA, and more specifically the localization and delocalization of aromatic electrons in purines, is the mechanism for our thinking process.

Intuition Versus Logic

Furthermore, consider a connection between the proton jumps in basepairs with conscious choice-making and between the delocalization with electrons and intuition. Terrence McKenna and others suggested that it was the evolution of hominids from gathering to hunting that forced us to develop our logical mind [33, 34]. They reasoned that it is our predatory nature that requires us to logically create plans and execute them, otherwise, we would not survive. This leads our modern civilization to prefer logic over intuition and action over passivity. McKenna noted that traditional tribes had a different mindset which was much more passive and intuitive than modern-day humans. Consider also that such qualities as ego, logic, and choice-making (popularly called masculine and left-brain) may be mechanistically mediated by localization of aromatic electrons and temporary loss of aromaticity in DNA. Conversely, such qualities as selflessness, intuition, and passivity (popularly called feninine and right-brain) may be mechanistically mediated by delocalization of aromatic electrons and gain of aromaticity in DNA.

Collective Behavior of Purines

When multiple purines follow each other in the DNA sequence, Fig. 3, their aromatic rings are attracted by the stacking forces and their magnetic moments should face in one direction and add together thus stabilizing their collective orientation. This would likely create a delocalized cloud of aromatic electrons spanning this stretch of purines and thus create a structure prone to charge oscillations. Since stretches of purines are frequent in the genome, they would create antennas allowing for wireless communication between the parts of the genome and between the genome copies of all the cells in the body. Thus, the delocalized state of electrons in purine stretches would allow for organism-wide resonances and nonlocal communications that nicely produce the intuitive state of mind. Conversely, the collapse of the wave function and the loss of aromaticity in purine stretches would correspond to the logical way of thinking and choice making.

Schrödinger's Mice

One peculiarity of the delocalization of charges in the basepair shown in Fig. 1 is that it is not only electrons that are delocalized, but also protons. Quantum delocalization of protons in basepairs is known from molecular dynamic calculations [25]. Since protons are 800 times heavier than electrons, their delocalization, while less pronounced, is still real. Both protons and electrons exist in the state of delocalization, quantum superposition, and obey Heisenberg's uncertainty principle. In the basepair's natural state, an outside observer not only cannot know the position of electrons of the central ring but also cannot know the position of the protons of the

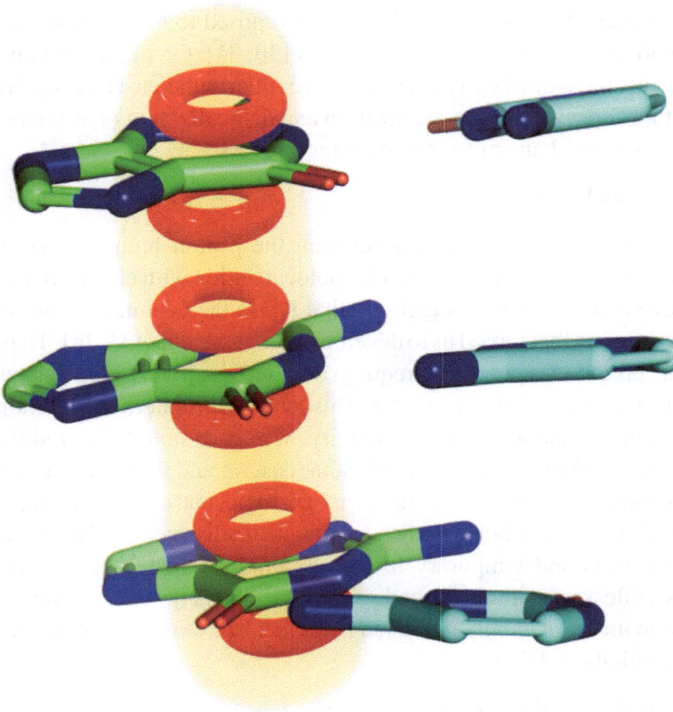

Fig. 3 Aromatic electron rings are merged in a purine stretch. Three basepairs of the basestack are shown. The backbone is hidden. Aromatic pi electrons of the central rings are shown in red. A cloud of delocalized aromatic electrons is shown in yellow

hydrogen bonds. The electrons are fused together in a double electron ring above and below the central ring (Fig. 3) and protons are delocalized into a probability cloud spreading along the hydrogen bond. When the basepair has a moment of quiet and is not bumped by infrared photons or water molecules, its electrons and protons delocalize and when it is disturbed by the bumps from outside, its delocalization collapses and protons and electrons take positions or in other words make a choice, as in Fig. 1.

In this respect, the dependence of the basepair aromaticity on its tautomeric form is very interesting. Only tautomer GC1 enables the central ring to become aromatic, while GC2 and GC3 disable the aromaticity of the central ring. Yet, much of the time, all three tautomers coexist in a state of quantum superposition. Quantum superposition is incompatible with the deterministic logic of the macroworld, and it only can be embraced as a gimmick of the quantum world. DNA is auspiciously located at the mesoscopic scale, the shadowland between the macroscopic and microscopic scale.

Another level of complexity is added by the fact that electrons are much more prone to delocalization than protons. To illustrate the resulting intrigue, let's expand the analogy of Schrödinger's cat. Let's have the GC basepair symbolized by a

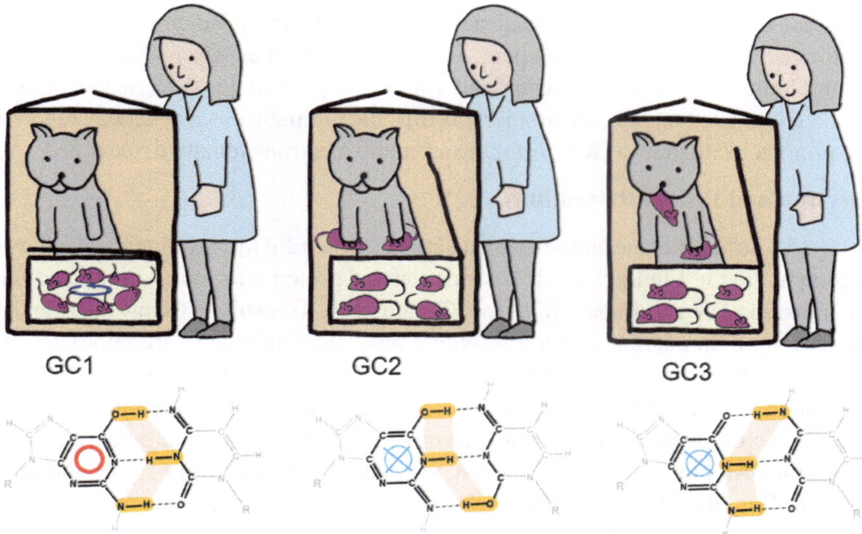

Fig. 4 Schrödinger's mice as an analogy of tautomeric states of GC basepair. The human observer is observing a closed box with a Schrödinger's cat. The cat is alive in all three superimposed states: GC1, GC2, and GC3. In GC1, the cat is observing a closed smaller box containing 6 mice which are numbered and spinning in a circle. In the state GC2, the cat has opened the box and grabbed 2 mice with two paws. Only at that point do the identifiers of the mice become visible to the cat. The state GC3 is the same as GC2 except the cat grabbed the mice with one paw and its mouth

Schrödinger's cat in a box, although without a threat to its life, Fig. 4. The cat can exist in three positions GC1, GC2, and GC3 corresponding to three tautomers of the GC basepair. For the outside observer, the position of the cat is unknown, until the observer opens the box. The cat also is an observer. It observes a smaller closed box containing a self-spinning carousel with 6 mice representing 6 electrons of the central ring, this position of the cat represents the tautomer GC1. The mice have numbers on *their t-shirts (not shown). The spinning of the wheel represents the delocalization of the electrons. Occasionally, the cat opens the smaller box and grabs two mice from the carousel and the remaining 4 mice hide in the corners. This represents the loss of aromaticity. If the cat grabs the 2 mice with two paws, this represents his second position and the tautomer GC2. If the cat grabs the 2 mice with the right paw and the mouth, this represents his third position and the tautomer GC3. The cat reads the identifiers of the mice and lets them go, switching back to the first position GC1, and the 6 mice start spinning on the carousel again.

Therefore, we can see that the quantum effects (uncertainty, delocalization, and superposition) are nested. The human observer observes a cat, which observes the mice. What is fascinating in this model is that the mice are delocalized (superimposed and uncertain) only when the cat is in the position GC1. In the other two positions, the mice are fixed, localized, their positions are determined. Therefore the human observer observes a part-time delocalized cat that observes the delocalization of mice only part-time. This illustrates the idea that the aromatic electrons and hydrogen bond protons delocalize when the basepairs are undisturbed; when basepairs are disturbed by the bumps of infrared photons and water, the protons localize; if the protons localize into GC2 and GC3 tautomeric forms, the aromatic electrons localize too; if the protons localize into GC1 form, the aromatic electrons remain delocalized.

Electron and Proton Relocations

To detail a little better the tautomeric transition, it is helpful to trace charge relocations accompanying it. Figure 5 illustrates electron and proton relocations accompanying the transition from tautomer GC1a to GC2 and back. Transitions follow the rule that charges of all atoms remain neutral. Figure 5a–c show that there are relocations of electrons in both directions. Figure 5d summarizes the relocations to illustrate that as a result, in the GC1a to GC2 transition, the protons move one step counterclockwise and a chain of electrons moves one step clockwise, thus neutralizing the relocation of protons. All other classical Watson–Crick tautomeric transitions follow the same pattern. This also illustrates that the basic chemistry of tautomeric transitions is simple and can be formalized.

Proton Clouds

In addition to electron clouds in purine stretches, we previously predicted the existence of delocalized proton clouds spanning multiple nucleotides in the DNA chain and produced preliminary evidence for their existence, Fig. 6 [23]. Proton clouds are known from protein research where they are sometimes called proton wires [35].

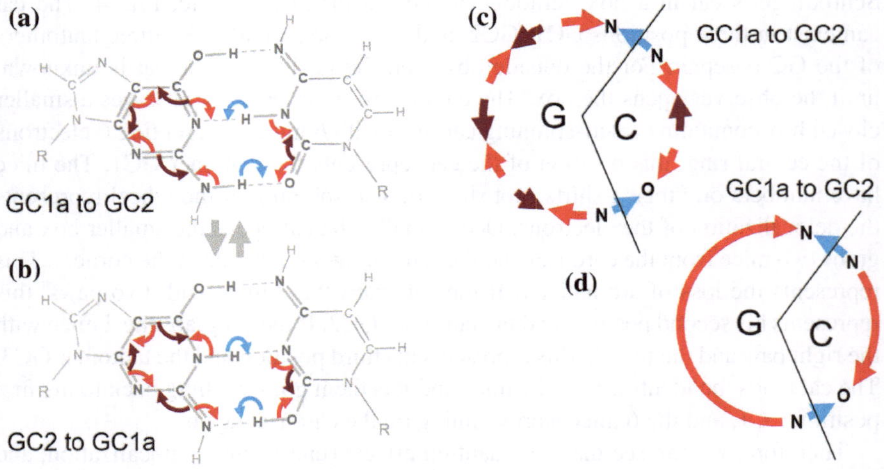

Fig. 5 Tautomeric transition GC1a to GC2. Proton relocations are in blue. Relocations of individual electrons in the same direction as protons are in brown and relocations in the opposite direction are in red. **a** Transition GC1a to GC2. **b** Transition GC2 to GC1a. **c** Simplified summary of charge relocations GC1a to GC2. **d** Even more simplified summary of charge relocations GC1a to GC2

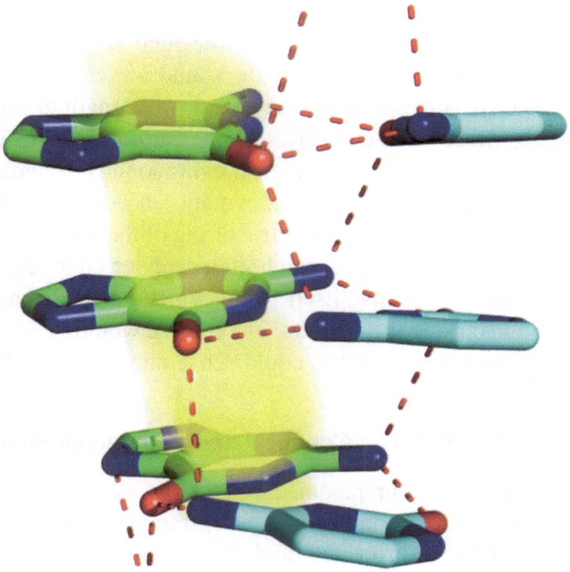

Fig. 6 Electron clouds and proton wires in DNA. A stretch of purines in the double helix is shown. It is a GGG sequence, GGG is shown in green and the complementary strand CCC in aqua color. The DNA backbone is not shown. The electron cloud is shown in yellow. Proton wires are shown in dotted lines

In our model of DNA resonance signaling, proton clouds also serve as antennas for wireless communication alongside with electron clouds. This way, there is an interplay of partially overlapping delocalized positive proton and negative electron clouds that are attracted to each other and oscillate in coordination with each other [22, 23, 29]. Their oscillations would only partly overlap in frequency, since protons are 800 times heavier than electrons. Consider also that oscillations of delocalized charge clouds spanning multiple basepairs will be affected by the tautomeric transitions at each basepair. This simplified model gives us a glimpse into the sophisticated machinery that we believe underlies the intuitive and logical thinking processes in our DNA and in our minds.

Function

Until this point, we have been looking at resonance signaling mainly within DNA. Let us now consider the ways in which oscillations of aromaticity in basepairs can communicate with the biochemical processes outside of DNA. Consider collective delocalization of electrons in the stretch of purines. This would turn on their aromaticity, increase base stacking, increase the attraction of the bases to each other, and shorten the base stack. Since the phosphates are charged, they repel each other, keeping the length of the DNA backbone from shortening. This should increase DNA winding and bending. Conversely, tautomeric transitions leading delocalization and loss of aromaticity will cause DNA unwinding and straightening. These changes (winding and unwinding, bending and straightening) are known to affect chromatin structure and gene expression.

Another way aromaticity oscillations can affect biochemistry is via electromagnetic oscillations. Charge oscillations that we suggest occur in electron and proton clouds spanning multiple bases can add together and their lower harmonics in the MHz–GHz range can induce ultrasound waves in the nucleoplasm. The frequency of 214 MHz corresponds to the sound wavelength of 7 um, the size of the nucleus. 750 GHz corresponds to the sound wavelength of 2 nm, the diameter of the double helix. Since DNA comprises a large part (about 1.5% of the nucleus mass) its harmonized oscillations could create moving sound interference patterns within the nucleus. This brings us to the ideas of cymatics, according to which moving sound patterns in tissues are responsible for structuring of the organism and driving organized motility of cellular components and proteins, reviewed in references [36, 37]. In this way, the genome could move itself using cymatic propulsion and control the movements of proteins inside the nucleus. The reverse would be also possible—the interaction can be bidirectional—DNA could sense and influence the environment by interacting with the wave patterns.

Interface Between Chemical and Vibrational Signaling

According to our model, DNA serves as an interface between the chemical and vibrational signaling. Binding of proteins to a specific location in DNA will change its vibrational properties and thus biochemical information would be converted to wave information that DNA is sending and receiving. In other words, binding of proteins (and nucleosomes) to a genomic locus tunes it to different frequencies. Conversely, a resonance signal received by a locus can excite it and thus lead to a structural change, which in turn would attract or repel specific proteins, which in turn would affect its biochemical function. Moreover, charge oscillations could drive the opening and closing of chromatin, thus directly controlling gene transcription. Both conversion of the chemical to a resonance signal and back could be explicitly driven by chemical energy, such as the energy of ATP hydrolysis by proteins bound to DNA. Also, possibly, the energy of Brownian motion [38] and infrared background radiation [39] could be harvested by DNA for the purposes of signal conversion.

Aromaticity

The idea that aromaticity of biopolymers is essential for consciousness is not new. The importance of aromaticity in microtubules for consciousness was developed by Hameroff during the past several decades [40]. Classically, the psychoactive effects of drugs are explained via binding of drugs to proteins and blocking neurotransmitter reuptake, inhibiting neurotransmitter synthesis, and inhibiting enzymes involved in molecular signaling pathways. Yet, the idea of DNA resonance signaling has been around at least since the early 1970s. Thus, 50 years ago, it was proposed that psychoactive substances, being predominantly aromatic, work by binding to DNA and changing its aromaticity and quantum delocalization of electrons [41]. Smythies pointed out that most of the psychoactive drugs contain aromatic groups similar to nucleobases, easily penetrate via cellular and nuclear membranes, and can bind to DNA either via intercalation or via hydrogen bonds [41]. Miller also emphasized the significance of aromaticity in DNA and electron delocalization for the phenomena of life [16]. Hameroff pointed out the correlation between the aromaticity correlates of anesthetic compounds and their potency [40].

To further highlight the importance of aromaticity for consciousness, we illustrated the aromaticity of the main psychoactive substances, their similarity to nucleobases and listed them according to the types of aromatic groups they contain, Fig. 7. Also, to illustrate intercalating structures, included are two intercalating substances: ethidium and SYBR Green, for which psychoactive effects are unknown. Although the list of substances in Fig. 7 is far from exhaustive, this exercise strongly confirms the observation that the majority of psychoactive substances are aromatic and this further supports the model of the mind as a DNA resonance-based fiberoptic network.

Binding of Aromatic Drugs to DNA

Small aromatic molecules such as the psychoactive substances listed in Fig. 7 easily penetrate the cell and nuclear membranes [42]. Most of them bind to DNA [43]. Indole derivatives such as melatonin, harmine [44], and ibogaine migrate into the nucleus and bind to DNA. Other indole derivatives also bind to DNA [42]. Hallucinogen ibogaine enters the nucleus and regulates gene expression [45]. Caffeine

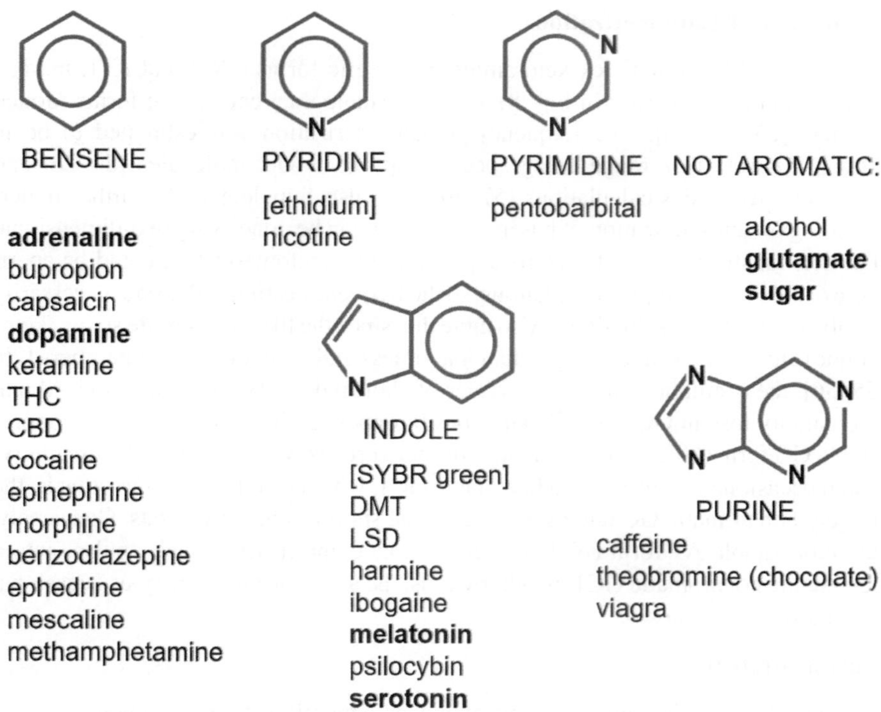

Fig. 7 Aromatic groups and exogenous and endogenous psychoactive substances classified by the aromatic groups they contain. In bold are endogenous psychoactive substances. In square brackets are intercalating substances widely used in DNA research, for which psychoactive properties are unknown. Only the benzene, pyridine, pyrimidine, indole, and purine groups are shown. All the psychoactive substances listed underneath contain additional, sometimes also aromatic, radicals which are not shown

and chocolate's theobromine bind to DNA via hydrogen bonds [46, 47]. Cannabinol (CBN) from cannabis binds in the major groove of DNA and does not intercalate into it [48]. Prozac is a DNA groove binder [49].

Intercalation

When an aromatic small molecule intercalates into DNA, it inserts itself into the base stack as if it is an additional basepair in the DNA and its aromatic ring of pi-electrons is fuzed into the periodic set of pi-electron rings of the nucleobases [43]. Morphine binds and intercalates into DNA [50, 51]. Adrenaline binds to DNA and may intercalate into DNA [52]. Hallucinogen harmine penetrates into the nucleus and binds to DNA [44] via intercalation [53]. Serotonin and tryptamine intercalate into DNA [54]. In summary, although the researchers of aromatic psychoactive drugs were primarily testing psychoactive drugs for the lack of mutagenic properties, they unwittingly produced supportive evidence for the hypothesis of Smythies from 50 years ago [41] that psychoactive drugs are expanding consciousness by boosting the aromaticity of DNA.

Frequency of Tautomerization

The classical Watson–Crick keto-amine tautomeric forms (GC3 and AT1, marked with a continuous border in Fig. 8) are more stable than enol-imine forms (dotted border) [25, 55, 56]. The frequency of tautomerization was estimated to be in 100 MHz–100 THz range using spectroscopy of model molecules [24–26] and molecular dynamics calculations [57–59]. See also Supplement 1. Further understanding of tautomerization of basepairs in DNA can be done using two-dimensional Fourier-transform infrared spectroscopy. Note that tautomerization could be aperiodic or subject to complex oscillations, so the frequency estimate does not necessarily imply regularity in oscillations. Also, note that since the lifetime of more stable (keto-amine) tautomers is much longer than that of less-stable (enol-imine) tautomers [25, 55, 56], the oscillations have a character of short pulses. We have also noticed that aromaticity loss in GC and AT pairs goes in opposite directions, Fig. 8. The more-stable GC form (GC3) is nonaromatic (or more precisely, has a lowered aromaticity) and it occasionally pulses into a less-stable GC1, which is fully aromatic—that is, the largely nonaromatic GC undergoes occasional short aromaticity bursts. Conversely, the more-stable AT form (AT1) is fully aromatic and it occasionally pulses into a less-stable nonaromatic GC1, in other words, the aromatic GC undergoes occasional short aromaticity lapses.

Purine Stretches

Among functionally important and abundant genomic elements, genomic polyA tracts and CpG islands stand out. PolyA tracts are important for viruses and transposons and often a deletion of a polyA tract impairs gene function [60]. CpG islands are typically located in genes and gene promoters and are involved in the activation of gene transcription [61]. Based on the above observation of the opposite character of aromaticity oscillations between GC and AT basepairs, polyA tracts having a stretch of pi-electron rings of adenines, should spend most of the duty cycle in an aromatic

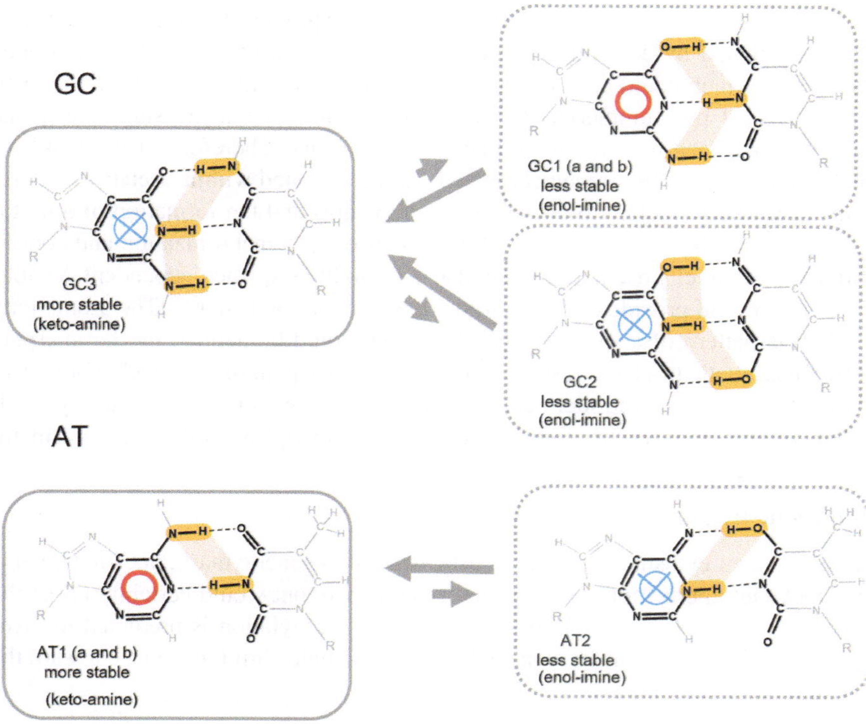

Fig. 8 Stability and aromaticity oscillations of purine tautomers in basepairs

state and have occasional lapses of aromaticity loss. Since the pi electrons in the basestack are organized in a periodic structure, they very likely exist as a synchronized electron cloud and their aromaticity loss is coordinated within the entire polyA tract. Both high aromaticity of the uniformly periodic basestack and occasional coordinated loss of aromaticity might affect their oscillatory and biomolecular function due to possible effects on DNA structure, packing of chromatin, binding of nucleosomes, and protein factors. Similar aromaticity oscillation patterns would also be observed in $(AT)_n$ tandem repeats.

Just the opposite should happen to CpG islands made exclusively of GC basepairs. They should exist in a reduced aromaticity state for most of their duty cycle and collectively produce short aromatic bursts. This could also affect their DNA resonance signaling and biomolecular functions.

Coordination of Aromaticity Oscillations

There are several mechanisms that would synchronize aromaticity oscillations within stretches of basepairs in DNA. First, aromatic pi-electron rings of purines unite into a synchronized periodic pattern, especially when the sequence is periodic, such as in tandem genomic repeats. Such periodic stacking of pi-electron rings is thought to be responsible in part for the experimentally observed high electrical conductivity of

DNA in physiological conditions [62]. Second, as we previously published, basepairs are likely bound by delocalized proton wires composed of longitudinal hydrogen bonds [23] which could also synchronize aromaticity oscillations. Third, the excitations caused by tautomerization could be transmitted via the sugar-phosphate backbone and lead to coordination between basepairs. Therefore, it is likely that aromaticity oscillations are synchronized and coordinated within stretches of basepairs. Since both stacking of aromatic electron rings and the formation of longitudinal hydrogen bonds depends on DNA sequence, the synchronization and coordination of aromaticity oscillations would also be highly sequence-dependent. Various sequences would provide different aromaticity oscillation patterns. The aromaticity oscillation pattern of a specific DNA fragment would be defined by the interplay between aromatic pi-electron stacks and proton wires, patterns of which would vary with DNA sequence. Yet, identical sequences will have identical aromaticity oscillation patterns and synchronize with each other, thus providing a mechanism for resonance signaling.

Epigenetic Regulation

Methylation of cytosine bases, the most frequent epigenetic mark, does not change the DNA tautomerization formulas described above, but would certainly affect the tautomerization rates and stability. In particular, methylation is predicted to favor aromatic ring stacking interactions [63] and thus should shift the balance toward the aromatic tautomers.

3 Discussion

ORCH-OR Theory

The delocalized state of aromatic electrons and protons in biological systems is described by Schrödinger's wave function. The loss of delocalization results in the collapse of Schrödinger's wave function and, according to Objective Reduction (OR) of the quantum state, this collapse is a choice and collectively these choices produce conscious awareness [64]. This was expanded to the Orchestrated Objective Reduction (ORCH-OR) theory of Penrose and Hameroff [65] which proposed the key role of microtubules. There, the aromatic rings of aromatic amino acids tyrosine, phenylalanine, and tryptophan of tubulin were suggested to periodically collapse and expand, producing choices and thus creating conscious awareness. Hameroff also posted online an unfinished paper suggesting the role of DNA in the process.

Here, we expand the ORCH-OR theory to include DNA in better detail. DNA and microtubules share aromatic and helical natures and their dimensions are comparable. DNA is plausible as a thinking molecular machine since it carries the genetic code and has an efficient addressing system: it is often sufficient to know only 15 bases of the code to find a specific spot among the 3.2 billion bases of the genome. ORCH-OR theory proposes that the periodic collapse of the wave function of the aromatic aminoacids produces consciousness. Here, we propose the same for the aromaticity in DNA. In this process, basepairs oscillate between their aromatic and nonaromatic tautomeric forms, Fig. 1, the aromatic electrons oscillate between delocalized and localized states and their wave functions collapse and expand. According to our model, this takes place in each of the 6.4 billion purines in the cell. This number can be multiplied by 80 billion neurons in the brain or up to 30 trillion cells of our body considering that not only brain neurons are involved in the thinking process. As we proposed previously [1], the genomes of the body located in the nuclei are informationally coupled into one fiberoptic network and thus all DNA and microtubules of the body are united into one thinking network.

Hameroff also proposed that occasional wave function collapses produce time as a byproduct of creating consciousness [32]. Here, we expand this by suggesting that it is the experience of time and self-awareness that are produced by the wave function collapses in the DNA of the body. Nonbiological objects and unidirectional processes also exist in the space–time of our universe, but we suggest that it is the wave function collapses and expansions of aromatic electrons in DNA that produce the experience of conscious awareness and sliding unidirectionally through time.

Decoherence

Understanding decoherence is one of the key developments in the quantum mechanics of recent decades [66]. This concept allows modeling the biological processes in mesoscopic scale—the scale of DNA. When purines transition into their aromatic forms, their pi electrons are united into an aromatic ring and delocalize. This results in the quantum entanglement of these electrons and increases the coherence of their union. The loss of aromaticity could be caused when the base stack is bumped by the water molecules or infrared photons. The loss of aromaticity is accompanied by localization (or de-delocalization) of electrons of the aromatic ring, loss of coherence, and collapse of Schrödinger's wave function. Thus, purines oscillate between two worlds—that is, between the quantum world of coherence and delocalization and the macroscopic world of decoherence and localization. The quantum delocalized coherent state occurs spontaneously whenever the electrons are left to themselves, which is possible because purines are protected from the outer nucleoplasm by the highly charged backbone of DNA. The macroscopic localized decoherent state is created when Brownian motion or infrared irradiation causes double proton transfer which pulls out an electron from the aromatic ring and causes the ring to fall apart. This way, oscillations of aromaticity in DNA provide an interface between the quantum world and the macroscopic world. DNA can be considered a natural quantum computer and possibly a receiver and transmitter of nonlocal quantum information.

Nonlocality

Nonlocality, or Einstein's "spooky action at a distance", is a quantum world phenomenon arising from the entanglement of elementary particles. Entanglement and nonlocality were demonstrated in experiments with electrons and photons. Although DNA, being of mesoscopic scale, is a few orders of magnitude larger than particles for which quantum effects were demonstrated, it still retains some of the properties of the microscopic world: delocalization of electrons in aromatic rings of purines is well known and delocalization of protons in hydrogen bonds has also been shown. Another quantum property in DNA is known from the experiments on its electrical conductivity. It was experimentally shown that, in short tandem DNA repeats, electrons tunnel (same as jump or hop) through more than one base [67].

Nonlocality, or action at a distance, was also experimentally studied in biology. Thorough and well-controlled studies of Radin, Sheldrake, and others demonstrate that consciousness has a nonlocal component [68–72]. Sheldrake proposed that a substantial part of the human consciousness is located outside of the body in a nonlocal "morphic field" [69]. To expand this, we suggest that oscillations of aromaticity in stretches of DNA could serve as an interface between the local macroscopic world and the nonlocal "morphic field" governed by the laws of quantum physics. This nonlocality would also correspond to Bohm's implicate order of the De Broglie–Bohm interpretation of quantum mechanics [73].

Sheldrake also convincingly argues that the work of the mind is not limited to the brain and the rest of the body is involved in the work of the mind [69]. For example, there are documented cases in which organ transplants transferred memories and character traits of transplant donors to recipients [69, 74–76]. On the same note, the fish having no visual cortex was able to perceive a visual illusion, the function traditionally ascribed to the visual cortex. The above observations are in agreement with our model that unites the DNA of all the cells of the brain and the rest of the body into a fiberoptic network.

Genome as a Quantum Computer

It has been previously proposed that the genome works as a quantum computer [77–79]. Here, we expanded this by adding a specific mechanism for quantum computation. The aromaticity oscillations are coordinated in stretches of DNA and are coupled to the oscillations of delocalized protons. Sequence-specificity of the patterns of the electron and proton clouds allows the DNA code to directly define the oscillations and thus serve as a program for the quantum computer. According to our model, the key intuitive/logical unit in this computer is the basepair, which oscillates between intuitive uncertainty and logical certainty, thus making a choice for every cycle that happens approximately with GHz–THz frequency. The number of intuitive-logical units in this computer can be obtained by multiplying 6.4 billion purines per cell and by 30 trillion cells per human body. Therefore, or body contains $2 * 10^{23}$ intuitive-logical units (basepairs) that make choices with GHz–THz frequency each.

An important role in the model of the genome as a quantum computer must be given to the balance between chaos and self-organization of chromatin. The theory of self-organization of chaos is gradually expanding from weather and financial forecasts to biology [80]. Fractality of chromatin is experimentally demonstrated in living cells [81]. High resolution spatial mapping of chromatin in cells demonstrates that chromatin oscillates between chaos and organization: it organizes itself into fractal structures which are then destroyed by the randomnicity of Brownian motion [82]. Therefore the key features of the genome as a quantum computer are the interplay between order and randomnicity, coherence and decoherence, delocalization and localization, quantum and classical mechanics. The results of the genomic computation inside the nucleus are integrated by the nuclear membrane and then the information is integrated by the fiberoptic network, which unites all genome copies of the body. In this way, the intuitive and logical computation on the molecular level is integrated in the work of our mind.

4 Conclusions

In summary, we started from searching for sequence-specific oscillations in DNA and this brought our attention to sequence-dependent stacking of aromatic rings and proton wires stretched along the base stack. We then noticed that tautomerization must be sequence-dependent and should affect oscillations in delocalized charge clouds. Then we noticed that purines would oscillate between aromatic and nonaromatic states. Further, we incorporated these observations with the idea of fiberoptic signal transmission and the work of the mind. Verification of this model will require further studies including spectroscopic studies and quantum mechanical molecular modeling.

Acknowledgements We thank Richard Alan Miller, Rupert Sheldrake, Dean Radin, Stanley Krippner, Glen Rein, Peter Gariayev, Dirk K. F. Meijer, Alexey Melkikh, Erico Azevedo, Tobias A. Knoch, Erico Azevedo, and Anirban Bandyopadhyay for the feedback on the manuscript. We thank Himansu S. Biswal, Rajesh Vadgama, Cina Foroutan-Nejad, Judy Wu, Borys Osmialowski, Salvatore Profeta Jr, Hashim Al-Hashimi, Atul Kaushik Rangadurai, Dominik Lungerich, and Tatiana S Oretskaya for help with understanding tautomerization and/or aromaticity loss in purines. The work was funded solely by MMR.

Author Contributions MMR developed the hypothesis and wrote the manuscript. IVS did the literature work and contributed to the discussion.

References

1. Savelyev I, Zyryanova NV, Polesskaya OO, Myakishev-Rempel, M (2019) On the existence of the DNA resonance code and its possible mechanistic connection to the neural code. Neuroquantology 17
2. Dexter JP, Prabakaran S, Gunawardena J (2019) A complex hierarchy of avoidance behaviors in a single-cell eukaryote. Curr Biol 29:4323–4329.e2. https://doi.org/10.1016/j.cub.2019.10.059
3. Dussutour A, Latty T, Beekman M, Simpson SJ (2010) Amoeboid organism solves complex nutritional challenges. Proc Natl Acad Sci U S A 107:4607–4611. https://doi.org/10.1073/pnas.0912198107
4. Maegawa S (2017) Molecular characteristics of neuron-like functions in single-cell organisms. In: Brain evolution by design: from neural origin to cognitive architecture. Springer, Japan, Tokyo, pp 25–44. https://doi.org/10.1007/978-4-431-56469-0_2

5. Mathews AP (1903) Electrical polarity in the hydroids. Am J Physiol-Legacy Content. Am Physiol Soc 8:294–299. https://journals.physiology.org/doi/pdf/10.1152/ajplegacy.1903.8.4.294
6. Morgan TH, Dimon AC (1904) An examination of the problems of physiological "polarity" and of electrical polarity in the earthworm. J Exp Zool 1:331–347. https://doi.org/10.1002/jez.1400010206
7. Lund EJ (1917) Reversibility of morphogenetic processes in Bursaria. J Exp Zool 24:1–33. http://doi.wiley.com/10.1002/jez.1400240102
8. Gurwitsch A (1922) Über den Begriff des Embryonalen feldes. Wilhelm Roux Arch Entwickl Mech Org 51:383–415. https://doi.org/10.1007/BF02554452
9. Gurwitsch AA (1988) A historical review of the problem of mitogenetic radiation. Experientia 44:545–550. https://doi.org/10.1007/BF01953301
10. Volodyaev I, Beloussov LV (2015) Revisiting the mitogenetic effect of ultra-weak photon emission. Front Physiol frontiersin.org 6:241. https://doi.org/10.3389/fphys.2015.00241
11. Cifra M, Fields JZ, Farhadi A (2011) Electromagnetic cellular interactions. Prog Biophys Mol Biol Elsevier 105:223–246. https://doi.org/10.1016/j.pbiomolbio.2010.07.003
12. Scholkmann F, Fels D, Cifra M (2013) Non-chemical and non-contact cell-to-cell communication: a short review. Am J Transl Res ncbi.nlm.nih.gov 5:586–593. https://www.ncbi.nlm.nih.gov/pubmed/24093056
13. Trushin MV (2004) Distant non-chemical communication in various biological systems. Riv Biol researchgate.net 97:409–442. https://www.ncbi.nlm.nih.gov/pubmed/15754593
14. Xu J, Yang F, Han D, Xu S (2017) Wireless communication in biosystems. arXiv [physics.bio-ph]. http://arxiv.org/abs/1708.02467
15. Frohlich H (1988) Theoretical physics and biology. In: Biological coherence and response to external stimuli. Springer-Verlag, Berlin, pp 1–24. http://link.springer.com/content/pdf/10.1007/978-3-642-73309-3.pdf#page=9
16. Miller RA, Webb B, Dickson D (1975) A holographic concept of reality. Psychoenergetic Syst (Gordon and Breach Science Pub Ltd.) 1:55–62. http://www.nwbotanicals.org/events/real.pdf
17. Hameroff SR (1974) Ch'i: a neural hologram? Microtubules, bioholography, and acupuncture. Am J Chin Med (World Scientific Publishing Co.) 02:163–170. https://doi.org/10.1142/S0192415X74000213
18. Maurer N, Nissel H, Egerbacher M, Gornik E, Schuller P, Traxler H (2019) Anatomical evidence of acupuncture meridians in the human extracellular matrix: results from a macroscopic and microscopic interdisciplinary multicentre study on human corpses. Evid-Based Complement Alternat Med 2019:1–8. https://doi.org/10.1155/2019/6976892
19. Bai Y, Wang J, Wu J-P, Dai J-X, Sha O, Tai Wai Yew D, Yuan L, Liang Q-N (2011) Review of evidence suggesting that the fascia network could be the anatomical basis for acupoints and meridians in the human body. Evid Based Complement Alternat Med 2011:260510. http://dx.doi.org/10.1155/2011/260510
20. Scott AC (1985) Soliton oscillations in DNA. Phys Rev A Gen Phys 31:3518–3519. https://www.ncbi.nlm.nih.gov/pubmed/9895924
21. Volkov SN, Kosevich AM (1987) Conformation oscillations of DNA. Mol Biol (europepmc.org.) 21:797–806. https://europepmc.org/abstract/med/3657778
22. Savelyev I, Myakishev-Rempel M (2019) Possible traces of resonance signaling in the genome. https://www.researchgate.net/profile/Max_Myakishev-Rempel/publication/335149527_Possible_traces_of_resonance_signaling_in_the_genome/links/5d5310c4299bf16f073695fd/Possible-traces-of-resonance-signaling-in-the-genome.pdf
23. Savelev I, Myakishev-Rempel M (2020) Evidence for DNA resonance signaling via longitudinal hydrogen bonds. Prog Biophys Mol Biol. https://doi.org/10.1016/j.pbiomolbio.2020.07.005
24. Abou-Zied OK, Jimenez R, Romesberg FE (2001) Tautomerization dynamics of a model base pair in DNA. J Am Chem Soc 123:4613–4614. https://doi.org/10.1021/ja003647s
25. Pérez A, Tuckerman ME, Hjalmarson HP, Von Lilienfeld OA (2010) Enol tautomers of Watson–Crick base pair models are metastable because of nuclear quantum effects. J Am Chem Soc (ACS Publications) 132:11510–11515. https://pubs.acs.org/doi/abs/10.1021/ja102004b

26. Abo-Riziq A, Grace L, Nir E, Kabelac M, Hobza P, de Vries MS (2005) Photochemical selectivity in guanine–cytosine base-pair structures. Proc Natl Acad Sci USA (National Academy of Sciences) 102:20–23. https://www.pnas.org/content/102/1/20.short
27. Katritzky AR, Barczynski P, Musumarra G, Pisano D, Szafran M (1989) Aromaticity as a quantitative concept. 1. A statistical demonstration of the orthogonality of classical and magnetic aromaticity in five- and six-membered heterocycles. J Am Chem Soc (American Chemical Society) 111:7–15. https://doi.org/10.1021/ja00183a002
28. Gershoni-Poranne R, Stanger A (2015) Magnetic criteria of aromaticity. Chem Soc Rev (pubs.rsc.org.) 44:6597–6615. http://dx.doi.org/10.1039/c5cs00114e
29. Polesskaya O, Guschin V, Kondratev N, Garanina I, Nazarenko O, Zyryanova N, Tovmash A, Mara A, Shapiro T, Erdyneeva E, Zhao Y, Kananykhina E, Myakishev-Rempel M (2018) On possible role of DNA electrodynamics in chromatin regulation. Prog Biophys Mol Biol 134:50–54. https://doi.org/10.1016/j.pbiomolbio.2017.12.006
30. Shimony A, Cushing JT (1994) Search for a naturalistic world view, vol. 1: scientific method and epistemology, vol. 2: natural science and metaphysics. Am J Phys 62:286–288. https://doi.org/10.1119/1.17617
31. Shimony A (1997) On mentality, quantum mechanics and the actualization of potentialities. na
32. Hameroff S (2003) Time, consciousness and quantum events in fundamental spacetime geometry. In: The nature of time: geometry, physics and perception. Springer Netherlands, Dordrecht, pp 77–89. https://doi.org/10.1007/978-94-010-0155-7_9
33. Thagard P, Shelley C (1997) Abductive reasoning: logic, visual thinking, and coherence. In: Logic and scientific methods: volume one of the tenth international congress of logic, methodology and philosophy of science, florence, August 1995. Springer Netherlands, Dordrecht, pp 413–427. https://doi.org/10.1007/978-94-017-0487-8_22
34. Sheldrake R, McKenna TK, Abraham R (1998) The evolutionary mind: trialogues at the edge of the unthinkable. Trialogue Press
35. Shinobu A, Agmon N (2009) Mapping proton wires in proteins: carbonic anhydrase and GFP chromophore biosynthesis. J Phys Chem A (ACS Publications) 113:7253–7266. https://doi.org/10.1021/jp8102047
36. Meijer DKF, Geesink HJH (2016) Phonon guided biology: architecture of life and conscious perception are mediated by toroidal coupling of phonon, photon and electron information fluxes at discrete eigenfrequencies. Neuroquantology 14. https://www.academia.edu/download/52162519/Phonon_guided_biology_NQ.pdf
37. Meijer DKF, Jerman I, Melkikh AV, Sbitnev VI (2020) Consciousness in the universe is tuned by a musical master code, Part 3: A hydrodynamic superfluid quantum space guides a conformal mental attribute of reality. Quantum (academia.edu.) 11:72–107. http://www.academia.edu/download/63283954/QBS_part_3_f152fa_74f949c7d405405789a7637d161201b420200512-121026-iu7xdy.pdf
38. Minchew CL, Didenko VV (2014) Nanoblinker: Brownian motion powered bio-nanomachine for FRET detection of phagocytic phase of apoptosis. PLoS ONE 9:e108734. https://doi.org/10.1371/journal.pone.0108734
39. Woller JG, Hannestad JK, Albinsson B (2013) Self-assembled nanoscale DNA-porphyrin complex for artificial light harvesting. J Am Chem Soc 135:2759–2768. https://doi.org/10.1021/ja311828v
40. Hameroff SR, Craddock TJA, Tuszynski JA (2014) Quantum effects in the understanding of consciousness. J Integr Neurosci (World Scientific) 13:229–252. https://www.worldscientific.com/doi/abs/10.1142/S0219635214400093
41. Smythies JR, Benington F, Morin RD (1970) The mechanism of action of hallucinogenic drugs on a possible serotonin receptor in the brain. Int Rev Neurobiol 12:207–233. https://doi.org/10.1016/s0074-7742(08)60062-8
42. Lafayette EA, de Almeida SMV, Cavalcanti Santos RV, de Oliveira JF, Amorim CA da C, da Silva RMF, Pitta MG da R, Pitta I da R, de Moura RO, de Carvalho Júnior LB, de Melo Rêgo MJB, de Lima M do CA (2017) Synthesis of novel indole derivatives as promising DNA-binding agents and evaluation of antitumor and antitopoisomerase I activities. Eur J Med Chem 136:511–522. http://dx.doi.org/10.1016/j.ejmech.2017.05.012

43. Rescifina A, Zagni C, Varrica MG, Pistarà V, Corsaro A (2014) Recent advances in small organic molecules as DNA intercalating agents: synthesis, activity, and modeling. Eur J Med Chem 74:95–115. https://doi.org/10.1016/j.ejmech.2013.11.029
44. Vignoni M, Erra-Balsells R, Epe B, Cabrerizo FM (2014) Intra- and extra-cellular DNA damage by harmine and 9-methyl-harmine. J Photochem Photobiol B 132:66–71. https://doi.org/10.1016/j.jphotobiol.2014.01.020
45. Marton S, González B, Rodríguez-Bottero S, Miquel E, Martínez-Palma L, Pazos M, Prieto JP, Rodríguez P, Sames D, Seoane G, Scorza C, Cassina P, Carrera I (2019) Ibogaine administration modifies GDNF and BDNF expression in brain regions involved in mesocorticolimbic and nigral dopaminergic circuits. Front Pharmacol 10:193. https://doi.org/10.3389/fphar.2019.00193
46. Johnson IM, Prakash H, Prathiba J, Raghunathan R, Malathi R (2012) Spectral analysis of naturally occurring methylxanthines (theophylline, theobromine and caffeine) binding with DNA. PLoS ONE 7:e50019. https://doi.org/10.1371/journal.pone.0050019
47. Nafisi S, Manouchehri F, Tajmir-Riahi H-A, Varavipour M (2008) Structural features of DNA interaction with caffeine and theophylline. J Mol Struct 875:392–399. http://www.sciencedirect.com/science/article/pii/S0022286007003791
48. Tian Z, Wang Z, Han X, Wang N, Wang R (2018) Study on the interaction between cannabinol and DNA using acridine orange as a fluorescence probe. J Mol Recognit 31. https://doi.org/10.1002/jmr.2682
49. Kashanian S, Javanmardi S, Chitsazan A, Paknejad M, Omidfar K (2012) Fluorometric study of fluoxetine DNA binding. J Photochem Photobiol B 113:1–6. https://doi.org/10.1016/j.jphotobiol.2012.04.002
50. Li JF, Dong C (2009) Study on the interaction of morphine chloride with deoxyribonucleic acid by fluorescence method. Spectrochim Acta A Mol Biomol Spectrosc 71:1938–1943. https://doi.org/10.1016/j.saa.2008.07.033
51. Talemi RP, Mashhadizadeh MH (2015) A novel morphine electrochemical biosensor based on intercalative and electrostatic interaction of morphine with double strand DNA immobilized onto a modified Au electrode. Talanta 131:460–466. https://doi.org/10.1016/j.talanta.2014.08.009
52. Zheng S-J, Lin X-Q (2003) Spectral and electrochemical investigation of intercalations of adrenaline and CT-DNA. Chin J Chem (Wiley) 21:767–771. https://doi.org/10.1002/cjoc.20030210712
53. Wink M, Schmeller T, Latz-Brüning B (1998) Modes of action of allelochemical alkaloids: interaction with neuroreceptors, DNA, and other molecular targets. J Chem Ecol (Springer) 24:1881–1937. https://idp.springer.com/authorize/casa?redirect_uri=https://link.springer.com/article/10.1023/A:1022315802264&casa_token=hcRaCBPWqJgAAAAA:en7s9Wb0_tvtPxjCJLoDqTfpMS06NPph2E_eWpqh5L2RQ8FE9VjuWSbNJDW0eV1sjLyk17Veddg KDmvnNA
54. Hélène C, Dimicoli JL, Brun F (1971) Binding of tryptamine and 5-hydroxytryptamine (serotonin) to nucleic acids. Fluorescence and proton magnetic resonance studies. Biochemistry 10:3802–3809. https://doi.org/10.1021/bi00796a025
55. Brovarets OO, Hovorun DM (2015) The physicochemical essence of the purine pyrimidine transition mismatches with Watson-Crick geometry in DNA: AC* versa A C. A QM and QTAIM atomistic understanding. J Biomol Struct Dyn (Taylor & Francis) 33:28–55. https://www.tandfonline.com/doi/abs/10.1080/07391102.2013.852133
56. Brovarets' OO, Oliynyk TA, Hovorun DM (2019) Novel tautomerisation mechanisms of the biologically important conformers of the reverse Löwdin, Hoogsteen, and Reverse Hoogsteen G*·C* DNA base pairs via proton transfer: a quantum-mechanical survey. Front Chem 7. https://dx.doi.org/10.3389/fchem.2019.00597
57. Ol'ha OB, Hovorun DM (2018) Renaissance of the tautomeric hypothesis of the spontaneous point mutations in DNA: new ideas and computational approaches. Mitochondrial DNA: New Insights (BoD--Books on Demand) 31. https://books.google.com/books?hl=en&lr=&id=Ui-RDwAAQBAJ&oi=fnd&pg=PA31&dq=Renaissance+of+the+Tautomeric+Hypothesis+

of+the+Spontaneous+Point+Mutations+in+DNA+New+Ideas+and+Computational+Approaches+Ol+ha+O+Brovarets+and+Dmytro+M+Hovorun&ots=1OVPDNNqDF&sig=eFIC-GWVP5xhbezRKKCSRX47RPw
58. Brovarets' OO, Hovorun DM (2014) Why the tautomerization of the G·C Watson–Crick base pair via the DPT does not cause point mutations during DNA replication? QM and QTAIM comprehensive analysis. J Biomol Struct Dyn (Taylor & Francis) 32:1474–1499. https://doi.org/10.1080/07391102.2013.822829
59. Brovarets (2015) Proton tunneling in the AT Watson-Crick DNA base pair myth or reality. J Biomol Struct Dyn (Taylor & Francis) 33:2716–2720
60. Guerrini R, Moro F, Kato M, Barkovich AJ, Shiihara T, McShane MA, Hurst J, Loi M, Tohyama J, Norci V, Hayasaka K, Kang UJ, Das S, Dobyns WB (2007) Expansion of the first PolyA tract of ARX causes infantile spasms and status dystonicus. Neurology 69:427–433. https://doi.org/10.1212/01.wnl.0000266594.16202.c1
61. Deaton AM, Bird A (2011) CpG islands and the regulation of transcription. Genes Dev 25:1010–1022. https://doi.org/10.1101/gad.2037511
62. Kratochvílová I, Vala M, Weiter M, Špérová M, Schneider B, Páv O, Šebera J, Rosenberg I, Sychrovský V (2013) Charge transfer through DNA/DNA duplexes and DNA/RNA hybrids: complex theoretical and experimental studies. Biophys Chem 180–181:127–134. https://doi.org/10.1016/j.bpc.2013.07.009
63. Kabeláč M, Hobza P (2007) Hydration and stability of nucleic acid bases and base pairs. Phys Chem Chem Phys (The Royal Society of Chemistry) 9:903–917. https://pubs.rsc.org/en/content/articlehtml/2007/cp/b614420a
64. Penrose R (1994) Shadows of the mind: a search for the missing science of consciousness. Oxford University Press. https://play.google.com/store/books/details?id=gDbOAK89tmcC
65. Hameroff (1997) Models of classical and quantum computation in microtubules: implications for consciousness. BCEC 193–217. https://books.google.com/books?hl=en&lr=&id=TR7QDgAAQBAJ&oi=fnd&pg=PA193&dq=Orchestrated+reduction+of+quantum+coherence+in+brain+microtubules+A+model+for+consciousnessS+Hameroff+R+Penrose+Mathematics+and+computers+in+simulation+40+(3-4)+453-480&ots=vsdTvgZNoW&sig=iueSxW9yWgoH0AwnctrWRMxW168
66. Ball P (2018) Beyond weird: why everything you thought you knew about quantum physics is different. University of Chicago Press. https://play.google.com/store/books/details?id=Nk1vDwAAQBAJ
67. Lewis FD, Liu J, Weigel W, Rettig W, Kurnikov IV, Beratan DN (2002) Donor-bridge-acceptor energetics determine the distance dependence of electron tunneling in DNA. Proc Natl Acad Sci U S A 99:12536–12541. https://doi.org/10.1073/pnas.192432899
68. Radin D (2009) Entangled minds: extrasensory experiences in a quantum reality. Simon and Schuster. https://play.google.com/store/books/details?id=sUM1Hc-KwJQC
69. Sheldrake R (2009) Morphic resonance: the nature of formative causation. Inner Traditions/Bear & Co. https://play.google.com/store/books/details?id=F75gxeTBgocC
70. Storm L, Sherwood SJ, Roe CA, Tressoldi PE, Rock AJ, Di Risio L (2017) On the correspondence between dream content and target material under laboratory conditions: a meta-analysis of dream-ESP studies, 1966–2016. Int J Dream Res (psycnet.apa.org) 10:120–140. https://psycnet.apa.org/fulltext/2018-14483-003.pdf
71. Bem D, Tressoldi P, Rabeyron T, Duggan M (2015) Feeling the future: a meta-analysis of 90 experiments on the anomalous anticipation of random future events. F1000Res 4:1188. http://dx.doi.org/10.12688/f1000research.7177.2
72. Mossbridge JA, Radin D (2018) Precognition as a form of prospection: a review of the evidence. Psychol Conscious Theory Res Pract (Educational Publishing Foundation) 5:78. https://psycnet.apa.org/journals/cns/5/1/78/
73. Bohm D (1980) Wholeness and the implicate order. Routledge & Kegan Paul Ltd., London & Boston
74. Pearsall P, Schwartz GER, Russek LGS (2002) Changes in heart transplant recipients that parallel the personalities of their donors. J Near-Death Stud (Springer) 20:191–206. https://

idp.springer.com/authorize/casa?redirect_uri=https://link.springer.com/article/10.1023/A:1013009425905&casa_token=uGxsEA-qtPcAAAAA:ENjBkxx_8n2RNKu9Kp6IH1h_0jGgELODSNQDCH0AiXT1dpvra1bT3qfXnPBR6hMmJItyTq3JQmrTfcZBNg

75. Joshi S (2011) Memory transference in organ transplant recipients. J New Approaches Med Health (namahjournal.com) 19. https://www.namahjournal.com/doc/Actual/Memory-transference-in-organ-transplant-recipients-vol-19-iss-1.html
76. Liester MB (2020) Personality changes following heart transplantation: the role of cellular memory. Med Hypotheses (Elsevier) 135:109468. https://doi.org/10.1016/j.mehy.2019.109468
77. Miller RA, Webb B (1973) Embryonic holography: an application of the holographic concept of reality. DNA Decipher J 2. http://www.dnadecipher.com/index.php/ddj/article/view/26
78. Gariaev P, Birshtein BI, Iarochenko AM, Marcer PJ, Tertishny GG, Leonova KA, Kaempf U (2001) The DNA-wave biocomputer. Int J Comput Anticip Syst (Ed. Daniel Dubois, Published by CHAOS) 10. http://www.curealternative.net/bioelettr/memoria_acqua_DNA_wave_computer.pdf
79. Pitkänen M (2010) DNA as topological quantum computer. In: TGD as a generalized number theory (Matti Pitkanen, tgd.wippiespace.com). http://scireprints.lu.lv/66/
80. Gleick J (1997) Chaos: making a new science. Vintage. https://play.google.com/store/books/details?id=FZmSJPfTQVwC
81. Bancaud A, Huet S, Daigle N, Mozziconacci J, Beaudouin J, Ellenberg J (2009) Molecular crowding affects diffusion and binding of nuclear proteins in heterochromatin and reveals the fractal organization of chromatin. EMBO J (embopress.org) 28:3785–3798. http://dx.doi.org/10.1038/emboj.2009.340
82. Knoch TA (2020) A consistent systems mechanics model of the 3D architecture and dynamics of Genomes. Chromatin and Epigenetics. IntechOpen, Rijeka. https://doi.org/10.5772/intechopen.89836

Chapter 9
QED Coherence and Super-Coherence of Water in Brain Microtubules and Quantum Hypercomputation

Luigi Maxmilian Caligiuri

1 Introduction

During the past years, a lot of attempts have been carried on in order to understand human brain superior capabilities in the light of Quantum Mechanics (QM) or, more correctly, of Quantum Field Theory (QFT). Nevertheless, the most part of these studies has been based on the erroneous assumption that quantum theory would be limited to describe microscopic world. This assumption is deeply wrong as it is clearly shown by the existence of macroscopic quantum phenomena like, for example, superfluidity, superconductivity (both "cold" and "hot"), not to mention the astonishing evidence of macroscopic quantum behavior recently shown by Zeilinger [1] about a 60 carbon atoms molecule known as *"buckminsterfullurenes"*.

These findings open the very fascinating possibility of applying QM and QFT to explain some fundamental, but still ununderstood, phenomena as the emergence of consciousness, the formation of memory and, in general, the higher capabilities of human brain as, first of all, its computational abilities.

All these processes, in fact, seem to be characterized by typical quantum features as, for instance, the unity and non-local manifestation of consciousness [2–4] as well as the apparent non—algorithmic and non—computational features, at least in the conventional and classical framework, of certain brain functions [5, 6]. Such features are also at the basis of Pribram's model of "holographic brain" [7] according to which some of the main brain's functions, like memory and perceptual processing, can be traced back to the associative features of typical parallel distributed processes, like those implemented in the optical holographic techniques, as well as non-local and cooperative processes occurring in brain.

L. M. Caligiuri (✉)
Foundation of Physics Research Center (FoPRC), Cosenza, Italy
e-mail: luigimaxmilian.caligiuri@foprc.org

A. Bandyopadhyay and K. Ray (eds.), *Rhythmic Advantages in Big Data and Machine Learning* , Studies in Rhythm Engineering,
https://doi.org/10.1007/978-981-16-5723-8_9

Of the most ambitious attempt to apply QM and QFT to the living system refers, without doubt, to neurosciences. The first rigorous proposal in this sense dates back to 1990 when Hameroff and Penrose [8] suggested that quantum effects could play a primary role in brain functioning and, in particular, in the emergence of consciousness. Their model considered the dynamics of microtubules (MTs) of brain cells modeling their tubulin dimer units as quantum systems described by a coherent superposition of two-levels quantum states, respectively corresponding to two configurations of tubulin proteins (α- and β-tubulin) related to the position of the unpaired electrical charge of 18 e relative to pocket.

Very interestingly, Hameroff also suggested [9], by considering the periodic lattice structure of microtubules (providing "periodically arrayed slits"), that they would behave like waveguides for the photons simulating the behavior of holographic information processors. This possibility supposes the occurrence, under suitable conditions, of a macroscopic quantum phenomenon (also known as "superradiance") in the water molecules contained inside brain's MTs consisting in a quantum collective behavior of water molecules and electromagnetic field modes able to convert the perturbative thermal and molecular disordered oscillations into coherent photon modes inside MTs.

The first rigorous description of such phenomenon was given only by Del Giudice, Preparata and Vitiello [10] by their model of water as a free electric dipole laser. The theoretical background on which it is based was formalized by Preparata [11] who considered the coherent interaction between the quantum "matter field" of water molecules and a self-generated quantized electromagnetic field, arising from QED quantum vacuum, spontaneously occurring when suitable conditions about matter density and temperature have satisfied.

In particular, such interaction determines an exponential increase, due to a mutual coherent amplification, of the field's amplitude within macroscopic quantum domains, for this reason called "Coherent Domains" (CDs), whose extension L is related to the energy associated to the transition between a given couple of levels of the matter quantum systems.

In this model [10] these levels respectively correspond to the ground state and to some of the low-lying states of the rotational energy spectrum of water molecules (of the order of 4 meV), so giving a value of L of few microns and a typical period of coherent oscillations of the order of 10^{-14} s, that results much shorter than the typical value characterizing the Coulomb short-range interactions. The related coherent dynamics are characterized by a spatially extended oscillating polarization field able to correlate a very high number of water electric dipoles and to create stable and ordered macroscopic structures.

The need for a QFT-based approach (as those conducing to above described QED coherence in matter), in order to correctly describe living structures, was also recognized by Umezawa [12] that suggested the presence, in brain's cells, of a spatially distributed system characterized by mechanical degrees of freedom subjected to quantum phenomena.

Jibu et al. [13] suggested the photons composing the coherent e.m. field inside MT entangle the cytoskeletal protein and the MT quantum states of a given neuron link to those of other neurons by the tunneling of such photons through biological membranes. In this way, according to these authors, consciousness would could arise as the result of creation-annihilation dynamics of a finite number of superradiant photons in brain.

Nevertheless, despite the very interesting picture emerging from these models, different studies put in evidence some critical issues related to them [14, 15].

In particular, no convincing physical mechanism has been suggested so far, able to thoroughly explain a certain number of important aspects related to MTs quantum dynamics as, primarily: the generation of superradiant evanescent photons of the "right" frequency inside brain's MTs; the required stability against environmental decoherence to ensure the quantum dynamics to take place; how the coherent interaction between MTs would occur and, in particular, how this interaction could "manipulate" qubits inside the brain in order to generate the human brain higher computational activities.

In order to suggest a possible answer to these questions, we have already shown [16–19] that in brain's MTs there exist the right conditions for a spontaneous quantum phase transition of water molecules to occur from perturbative ground state to a coherent one in which water molecules coherently oscillate in phase with an auto-generated e.m. field associated to a two-levels transition belonging to the electronic spectrum of water molecule rather than to the rotational one as so far considered in the above mentioned studies.

The coherent dynamics here proposed implies the formation of macroscopic quantum CDs (having a size of about 0.1 μm), in which the oscillating e.m. field, due to the rescaling of oscillation frequency, is trapped and whose frequency of oscillation is in turn determined by the number of molecules participating in this oscillation at a given temperature. As we'll see, a very important feature of such e.m. coherent field is that it shows, at the CD's boundaries, an evanescent-wave-like behavior [11] characterized by an exponential decreasing tail leaking out from the CDs. For this reason, these macroscopic quantum domains behave, with respect to coherent photons generated inside it, as e.m. cavities at whose boundaries a "total frustrated reflection" occurs and that resemble a naturally occuring laser without the need for a pump.

As we have seen above, one of the most critical point in all the theoretical models about quantum computation in brain is the physical mechanism according to which such calculation would occur. We propose it could be possible by means of a quantum interaction between the CDs due to the coherent dynamics of water contained in brain's MT. Such interaction could occur through the exchange of photons from one CD to another by a quantum tunneling coupling between them. Nevertheless, differently from any other theoretical proposals, we show this interaction give rises to a different and extra QED coherent dynamics, called "supercoherence", namely a coherence between CDs in which many coherent regions, each having a size of about 0.1 μm, could form much more extended regions in which a high number of CDs oscillate in tune.

A fundamental role in such tunneling process is played by the evanescent tails associated to the e.m. coherent field produced by each CD and whose overlapping determines the quantum coupling. Furthermore, as we have recently shown [18, 19], if we suppose the brain's MTs to have properties similar to that of such metamaterials (MTMs), MTs are able to stabilize and preserve the coherent state of water molecules inside them also amplifying the evanescent field generated by each water CD contained in the nearby interacting MTs so ensuring a persistent and efficient tunneling of virtual photons.

In this chapter, also taking into account our recent results on this topic [18], we'll discuss a novel theoretical model describing the interaction between brain MTs, which considers the excited spectrum of the coherent state of water molecules enclosed in them, as well as the tunneling of virtual photons among them. Such dynamics give rise to a sort of "phase interaction" between MTs, substantially instantaneous, because of the zero-time tunneling of the virtual photons across the barriers represented by the CDs e.m. "cavities". This special kind of interaction would be able to correlate, in a stable manner, a very high number of MTs, so creating a "network" of coherently oscillating CDs, extending over macroscopic regions in brain.

As a very important result, we show this dynamics is able to simulate a universal set of quantum gate so realizing a quantum hypercomputing system in which a locally generated information at a point in the brain could be instantaneously transferred to another point. This model could then give a novel and very promising explanation of some of the extraordinary capabilities of human brain on a well-founded QFT basis without incurring in most of the difficulties affecting the other theoretical models so far proposed.

2 A Brief Overview of QED Coherence in Condensed Matter and Its Emergence in Liquid Water

For many decades, condensed matter has been considered as a collection of microscopic components, atoms or molecules, keep together by a set of intermolecular short-range forces when temperature T is below a defined value and kinetic energy of corpuscles is lower than potential energy between them. In this framework, that represents the commonly accepted one, the acting forces, as the Van Der Waals or Lennard–Jones forces, are supposed to have a range of action no longer than two or three molecular diameters so it is an unsolved dynamical issue the emergence of long range order in macroscopic condensed matter systems, as for example, in crystals and living organisms.

Yet in 1916 physicist W. Nernst proposed a very different point of view according to which the quantum fluctuations of microscopic components of matter, predicted by the Heisenberg uncertainty principle, could synchronize establishing a phased common oscillation over extended space–time regions that completely characterizes the condensed system as a whole. This fundamental principle of macroscopic coherence has been reinterpreted only recently in the more correct framework of (QFT) by Umezawa [12].

The possibility to admit, for a given system, the contemporary existence of matter fluctuations on one side and a non-fluctuating and well-defining evolution on the other is based on the required invariance of the system's Lagrangian under local phase transformation of matter field, due to the action of the gauge field $A_\mu(\vec{x}, t)$ that, at atomic and molecular scale, is just the electromagnetic field. This gauge invariance allows the quantum fluctuation of matter and e.m. fields to tune together and, if some conditions are fulfilled, to giving rise to a common phased oscillation of the matter system spanning a macroscopic spatial region covered by e.m. field. This process is, in some respect, conceptually similar to that determining the Lamb-shift of the energy levels of the hydrogen atom due to the coupling between the quantum fluctuations of vacuum (QV), the so-called Zero-Point (ZP) fluctuations, and the electron orbital density current through the interaction term $\vec{j} \times \vec{A}$.

The above dynamics arises from a fundamental principle, namely that any quantum system (either composed by particles of fields) is always subject to fluctuations as well as the vacuum. The occurrence of the Lamb shift actually shows a fluctuating field is always present as the result of the quantum fluctuations of vacuum oscillators and that we should always consider that atoms and molecules are coupled to a quantized e.m. field.

If we consider a matter system composed of electrical charged particles (electrons and nuclei) and characterized by a discrete energy spectrum of values $\{E_i\}$, a QV (or thermal bath)fluctuation, able to excite one of it from the fundamental state (with energy $E_0 = \omega_0$) to the state k (with $E_k = \omega_k$), must have a wavelength $\lambda \simeq 1/\delta E$ where $\delta E = E_k - E_0$. In the case of water molecules the typical energy difference between two levels can be estimated as $\Delta E_{\text{avg}} \simeq 10\,\text{eV}$ corresponding to a wavelength $\lambda = 1200$ A that is much greater than the typical size of water molecule $d \simeq 1$ A. This means that an e.m. vacuum fluctuation, able to excite a molecule, is composed by a photon "extending" over a length more than one thousand greater than the dimension a single molecule itself and includes, at the usual density, in its volume λ^3, more than 20.000 water molecules.

If $P = f_{1k}$ is the Lamb-shit type probability that a photon of the vacuum fluctuations excites the level k (quantified by the "oscillator strength" for the transition $0 \rightarrow k$), for an atom or a molecule with n electrons, we have

$$\sum_k f_{0k} = n \tag{1}$$

and the probability that a photon, due to its subsequent decay, could in turn excite one of the N molecules contained in $V = \lambda^3$ is given by

$$P_N = PN = PV\left(\frac{N}{V}\right) = P\lambda^3\rho \tag{2}$$

where ρ is the water density.

If $P_N \ll 1$ the emitted photon will probably come back to QV or fly away, but when $P_N = 1$, it will surely excite another molecule in the V volume and the process will repeat; other photons from quantum fluctuations then will undergo the same phenomenon and, if $\rho \geq \rho_{\text{crit}} = \left(P\lambda^3\right)^{-1}$, after a suitable time interval, a considerable e.m. field develops in the region of extension λ by attracting more and more molecules able to resonate with the growing e.m. As we'll see in the following, this attractive process will go on, increasing the system density, until it reaches a defined value when the autogenerated e.m. field exponentially increases, so spontaneously driving the system toward a more stable state (because characterized by lower energy and so strongly favored), where the further density increase is counterbalanced by the repulsive intermolecular forces and in which the above quantum fluctuations become strongly amplified and phase correlated.

These coherent oscillations are then confined within defined spatial regions, called "Coherence Domains" (CDs), associated to the wavelength of the tuning electromagnetic field, whose extension is of the order of

$$L \sim \frac{2\pi}{\omega_0} \tag{3}$$

where ω_0 (from now on in this chapter, unless otherwise specified, we'll adopt the natural units system $\hbar = c = 1$) is the energy associated to the transition between a given couple of levels of energy of the quantized matter system driving the coherent evolution of the whole system.

Without entering into a detailed mathematical analysis of the coherent dynamics of a general two-levels system, already discussed in other works [10, 11, 18, 20, 21], we'll limit ourselves to its application to the water molecules. Furthermore, since we are interested in the coherent evolution of a system oscillating between two discrete energy levels, we'll consider a two-components matter field, respectively associated with the space–time distribution of the molecules among the considered energy levels.

In the case of multilevel atoms or molecules the contribution to the described dynamics of the other energetic levels, different than the two considered, takes place through second order radiative corrections that are suppressed by the dominant dynamics that drive the system toward the coherent state.

The mathematical description of dynamics evolution of matter + e.m. field interacting system is achieved by considering its Lagrangian function and the corresponding Eulero–Lagrange equations. Such equations contain, as dynamical variables, the matter field associated to the space–time distribution of matter components (water molecules) corresponding to the different energetic configurations and the e.m. field, described in terms of its potential vector $\vec{A}$.

The interesting physical evolution over long time period is that considering only the e.m. modes that couple with the matter oscillations, namely that defined by the condition

$$\omega_{\vec{k}} = \omega_{ab} = E_b - E_a \tag{4}$$

Among all the possible quantum transitions $a \to b$ of the matter fields, it will be considered the transition between the ground state $|0\rangle$ and a particular excited state $|1\rangle$ such as $\omega_0 = E_1 - E_0$ and $\left|\vec{k}\right| = \omega_0$.

Furthermore, by limiting our description to a defined region whose size is the of order of the wavelength of the coupling mode (corresponding to the CD "extension"), $\lambda \simeq 2\pi/\omega_0$, we can neglect the spatial dependence of both the matter and e.m. fields inside it (whose volume is V_{CD}). The dynamic equations, describing the time-evolution of the electromagnetic field + matter interacting ensemble, can be written as

$$\begin{gathered} i\dot{\xi}_0(\tau) = gA^*(\tau)\xi_1(\tau) \\ i\dot{\xi}_1(\tau) = gA(\tau)\xi_0(\tau) \\ -\tfrac{1}{2}\ddot{A}(\tau) + i\dot{A}(\tau) - \mu A(\tau) = g\xi_0^*(\tau)\xi_1(\tau) \end{gathered} \tag{5}$$

where $\xi = (\xi_0, \xi_1)$ is a two-components matter field describing the space–time distribution of water molecules, A the space-averaged e.m., vector potential and $\tau \equiv \omega_0 t$, with

$$g \equiv 2\pi\left(\frac{\omega_p}{\omega_0}\right) f_{01}^{\frac{1}{2}} \tag{6}$$

the term ω_p is the electron plasma frequency (m_e is the electron mass)

$$\omega_p = \frac{e}{\sqrt{m_e}}\sqrt{\frac{N}{V}} \tag{7}$$

and μ represents the so-called photon "mass" term that can be written as the sum of two terms:

$$\mu = \mu_d + \mu_{\text{cont}} \tag{8}$$

in which

$$\mu_d = -\frac{3}{2}\omega_p^2 \sum_{n \neq 1} \frac{f_{0n}}{\omega_n^2 - \omega_0^2} \tag{9}$$

Includes the contributions of the transitions from ground state to excited discrete states and,

$$\mu_{\text{cont}} = -\frac{3}{2}\omega_p^2 \int\limits_{\omega > \omega_{\text{ion}}}^{+\infty} \frac{f(\omega)}{\omega^2 - \omega_0^2} d\omega \tag{10}$$

where ω_{ion} is the ionization threshold and $f(\omega)$ the oscillator strength for unit of frequency, is the contribution of the transitions from ground state to excited states of the continuum spectrum (namely those for which the excitation energy is larger than ionization threshold) and f_{mn} is the oscillator strength for the electronic transition from the quantum state $|m\rangle$ to the excited state $|n\rangle$.

The short-time behavior of system can be studied by differentiating the third of (5) and substituting it into the second one, so obtaining the differential equation

$$-\frac{1}{2}\dddot{A}(\tau) + \ddot{A}(\tau) + i\mu\dot{A}(\tau) + gA^2(\tau) = 0 \tag{11}$$

to be solved assuming the "initial" conditions $A(0) \sim 0,\ \xi_0(0) \sim 1,\ \xi_1(0) \sim 0$ that define the "perturbative" initial ground state (the PGS) in which all the fields just perform their zero-point oscillations and the electromagnetic field $A(0)$ is too low to ensure the phased collective behavior of the matter + e.m. system.

The transition toward the CGS would then require the exponential increase of $A(\tau)$ able to overcome its nearly zero initial value and create the coherent tuning field. It can be shown [11] that this occurs, for a given μ, when

$$g^2 > g^2_{\text{crit}} \tag{12}$$

with

$$g^2_{\text{crit}} = \frac{8}{27} + \frac{2}{3}\mu + \left(\frac{4}{9} + \frac{2}{3}\mu\right)^{3/2} \tag{13}$$

Since the value of μ depends on the density $\rho = N/V$, the value of g_{crit} will then determine the corresponding critical value of density $\rho_{\text{crit}} = (N/V)_{\text{crit}}$ such as the (12) holds. As long as $\rho < \rho_{\text{crit}}$ the system remains in the vapor phase with all the molecules in PGS. But when $\rho > \rho_{\text{crit}}$ (and $g > g_{\text{crit}}$ as well), the system will undergo a "phase transition" from the PGS in which the electromagnetic and matter fields perform ZP very weak uncoupled fluctuations only, toward the CGS, the liquid state, in which almost every ZP fluctuation couples with the oscillations of atoms or molecules in the ensemble and then a strong electromagnetic field arises from QV tuning all the matter constituents to oscillate in phase with it and among themselves inside every CD. This results in the emergence of a macroscopic quantum state in which atoms and molecules lose their individuality and become part of a whole electromagnetic field + matter entangled system.

This condensation process happens within a time interval less than the typical period of oscillations of the coherent state, namely

$$\tau_{\text{coh}} \simeq \frac{2\pi}{\omega_0} \tag{14}$$

and goes on until the short-range electrostatic repulsion acts against the indefinitely increase of system density so determining the equilibrium state.

The stationary solution, corresponding to CGS, can be found by writing the coherent dynamical Eq. (10) in the phase representation by posing

$$\begin{aligned} \xi_0(\tau) &= \cos\alpha \cdot e^{i\theta_0(\tau)} \\ \xi_1(\tau) &= \sin\alpha \cdot e^{i\theta_1(\tau)} \\ A(\tau) &= A_0 \cdot e^{i\phi(\tau)} \end{aligned} \tag{15}$$

with $\theta \leq \alpha \leq \pi/2$. In the coherent Eqs. (15) the field ξ_0 describes the molecules in the ground state $|0\rangle$ whereas the field ξ_1 the molecules in the excited state $|1\rangle$ while the field A represents the photon absorption and the field A^* the photon emission.

The second term in (11) has a special meaning (as we'll see in the following) since it describes the "mass" get by the e. m. field when it is coupled with the water molecules and continuously exchanges photons with them.

The properties of water CGS, deriving from the Eqs. (15), have already been studied in a series of previous papers [11, 22, 23]. We note the parameters g,μ and ω_0 depending on the physical quantities defining the system, assume renormalized values due to the quantum transition (condensation) of each molecule from the non-coherent state $|0\rangle$ to coherent one $|\text{coh}\rangle$. The fields amplitudes defining such coherent state are the complete solution of the system (15) given by [23]

$$\begin{cases} A_0 = \frac{x\dot{\phi}}{2g} \\ \alpha = \arctan x \\ \dot{\theta}_0 = gA_0 \tan\alpha \\ \dot{\theta}_1 = \frac{gA_0}{\tan\alpha} \\ \omega_{re} = \omega_0\left|1-\dot{\phi}\right| \\ E_{\text{coh}} = \omega_0\left[A_0^2(1+2\mu_{re}) + \sin^2\alpha - \frac{3}{2}gA_0\sin 2\alpha\right] \equiv \omega_0 H \\ \dot{\phi} = \dot{\theta}_1 - \dot{\theta}_0 \end{cases} \tag{16}$$

where ω_{re} and μ_{re}, respectively indicate the re-normalized common value of e.m. and matter field oscillation ω_0 and of the mass term μ and E_{coh} the energy of the matter system in the coherent state $|\text{coh}\rangle$. The solution (16) also implies the condition $e^{i(\phi-\theta_1+\theta_0)} = \pm 1$ namely a "phase-locking" constraint between the fields.

It can be shown that the following condition holds [23]:

$$\omega_{re} \leq g^2\omega_0 \tag{17}$$

that implies, due to the smallness of $g^2 \sim 10^{-2}$ for water [22, 23], the common frequency of oscillation $\omega_{\text{coh}} = \omega_{re}$ of electromagnetic field and matter inside a CD is lower than the value characterizing the perturbative state of the incoherent phase ω_0, namely

$$\omega_{\text{coh}} = \left|1-\dot{\phi}\right|\omega_0 \ll \omega_0 \tag{18}$$

where ϕ is the phase factor ruling the behavior of the vector potential $A(\tau) = A_0 \exp[i\phi(\tau)]$. The frequency re-normalization $\omega_0 \to \omega_{re}$ has a very important consequence on the photon mass-term, whose value, as we'll see in the following, will be $\mu_{re} < 0$, showing that after the runaway water dynamics is completely changed with respect the non-coherent state.

The presence of a negative term in the expression of H in (16) also allows for E_{coh} to assume negative values, meaning the coherent state to have an energy lower than the non-coherent one, so resulting more stable and representing the true vacuum state of the system.

The quantum phase transition from gas-like incoherent state to coherent liquid state takes place spontaneously when the water density exceeds the smallest critical density corresponding to the various possible electronic transitions and, once the systems has "selected" the corresponding energy ω_0, all the other transitions are highly suppressed and don't have effect on the dynamical evolution of the matter + e. m. field system.

The level driving the transition toward the coherent state is just that characterized by the shortest time τ_{coh} required by the coherent oscillation to build up and consolidate and, in the case of water, is of the order of 10^{-14} s.

According to [22], the smallest value of critical density in water is associated to the electronic transition between the ground state and the 5d excited state of the molecule corresponding to the level at about $E_n = 12.06$ eV. The selection of this level as the excited partner of water coherent dynamics could also explain why the intermolecular distance in liquid water is larger than twice the molecular radius (respectively equal to $\simeq$3.1 A and $\simeq$1.0 A). Finally, the thermodynamics properties of liquid water, such as density and specific heat, obtained in this case also show a good agreement with experimental data.

The numerical solution of system (16), so far obtained for the coherent state, is shown in Table 1 [22, 23].

The energy gap per molecule, characterizing the coherent state with respect the non-coherent one, that ensures the stability of the former is then equal to

$$-\frac{\delta E}{N} = E_{GS} - E_0 \equiv |E_{coh}| \simeq \omega_{coh} \tag{19}$$

The frequency renormalization has an even more impressive effect since it determines the spontaneous creation of a resonating cavity whose e.m. mode coincides

Table 1 Calculated values of the parameters characterizing the coherent state of liquid water corresponding to oscillator strengths respectively used in: (a) [23] and (b) [22]. Energy units in eV, density units in g · cm^{-3}

x	$\dot{\phi}$	A_0	$\sin^2\alpha$	μ_{re}	g^2	E_{coh}	ρ_c	ω_{re}	ω_0	f_{mn}
0.76687	1.0136	2.748	0.103	−0.4842	0.020	−0.164	0.342	0.16	12.074	(a)
0.891	1.0215	2.40	0.127	−0.4734	0.036	−0.260	0.310	0.020	12.06	(b)

with the common coherent frequency ω_{re} of oscillation of matter and e.m. field inside it.

As we'll discuss in more detail in the following, according to the principles of Quantum Optics, this condition implies the self-generated coherent e.m. field is trapped within a well-defined macroscopic region coinciding with the CD defined above, being unable to be irradiated outwards, and whose dynamics is fundamental for our discussion.

3 Quantum Dynamics of Coherent Domains in Liquid Water

3.1 Spatial Behavior of the Coherent E.M. Field

The spatial behavior of the coherent field amplitude $A(\vec{x}, t)$ is one of the most interesting and meaningful features of the coherent state. Assuming spherical symmetry, we can define the radius $r_{\rm coh}$ as the radial dimension of the smallest CD in equilibrium with its own coherent e.m. that depends on the quantum "condensation" process. For a water CD we have [11], due to the condition $\omega_{\rm coh} \ll \omega_0$,

$$r_{\rm coh} \equiv R_{CD} \simeq \frac{3\pi}{4\omega_0} \tag{20}$$

so that, for $\omega_0 = 12.06$ eV, $R_{CD} \simeq 375$ A.

The radial profile of the coherent e.m. field then admits, for $r > r_0$, an evanescent-like solution given by

$$A(r) \simeq \frac{A(0)}{\sqrt{2}} \frac{\exp\left[-r\sqrt{\omega_0^2 - \omega_{\rm coh}^2}(r - r_0)\right]}{\omega_0 r} \tag{21}$$

where $A(0)$ is the vector potential at the CD's center ($r = 0$), whose spatial behavior is shown in Fig. 1. We clearly see the origin of such evanescent e.m. field is the notable energy shift $\omega_0 \to \omega_{\rm coh} \ll \omega_0$ due to the coherent dynamics.

For two closely packed CDs, the overall spatial profile of $A(r)$ is given by the function $A(x) = A(0)P(x)$, with $x = \omega_0 r/\pi (0 < x < 3/4)$ and

$$P(x) = \frac{\sin \pi x}{\pi x} + \frac{\sqrt{2}}{\pi} \frac{e^{-\pi(\frac{3}{4}-x)}}{3 - 2x} \tag{22}$$

The energy gap per molecule $\delta E/N$ given by (19) has the same spatial behavior as $A(r)$.

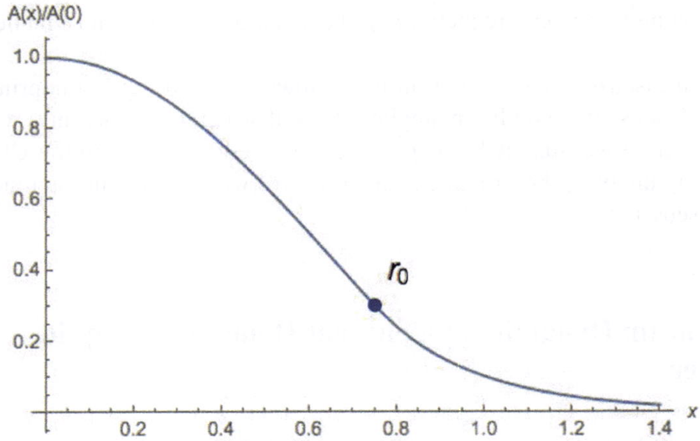

Fig. 1 Spatial profile of coherent e.m. field $A(r)$ of a CD

This can be easily viewed by noting that, within a CD, $\sin 2\gamma \sim A$, and then, since $\sin\gamma \sim \gamma$, we have $\sin^2\gamma \sim A^2$. Furthermore μ_{re} is practically constant since it represents a sort of "average photon mass" so we can write $\delta E/N \simeq -kA^2$(where k is a constant positive value within a CD)

$$\frac{\delta E}{N} = -k[A(0)P(x)]^2 = \left[-kA^2(0)\right]P^2(x) = \frac{\delta E}{N}(0)g(x) \tag{23}$$

where we have set $-kA^2(0) \equiv \delta E/N(0) = E_{\text{coh}}$ and $g(x) \equiv P^2(x)$.

The energy gap has then, as spatial profile, the function $P^2(x)$ shown in Fig. 2.

The coherent field generated inside a CD extends its influence outside the CD itself through its evanescent tail and allows two CDs of interacting each other by means of the superimposition of the evanescent tails of their respective coherent e.m. fields.

It is easy to see that for a pair of closely packed coherent domains (whose inter-centre distance is of order $2r_0 = 3\pi/2\omega_0$), the energy gap curve of the superimposed coherent e.m. fields due to the two CDs is lower than the curve of each isolated CD, determining an energy gaining up to four times larger than in the isolated domains at $r = r_{\text{coh}}$. At this distance, the difference between the two curves can be considered as the "binding" energy of the two CDs as shown in Fig. 2. This extremely important result shows that two or more nearby CDs interact each other through their respective coherent e.m. fields and, in particular, by means of their evanescent tails.

So, although the large frequency rescaling $\omega_0 \to \omega_{\text{coh}}$ keep the e.m. radiation field trapped inside every CD preventing it to escape to the vacuum (where $\left|\vec{k}\right| = \omega_0$), a coherent e.m. field, in the form of non-radiating evanescent wave, leaks out far beyond the CD's borders.

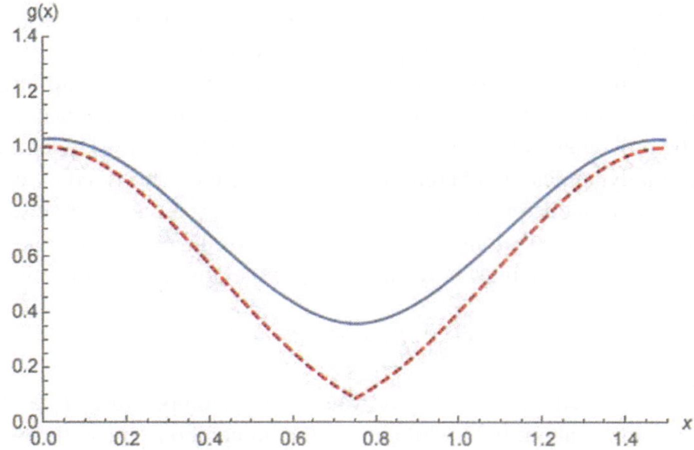

Fig. 2 Spatial profile of energy gap $\delta E/N$ in a coherence domain for two isolated CDs (dotted curve) and for two closely packed "interacting" CDs (dashed line)

As well shown in the following, it is just this evanescent field, acting as a "propagation channel" of virtual photons between two CDs, that would allow the realization of quantum hypercomputing in human brain.

3.2 *Thermodynamics Considerations and "Interstitial" Water in Living Tissues*

The above formulation of coherent dynamics doesn't consider the effect of thermal collision of water molecules on the stability of liquid coherent state and then it formally holds for $T = 0$. In this section, we briefly discuss this effect and review its consequences on the equilibrium between the incohernet (vapour-like) state and the coherent liquid state of water. The thermodynamics of coherent liquid water has been already examined [11]; in the following we'll limit our discussion to some features that will be relevant for our present work.

As we have seen, the coherent ground state of liquid water is characterized by an energy gap per molecule that prevents them to "escape" from the coherent phase whose value, at $T = 0$, is given by $\delta E = E_{\text{coh}}$ per water molecule. If the thermal fluctuations are able to transfer to a given CD an amount of energy $E_{\text{ther}} > E_{\text{coh}}$, they put some molecules out of phase with respect the coherent oscillations, pulling them out from the coherent domains. These molecules behave as a dense gas (a Van der Waals gas in which strong short-range attraction forces act), occupying the interstitial spaces between the CDs, whose sizes increase with the temperature. On the other hand, the energy minimization requirement keeps neighboring CDs closely packed so that the incoherent gas is trapped inside the spaces between an array of CDs. The pressure of this gas of incoherent molecules increases with the temperature until it is able to escape from CDs array, vaporizing to the open.

As soon as the thermal fluctuations will be able to transfer a sufficiently high number of molecules to the incoherent phase, a more and more increasing number of CDs will be broken and all the molecules will pass to the gaseous incoherent state.

The thermodynamics of coherent domains can be then described by considering the coherent and incoherent fractions $F_{\text{coh}}(T)$ and $F_{\text{coh}}(T)$, respectively indicating the number of molecules belonging, at each instant, to coherent and incoherent phase. In a given system we then must have, at a given temperature, the following constraint:

$$F_{\text{coh}}(T) + F_{\text{nc}}(T) = 1 \tag{24}$$

It can be calculated that [11, 18], for water at room temperature ($T \simeq 300\,\text{K}$), the coherent fraction is about 0.27 while inside the human body it can be even smaller due to the increase in temperature. In particular, if we assume the brain's average temperature to be $T \simeq 37\,^\circ\text{C}$ (corresponding to $T = 310\,\text{K}$), we have $F_{\text{coh}} \simeq 0.24$, namely about 25% of water molecules are in the coherent state.

From a statistical viewpoint, for every $T \neq 0$, the system oscillates between two states: the one in which some molecules are in the coherent state and the one in which they are in non-coherent state, according to whether the coherent electrodynamic attraction is overtaken or not by the thermal collisions. On average a molecule will then spend equal time between the two fractions and in order to reveal this two-phases structure, the observation time would require a resolution shorter than the oscillation time between the two fractions.

Nevertheless, in living systems, water can be almost always considered as interfacial water, since basically any point can be considered to be closer than a fraction of a micron to a surface. The special properties of interfacial water have been studied for a long time [24, 25] especially in connection with the so-called "EZ water" (i.e. "exclusion-zone" water). In particular, as studied by the Pollack group [26], water near a surface or a wall would be subjected to a new type of strongly attractive interaction able to counteract the effect of thermal collision. As a consequence, such interfacial water behaves as overcooled water (namely water at subzero temperature) that is characterized by glass-like structural and dynamical properties as the temperature decreases.

In this "glassy" state, water can be considered as a liquid possessing an enormously high viscosity in which perturbing just one molecule would perturb all of them. According to the above thermodynamics, this is just what happens in coherent water when we consider a sufficiently low temperature. In particular, when $T \to 135\,\text{K}$, we have $F_{\text{coh}}(T) \to 1$ ($F_{\text{unc}}(T) \simeq 1.5 \cdot 10^{-2}$)and to each decrease of about 30 K corresponds an increase of one order of magnitude of water density [27]. Thus the glassy state characterizing interfacial water can be considered as a fully coherent state whose enormous viscosity and rigid dynamics are due to strong phase-correlation between molecules, for it is impossible to move one molecule without moving the whole CD.

From a dynamic point of view, the arising of such fully coherent behavior can be explained by considering other coherent processes, occurring in liquid water close to a surface, in addition to that described above and concerning electron clouds [28]. The layer of fully coherent water close to a surface can then reach an extension up to about 500 μm [28] able to include some thousands of CDs as occurs in EZ water.

3.3 *Excited Spectrum of Coherent Oscillations in Water*

A very important consequence of the coherent dynamics of liquid water is the possibility to have "excited" coherent oscillations, that is excited states of CGS. This represents a very peculiar feature of liquid water due to the specific couple of levels involved in the coherent oscillation, namely the ground state and the excited level at 12.06 eV that lies just below the ionization threshold of 12.60 eV. For the water coherent state $|\text{coh}\rangle = \cos\gamma|0\rangle + \sin\gamma|1\rangle$ we have, from Table 1, $\sin^2\gamma \geq 0.103$, meaning than more than 0.10 of electron per water molecule can be considered as quasi-free particles.

A water CD can include, at the density of liquid water, a number of molecules of the order of $n \simeq 2 \cdot 10^5$, corresponding to a number of quasi-free electrons of about $n_e \simeq 2 \cdot 10^4$, meaning every water CD acts as an enormous reservoir of quasi-free electrons. These electrons can be easily excited by external, even weak, energy supply so originating metastable coherent excited levels in the form of "cold electron vortices" [29] as far as the energy supply is lower than the energy gap per molecule E_{coh} caractherizing the water CD. This energy, because of the coherence, is absorbed by the CD as whole. Very interestingly, such vortices cannot be excited nor decay thermically since they are coherent and possess a quantized magnetic moment that allows them to allin to external static magnetic field. For this reason, they can be characterized by a very long lifetime. The energy of an excited coherent vortex, associated to a single CD, is given by

$$E_n = \frac{\vec{L}_n^2}{2I} - g\vec{L}_n \cdot \vec{B} \tag{25}$$

where $\vec{L}_n$ is the angular momentum, I the momentum of inertia, $g = e/2m$ the gyromagnetic ratio and $\vec{B}$ the external magnetic field. It is easy to show that the moment of inertia I is given by:

$$I = \frac{8\pi}{15}\frac{N}{V}m_e R_{CD}^2 \tag{26}$$

showing the excited energy level depends on the size of the CD and on the number density of water molecules. The probability to excite each component of the spectrum $\{E_n\}$ is uniform, i.e. it is independent of the energy level to be excited.

The excited levels so obtained are associated to higher values of angular momentum $\vec{L}$ of the coherent vortex and to a shift in frequency ω_{coh} of the tuned oscillations.

It is important to observe that, being the spectrum $\{E_n\}$ a collective property of the entire CD, it has practically no upper limit since, although every molecule in the coherent state cannot absorb an energy higher than E_{coh}, all the N resonant molecules can account for a total energy of excitation of the order of NE_{coh} a very high quantity (including frequency up to visible and HV spectrum) since, at room temperature, $N \sim 10^{22}\,\text{cm}^{-3}$.

As we have seen, coherent states are characterized by a well-defined value of the common phase of oscillation of matter and e.m. field as follows from the phase-lock constraint relating the matter and e.m. field phases in the coherent state.

From a QFT point of view, this can be expressed by saying the relative wavefunctions are eigenstates of a suitable quantum phase operator whose features have been studied in detail elsewhere [30].

A change in phase ϕ then modifies the coherent oscillation frequency and vice versa. Even more interesting such a change would also affect the electromagnetic potentials $A_\mu = \left(V, \vec{A}\right)$ of the system.

If Ψ describes an incoherent state then its phase-shift due to a change in potentials has no effect on its evolution, on the contrary for a coherent state, Ψ being an autostate of the phase operator, a phase modification would determine a substantial change in its physical state and evolution.

As we have seen, in the coherent state of water, the wave function Ψ (both of particles and e.m. field) acquires a macroscopic meaning and a wave-like behavior, a situation quite similar to that occurring in laser or in superconductivity where the entire collection of microscopic objects (respectively photons and super-electrons) are described in terms of an electromagnetic field with a well-defined phase. By exploiting such analogy we can state a relationship between phase and potentials in a coherent state through Josephson's equations

$$\vec{\nabla}\Phi = \frac{e}{\hbar}\vec{A} \tag{27}$$

$$\frac{\partial \Phi}{\partial t} = \frac{e}{\hbar}V \tag{28}$$

where the quantum phase Φ includes, in the excited states, the information about rotational frequency of the associated cold vortices.

The above equations show that e.m. potentials are able to modify the phase of the coherent state, including the frequency of excited vortices, and vice versa even when the fields are not present as occurs in the Aharonov–Bohm effect [31]. The e. m. potentials can then establish a long range as well as superfast (actually faster than light being, driven by a phase velocity) interaction. We consider the consequences of this process from the standpoint of transmission and storage of quantum information in a forthcoming publication [30].

The existence of coherent excited levels in liquid water opens the possibility to considering the onset of a coherence between CDs, namely a “supercoherence” [21, 30, 32].

It is conceptually similar to the coherence between electron clouds oscillations and the autogenerated e.m. field just described above in the sense we now have a certain, generally high, number of CDs that become able to oscillate in phase with each other such as the resulting coherent oscillation could cover a much more extended region. Coherence among CDs doesn’t imply the achievement of a critical density to occur, as for the formation of single a CD, but is just limited to the possibility for every single isolated CD to oscillate in phase with each other, allowing them to be much less closely spaced. For the supercoherence to occur is yet needed the CDs could discharge their excitation energy outwards in order to give the required oscillations. As a consequence of energy discharge (or absorption) the CD changes its own frequency of oscillation. Supercoherence also further increases the stability of the involved CDs against the damaging effect of thermal fluctuations so further stabilizing the value of $F_{\text{coh}}(T)$.

In some previous papers [29, 32], a possible chemo-electrodynamical scheme for the energy discharging of a CD has been proposed. It requires the presence of suitable “guest” macromolecules able to “resonate” with the involved CDs.

In this chapter we instead discuss a novel and different model, based on a purely quantum-electrodynamic process, allowing a direct interaction between CDs (for this being more similar to that giving rise to the single CD itself), able to coherently couple them, that would take place inside brain’s MTs and could represent a new starting point for the explanation for the onset of quantum hypercomputing in human brain.

3.4 *Tunneling of Evanescent Modes (Virtual Photons) Between Water Coherent Domains*

The chemo-electrodynamics scheme proposed for supercoherence doesn’t consider any direct interaction between CDs but just the possibility for them to oscillate in phase with each other. Nevertheless, as we’ll see according to our model, it is possible for two or more CDs to directly interact through the e.m. vector potential of the evanescent fields leaking out from them by considering the quantum tunneling of virtual photons, associated to such evanescent modes.

3.4.1 Quantum Dynamics of “Evanescent” Electromagnetic Fields and Virtual Photons

From the standpoint of classical physics, evanescent waves usually originate by the “total reflection” of a light beam at the boundary between two media characterized by different values of refraction index (and correspondingly different densities) n_1 and n_2. If the transmitted wave is refracted at an angle $\vartheta \geq \pi/2$ in the medium with lower density, then we have, from Snell's law, for the wave number of the transmitted component of the incident wave orthogonal to the interface, the expression

$$k_{\perp} = \sqrt{\frac{\omega^2}{c^2}\left(1 - \frac{n_1^2}{n_2^2}\sin^2\alpha\right)} \tag{29}$$

where ω is the frequency of the incident wave and α is the reflection angle at the interface between two media. When the condition

$$1 - \frac{n_1^2}{n_2^2}\sin^2\alpha < 0 \tag{30}$$

is satisfied, $k_{\perp}$ becomes imaginary and the transmitted e.m. field will show an exponential decay with the distance z from the interface, so giving rise to an "evanescent" field.

The condition (30) also defines a "critical" angle $\alpha_c = \sin^{-1}(n_2/n_1)$ namely the angle for which, if $\alpha \geq \alpha_c$, the incident beam is totally reflected inside the first medium and doesn't propagate in the second one. Nevertheless, even in this case, a "transmitted" wave component is present in the second medium, namely just the evanescent wave. In fact, if a third medium (denser than the second) is placed after the second one at a distance d from the first interface, the wave number becomes real again in this medium and the total reflection amplitude is reduced, giving rise to a transmitted component, through the gap, that "escapes" into the last medium. This phenomenon is called "frustrated total internal reflection" (FTIR) in which the transmitted amplitude decreases exponentially with the gap d.

It is important to note this exponentially decreasing behavior makes the wave practically detectable within a limited distance from the propagation point so we can define a "propagation depth" d_p of evanescent wave as the distance where amplitude is reduced to the $1/e$ of its initial value, namely

$$d_p = \frac{1}{k} = \frac{\lambda}{2\pi\sqrt{n_1^2\sin^2\theta_1 - 1}} \tag{31}$$

where λ is the wavelength of the incident wave. It has been shown the optical refraction index plays the same role of the barrier potential in the quantum-tunneling phenomenon so suggesting the idea that evanescent waves could be actually tunneling modes of e.m. waves [18, 33, 34]. The spreading of e.m. waves across "impenetrable" opaque optical barriers is then analogous to the phenomenon of quantum tunneling experienced by particles through a potential barrier.

The tunneling of evanescent modes of e.m. waves are just interpreted, from a quantum viewpoint, as the tunneling of virtual photons through the corresponding

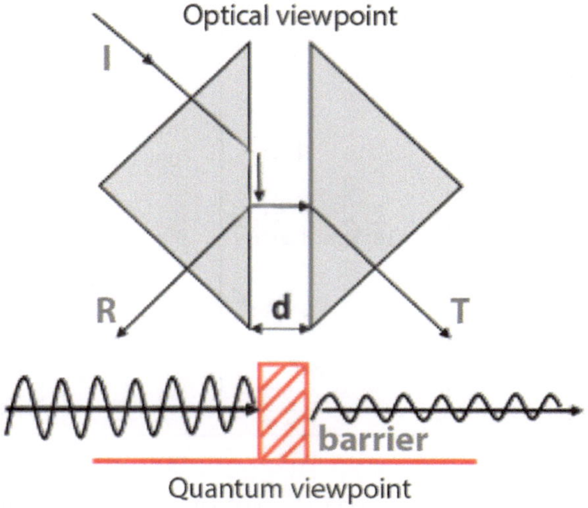

Fig. 3 Tunneling of e.m. modes through double prim configuration and its quantum analogous

potential barrier. Such tunneling photons are characterized by a negative square mass in the ordinary metric [18, 34, 35] or, equivalently, by a superluminal group velocity.

The tunneling time of evanescent e.m. waves (or virtual photons) through opaque optical barriers has been accurately measured in several experiments [18, 35], in particular involving the double prisms configuration (see Fig. 3) that, better than any other, can be considered as the optical analogous of the quantum tunneling through a potential barrier. In all the cases the experimental results confirm that the reflected and the transmitted (tunneled) waves (virtual photons) arrive at the detectors at the same time and this behavior holds for all fields (photons or massive particles) characterized by wave solutions with purely imaginary wave numbers.

This "universal" tunneling time τ of evanescent wave (or tunneling photons) is given by $\tau \simeq 1/f = 2\pi/\omega$ where f is the carrier frequency of the signal (or the associated quantum particle) undergoing the process and is substantially independent from the barrier length. An analogous relation can be obtained in the case of the tunneling of a particle wave packet for which the transmission time is given by $\tau \simeq \hbar/E$ where E is the energy of the particle.

This "universal" time relation implies a nearly "instantaneous" spreading of the signal or wave-packet across the barrier so that the measured tunneling time corresponds to a very high tunneling velocity, even greater than c.

This superluminal energy and signal velocity in quantum tunneling has been confirmed in many experiments [18, 34, 35] and its theoretical consequences have been already studied in deep in a number of previous publications [18].

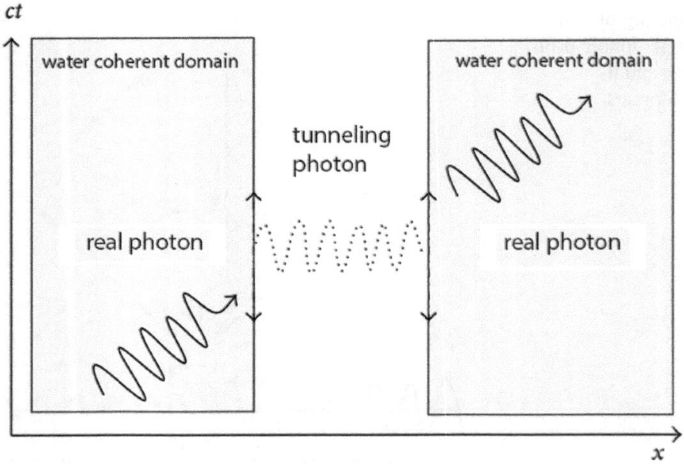

Fig. 4 Time–space diagram of evanescent mode or photonic tunneling

3.4.2 Quantum Coupling Between Coherent Domains and Their Interaction Through Virtual Photons

Evanescent modes can be considered as free e.m. field basing on the idea of refractive index of a passive, macroscopically continuous media [36]. From this standpoint such modes can be considered as the result of the spatial phase modulation, at the interface between incident and reflected wave. Evanescent modes, considered as classical c-number fields, can then interact with matter no differently from ordinary homogeneous e.m. waves.

On the other hand, as Feynman has shown, virtual particles appear as intermediate states of an interaction process. Then, despite they are not directly observable, virtual particles represent required intermediate states between real observable states having a direct influence on them (see Fig. 4). They also take part not only in the microscopic interaction processes but also in the macroscopic range.

In our model, the coupling among two CDs can be pictured in terms of the overlapping between their evanescent field tails [18] that allows the tunneling of virtual particles (tunneling photons) between them.

When two CDs are sufficiently close to each other, their respective e.m. evanescent fields, given by (36), could overlap. The width of the resulting overlapping zone, through which tunneling of virtual photons takes place, will depend on the distance between the two CDs and on the spreading of the two evanescent fields outside the respective CDs, whose "extension", described by their tales, is a function of ω_{coh}. The situation is schematically shown in Fig. 5

If the separation d between the two CDs is much greater than the average value of d_p characterizing the tails of evanescent e.m. field "escaping" from the respective CDs (Fig. 5), we can assume no useful overlapping exists between the two evanescent fields and consequently a negligible probability of interaction between them.

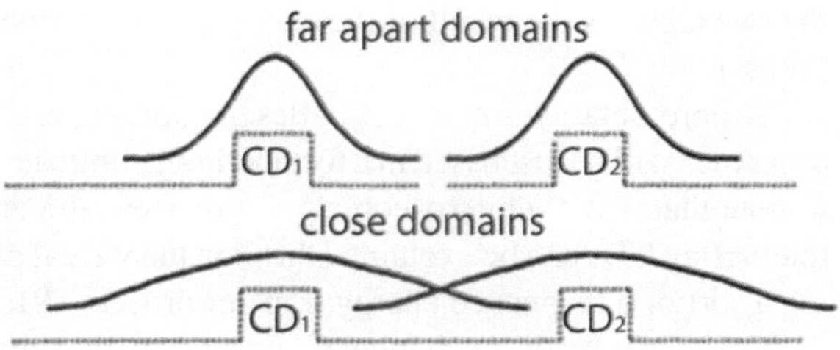

Fig. 5 Overlapping between the evanescent fields associated to two nearby CDs

When instead $d \sim d_p$ the overlap between these two evanescent e.m. fields takes place allowing a "non-local" interaction between the CDs.

As we have seen from (38), the closer are the two CDs the more stable is the system, since they increase their "binding" energy (being the largest when $d = 2r_{\text{coh}}$), so favoring in turn their interaction.

The value of d_p can be approximatively estimated by observing the interaction between CDs can be described as the tunneling of evanescent e.m. waves (or virtual photons) through opaque optical barriers. As already discussed in detail [18] the tunneling probability is inversely proportional to the barrier's length so it appears as a near-field phenomenon whose "detection" is limited to near-field zone. All the experimental evidence available about this process [33, 34, 37, 38] show this near field region has a spatial extension D up to

$$D \simeq 10^3\,\lambda_0 \tag{32}$$

where λ_0 is the wavelength of the incident field at the interface where the total reflection occurs. We can then assume in our case

$$d_p \simeq 10^3\,\lambda_0 = 10^3\left(\frac{2\pi}{\omega_0}\right) \tag{33}$$

where we have used the (3).

On the other hand, the tunneling time τ of the evanescent wave (or virtual photons) is related to the frequency f of the signal (or the associated quantum particle) undergoing the process, namely, for $\omega = \omega_{\text{coh}}$

$$\tau \simeq \frac{1}{f} = \frac{2\pi}{\omega} = \frac{2\pi}{\omega_{\text{coh}}} \tag{34}$$

that is independent from the barrier length for not too long barriers where evanescent field attenuation makes no interaction possible.

It is important to note this quantum-tunneling "interaction" depends on the existence of excited levels in the CDs whose excitations-deexcitations determine the tunneling of virtual coherent photons between the CDs, since in the fundamental state the CGS is stable. We could think this interaction to occur between the couples of quasi-free coherent electrons, respectively belonging to different excited levels in the two CDs, as the result of the exchange of a virtual photon through the tunneling process.

A more detailed microscopic description of this "tunneling" interaction and its dynamics will be discussed in a forthcoming publication [30]. Here we limit ourselves to note that, from a microscopic point of view, the energy transfer between the two interacting CDs can be explained through the virtual photons exchange between them via a sort of a resonance energy transfer process (RET).

If we consider the exchange of one photon, the "tunneling" coupling between two CDs can be described by the following total Hamiltonian

$$H = H_0 + H' \tag{35}$$

where H_0 and H' respectively indicate the "free" and the "interaction" Hamiltonian of the system. We further suppose the free system (the isolated water CD) can be considered as a two-levels system characterized by only two accessible eigenstates (normalized to unity), named $|1\rangle$ and $|2\rangle$, satisfying the stationary Schrodinger's equation

$$H_0|1\rangle = E_1|1\rangle; \quad H_0|2\rangle = E_2|2\rangle \tag{36}$$

with E_1 and E_2 energy eigenvalues of the system. According to our assumption, we can decompose the Hamiltonians as

$$H_0 = E_1|1\rangle\langle 1| + E_2|2\rangle\langle 2|; \qquad H' = \Gamma(|1\rangle\langle 1| + E_2|2\rangle\langle 2|) \tag{37}$$

where $\Gamma \in R$ represent the strength of the interaction. We then consider the complete eigenvalue problem

$$H|\alpha\rangle = E_\alpha|\alpha\rangle \tag{38}$$

where $\alpha = A, B$ and E_α is the eigenvalue of the eigenstate $|\alpha\rangle$. We can then write

$$|\alpha\rangle = c_{\alpha 1}|1\rangle + c_{\alpha 2}|2\rangle \tag{39}$$

satisfying the normalization condition $|c_{\alpha 1}|^2 + |c_{\alpha 2}|^2 = 1$. The eigenvalues problem (38) has the solution

$$\begin{aligned} E_A &= E_1(1+T)/2 + E_2(1-T)/2 \\ E_B &= E_1(1-T)/2 + E_2(1+T)/2 \end{aligned} \tag{40}$$

and $T = \sqrt{1 + 4\Gamma^2/(E_1 - E_2)^2}$ with eigenvalues satisfying $E_1 + E_2 = E_A + E_B$. The time-evolution of the system is described by

$$|\Psi(t)\rangle = U(t; t_0)|\Psi(t_0)\rangle = \exp[-iH(t - t_0)]|\Psi(t_0)\rangle \tag{41}$$

If we now suppose the initial state of the system is just $|\Psi(t=0)\rangle = |1\rangle$ we have

$$|\Psi(t)\rangle = \cos\vartheta e^{-iE_A t}|A\rangle - \sin\vartheta e^{-iE_B t}|B\rangle \tag{42}$$

By using (42) and after some algebra, we obtain the transition probability from the initial state $|1\rangle$ at $t = 0$ to the state $|2\rangle$ at time t as

$$P_{12} = |\langle 2|U(t)|1\rangle|^2 = \frac{\Gamma^2}{\Gamma^2 + (E_1 - E_2)^2}\sin^2\left(\sqrt{\frac{(E_1 - E_2)^2 + 4\Gamma^2}{4}}t\right) \tag{43}$$

namely the so-called "Rabi equation". This solution is equivalent to a Rabi oscillation in which a photon wave packet of given wave number is coupled back and forth between the two CDs.

A more interesting treatment of time evolution of the system can be formalized by using Pauli's matrices. In this notation the Hamiltonian is

$$H = n_0 I + \vec{n}\cdot\vec{\sigma} \tag{44}$$

where, I is the 2×2 identity matrix, $\vec{\sigma} = (\sigma_x, \sigma_y, \sigma_z)$ and

$$n_0 = \frac{1}{2}(E_1 + E_2),\ n_1 = \Gamma,\ n_2 = 0,\ n_3 = \frac{1}{2}(E_1 - E_2) \tag{45}$$

and the time-evolution operator becomes

$$U(t) = \exp\left[-i\frac{(E_1 + E_2)}{2}t\right]\left[\cos(|\vec{n}|t)I - i\sin(|\vec{n}|t)\frac{\vec{n}}{|\vec{n}|}\cdot\vec{\sigma}\right] \tag{46}$$

On the other hand, any arbitrary unitary operator U can be expressed as a rotation of an angle θ about an axis defined by the real unitary vector $\hat{u} = (u_x, u_y, u_z)$, namely

$$U = \exp(i\alpha)R_{\hat{u}}(\theta) = \exp(i\alpha)\exp\left(-i\theta\hat{u}\cdot\frac{\vec{\sigma}}{2}\right) \tag{47}$$

where the rotation operator $R_{\hat{u}}(\theta)$ can be expanded as

$$R_{\hat{u}}(\theta) = \cos\left(\frac{\theta}{2}\right)I - i\sin\left(\frac{\theta}{2}\right)\hat{u}\cdot\vec{\sigma} \tag{48}$$

The (47) is identical to (46) if we assume

$$\alpha = -\frac{E_1 + E_2}{2}t;\quad \theta = 2|\vec{n}|t;\quad \hat{u} = \frac{\vec{n}}{|\vec{n}|} \tag{49}$$

We have then proven a very important result namely the time-evolution operator of system of two interacting CDs through tunneling coupling can be expressed as the product of a rotation and an overall phase shift.

By the effect of the coupling, the two CDs cannot be further considered as independent entities but as two interconnected parts of the same system. We now remember the wavefunctions describing the coherent domains $\Psi_1 = \psi_1 \exp(i\Phi_1)$ and $\Psi_2 = \psi_2 \exp(i\Phi_2)$ are eigenstates of the phase operator (whose eigenvalue is related to the coherent frequency of oscillation including the angular frequency of the excited vortices) and also describe the state of tunneling photon as being in the one CD or in the other, through a shift $\Delta\Phi$ in the value of their relative phase.

The tunneling interaction is obviously not limited to the case of only two CDs but can be extended to a higher number of interacting domains.

4 Quantum Coherence in Water Inside Brain Microtubules Viewed as Metamaterials

Microtubules are rigid polymers consisting of groups of protofilaments, of length ranging between $1-30\,\mu\text{m}$ [15], cylindrically shaped with an outer and inner diameter, respectively of about 25 nm and 15 nm (Fig. 6).

They are composed by structural subunits, the tubulin heterodimers (of length about 8 nm), in turn containing the α- and β- tubulin having an high electric dipole moment (about 10^{-26} C m) [15] and determining the remarkable electric polarity of MTs that make them very sensitive to electromagnetic field. The mechanical properties of MTs have been studied in detail by means of many biophysical techniques such atomic force microscopy, thermal bending and single molecule.

Specifically, for the purpose of this chapter, we focus on the inner hollow volume of MT that can be assumed to be "filled" with (thermally) isolated water [18], showing in the following that, under the boundary conditions averagely satisfied inside the brain MTs, it undergoes a spontaneous quantum phase transition toward a coherent state in

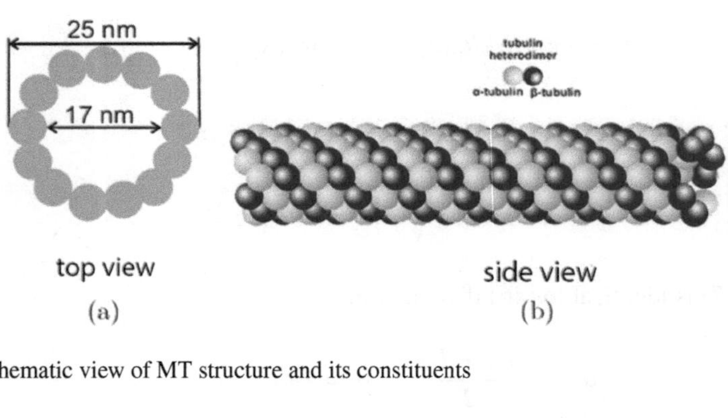

Fig. 6 Schematic view of MT structure and its constituents

which an electromagnetic field oscillates in tune with the water matter field between two energy levels corresponding to an electronic transition of water molecule.

Inside living systems, the average temperature is about 37 °C (≅310 K) and inside brain it can be even greater. As we have seen, at this temperature thermal fluctuations are able to meaningfully decrease the coherent fraction of water molecules. On the other hand, "biological" water can be considered as interfacial water and this particularly holds for the water contained in the inner hollow volume of brain's microtubules that we can assume to be "filled" with thermally isolated water. We can then assume water entrapped inside brain's MT as fully coherent. But are there inside them the conditions for the quantum phase transition toward coherent state to occur?

As we have seen, in a given thermodynamics environment, the runaway toward the GCS will happen, in the case of water (see Table 1), if $\rho > \rho_c \cong 0.3\,\mathrm{g\,cm^{-3}}$. Now, assuming for the brain an average temperature $T \sim 37\,°\mathrm{C}$ and a MT cavity volume [18] $V_{MT} \sim 5 \cdot 10^{-22}\,\mathrm{m}^3$ for a moderately long ($l \sim 10^{-6}$ m) MT, we have

$$\rho_{\mathrm{water,MT}} \sim 0,993\,\mathrm{g\,cm^{-3}} > \rho_{\mathrm{crit}} \tag{50}$$

showing the existence of the conditions, inside MT inner volume, required for the quantum phase transition of water toward the coherent state to occur.

An important point to consider is that the energy level $\omega_0 = 12.06\,\mathrm{eV}$ correspond to a CD whose "size" is, by (3), of order of

$$L_{CD} \sim 0.1\,\mu\mathrm{m} \tag{51}$$

that is about $1/10$ of the average length of a moderately sized MT and, in particular, of the order of magnitude of the MT dimers (~8 nm).

The rough estimate given by (51) shows that photons, belonging to the coherent electromagnetic field generated inside MT, according to our mechanism, are characterized by a wavelength much shorter than the typical length of MT so allowing the formation of node and antinodes within the MT inner cavity.

This result overcomes the trouble raised by Georgiev [14] with respect the infrared superradiant photons considered in the theoretical models of coherence in MT presented so far. The estimate of the radius of a single water CD, supposed as spherically symmetric, given by $r_{\mathrm{coh}} \sim 3.75 \cdot 10^{-8}$ m for the electronic transition $\omega_0 = 12.06\,\mathrm{eV}$, allows us to interpret the inner hollow cavity of a MT as "filled" by such CDs, since its value is practically comparable with that of the circular cross section of a MT ($\sim 2.5 \cdot 10^{-8}$ m) and less than a tenth of its average length ($\sim 10^{-6}$ m). If we now recall that each CD can be considered as a "cavity" for the coherent e.m. field "confined" within its neighborhood, this suggests the idea that the whole inner cavity of MT could be just considered as a waveguide for such electromagnetic fields. For a rectangular-shaped waveguide with length $L = a$ and width $W = b$, the propagation of an e.m. wave along the z-axis can be described by the stationary solutions of Maxwell's equations corresponding to the so-called TE_{mn} modes.

The corresponding wave number $\left|\vec{k}\right|$ is

$$k = \sqrt{\omega^2 - \omega_{mn}^2} \tag{53}$$

where

$$\omega_{mn} = \pi \sqrt{(m/a)^2 + \left(n/b\right)^2} \tag{54}$$

is the cutoff frequency whose minimum value, for a given waveguide, is given by

$$\omega_{10} = \pi / d \tag{55}$$

with $d = \max\{L, W\}$. In the case of a circular cross section, we can assume $d = 2r$, where r is the cavity radius and (53) can be also rewritten as

$$k = \omega_{\text{eff}} \tag{56}$$

with $\omega_{\text{eff}} \equiv \sqrt{\omega^2 - \omega_{mn}^2}$. This means the frequency of the e.m. wave propagating inside waveguide is smaller than its corresponding value in free space, being always $\omega_{\text{eff}} < \omega$. It is just what happens within a CD where $\omega_{\text{coh}} \ll \omega_0$, so preventing the "release" of the coherent electromagnetic energy toward the outside of a CD. Then in the inner hollow of brain MT, the coherent e.m. field generated in water behaves as it would be "confined" in an e.m. waveguide or cavity and characterized by a photon frequency $\omega_{\text{eff}} = \omega_{\text{coh}}$. This consideration needs further attention.

We see from (53) that, if $\omega < \omega_{mn} \equiv \omega_c$, the wave vector becomes imaginary, implying no wave propagates across the waveguide but we can only have an evanescent e.m. field and we say that the waveguide is in cutoff. More specifically, being ω_{10} the smallest value of ω_{mn}, a sufficient condition for a waveguide to be in cutoff is obviously $\omega < \omega_{10}$. For a water CD associated to the considered electron transition we have $\omega_{\text{coh}} \sim 10^{-2}\omega_0 = 0.1206\,\text{eV}$ while, from (55) and considering $r = R_{\text{in}}$, where $R_{\text{in}} \approx 17$ nm is the inner radius of MT, we obtain

$$\omega_{10,MT} \simeq 2.770 \cdot 10^{16}\,\text{Hz} \tag{57}$$

and

$$\omega_{\text{coh}} \simeq 1.832 \cdot 10^{14}\,\text{Hz} \tag{58}$$

namely

$$\omega_{\text{coh}} \ll \omega_{10,MT} \tag{59}$$

This important result proves a brain's MT is always in cutoff regime with respect to the coherent e.m. field generated in water molecules (at least if we don't consider too high excited energy levels), and then behave as "waveguide" for it. Consequently, we can think of water-filled brain's MT as a whole long cylindrical coherent domain in which, in turn, a lot of CDs are closely packed and whose overall evanescent field tunnel across the MT's boundaries. As we have seen, the interaction between couples of CDs can occur through the proposed mechanism of tunneling of evanescent modes (or virtual photons) between them. On the other hand, the interaction between two waveguides, by means of evanescent e.m. fields, has been already confirmed from the standpoint of classical physics and also applied in the construction of many devices (for example using optical fibers), just called "evanescent-field-optic-couplers".

In this case, the coupling between two waveguides can be expressed by the value of the rate of energy transfer between them which is proportional to the overlap of the modes of the two waveguides. The waveguides coupling then represents the classical counterpart of the quantum coupling through the exchange of virtual photons that further confirm the plausibility of our theory.

The brain's MTs, apart from stabilizing the coherent fraction of water enclosed in their inner volumes making it fully coherent, have the very important effect to amplify, as we'll see in the following, the evanescent field produced by each CDs, so meaningfully increasing the value of d_p. This aspect has been already studied in previous publications [18, 39] in which it has been shown that a "near-perfect" tunneling and amplification of evanescent e.m. field is theoretically possible and experimentally proven in a waveguide even in cutoff regime, provided that it is partially filled with a MTM.

The amplification process has been described by considering a MTM-filled waveguide section sandwiched between two empty sections operating in cut-off and two input/output waveguides above the cutoff.

We have shown [18] that in this way the evanescent e.m. field described by (49) would be amplified by passing through the MTs structure, provided that they could be considered as MTMs. The overall result of this amplification process will be then an important increase of the evanescent penetration depth d_p. We have also proven [18, 19] that MTs can be actually considered as MTMs by calculating the value of its refraction index n of the waveguide for $\omega = \omega_{\text{coh}}$, so obtaining, assuming $\omega_{mn} = \omega_{10}$, $n^2(\omega_{\text{coh}}) < 0$. A purely imaginary value of refraction index is a typical feature of well-known categories of MTMs as, for example, plasmas or ferromagnetic materials.

In this case, the amplification and the tunneling of an evanescent e.m. wave outside the waveguide is possible when [18, 19]

$$\omega = \omega_{\text{coh}} < \omega_c = (\pi/a)\big/\sqrt{\varepsilon\mu_0} \tag{60}$$

where a is the waveguide width, ε and μ_0, respectively, the dielectric permittivity of the MTM and the magnetic permittivity of vacuum. In this way, it would be sufficient to use an MTM with $\varepsilon > \varepsilon_0$ in order to allow the amplification and tunneling of the evanescent e.m. field into "free" space up to over four times the original field amplitude after the interaction with the MTM-filled portion of the waveguide. Brain's MTs can actually behave as a "natural" MTMs due to the positive and negative electric charges on the dimers causing them to align [40]. According to this idea, each tubulin can be then modeled like a split-ring resonator (SRR).On the other hand, MTMs just consist of recurring structures, like several nano-resonators, characterized by a period small if compared to the optical wavelength and able to excite electromagnetic oscillations when hit by photons of suitable frequency.

For frequencies different than the resonance one, this interaction can result in a negative value of permeability $\mu = \mu_0 \mu_r$ and then, due to the relation

$$k^2 = \varepsilon_r \mu_r \omega^2 / c^2 \tag{61}$$

where ε_r and μ_r, respectively indicate the relative electric and magnetic permittivity, it determines purely imaginary wave number which can produce evanescent wave enhancement.

5 The Coherent Interaction Between Microtubules and the Possibility of Hypercomputing in Human Brain

The previous discussion leads us to consider brain's MT as a network of cylindrical water CDs filling up their inner hollow volumes and coupled each other by the evanescent e.m fields produced by water coherent dynamics. In order to this evanescent coupling to take place the reciprocal distance between two CDs should allow a suitable overlapping between evanescent fields, according to the propagation depth d_p.

For our specific biosystem we have $\omega_0 = 12.06\,\text{eV}$ and $\omega_{\text{coh}} \simeq 0.12\,\text{eV}$ so we obtain, from (33) and (34), respectively

$$d_p \simeq 1.03 \cdot 10^{-4}\,\text{m}; \quad \tau \simeq 3.43 \cdot 10^{-14}\,\text{s} \tag{62}$$

If we now assume, as an average order of magnitude for the MTs spatial separation, the quantity

$$d_{MT} \sim 10^{-7}\,\text{m} \tag{63}$$

we see, by comparison with (62)

$$d_{MT} \ll d_p \tag{64}$$

so obtaining the very important result that brain's MTs are always in the overlapping case.

A very interesting result is that we can estimate the maximum tunneling "velocity" (the group velocity of evanescent waves or, equivalently, that of the virtual photons undergoing tunneling) as

$$v_{\max} = \frac{d_p}{\tau} \simeq 10^3 \left(\frac{\omega_{\text{coh}}}{\omega_0} \right) \tag{65}$$

that is superluminal if

$$\left(\frac{\omega_{\text{coh}}}{\omega_0} \right) > 10^{-3} \tag{66}$$

Now recalling that $\omega_{\text{coh}} \sim 10^{-2}\omega_0$ we see in our case the maximum value of tunneling photons is superluminal, i.e. $v_{\max} \simeq 10$, namely about ten times the value of light velocity in vacuum even for a non-excited CD.

According to this model, the coupling tunneling interaction could then occur between CDs even if they are placed far apart each other, since every CDs could interact with its relative first neighbors CDs and so on, to form a spatially extended network of interacting CDs.

The computational features of such tunneling interaction can be uncovered by going back to analyze the evanescent coupling between two MTs due to the water CDs. It is then appropriate to study the tunneling exchange of virtual photons between the two MT waveguides by using the dual-rail representation for the photon, according to which we would adopt the following encoding for the two logic states $|0\rangle_L$ and $|1\rangle_L$

$$|0\rangle_L \rightarrow |0\rangle_{MT_1}|1\rangle_{MT_2} = |01\rangle; \quad |1\rangle_L \rightarrow |1\rangle_{MT_1}|0\rangle_{MT_2} = |10\rangle \tag{67}$$

where $|0\rangle_{MT_j}$ and $|1\rangle_{MT_j}$ respectively represent the state in which a photon is absent and present as spatial mode in the *i*-th MT. If we consider the exchange of one photon, it can be shown the Hamiltonian of the system has the same form of (37) then if we assume $|0\rangle_L \rightarrow |1\rangle$ and $|1\rangle_L \rightarrow |2\rangle$, where the states $|1\rangle$ and $|2\rangle$ are those defined in (36), we have

$$P_{12} = |\langle 2|U(t)|1\rangle|^2 = |_L\langle 1|U(t)|0\rangle_L|^2 \equiv P(0_L \rightleftarrows 1_L) \tag{68}$$

In our theoretical framework, the propagation of a photon in an optical fiber is likened to the presence of a photonic spatial mode, due to tunneling, in one or the other MT as sketched in Fig. 7.

This can be shown if we consider, assuming a not too strong coupling, the total Hamiltonian for the system (we limit ourselves to the photon dynamics)

$$H = \sum_{\vec{k}} \left[\omega_{\vec{k}} \left(a^{\dagger}_{\vec{k},1} a_{\vec{k},1} + a^{\dagger}_{\vec{k},2} a_{\vec{k},2} \right) + \gamma \left(a^{\dagger}_{\vec{k},1} a_{\vec{k},1} + a^{\dagger}_{\vec{k},2} a_{\vec{k},2} \right) \right] \tag{69}$$

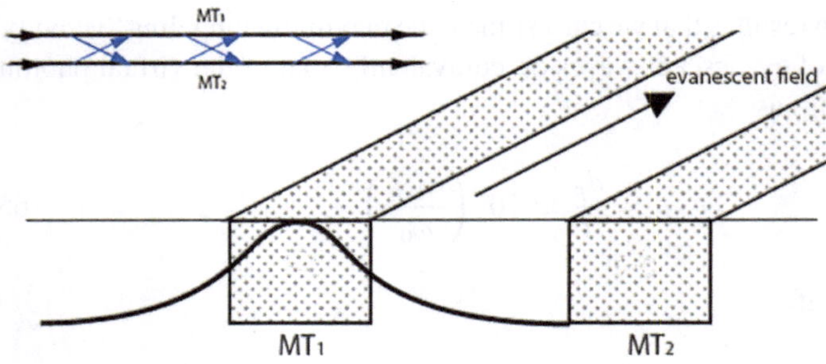

Fig. 7 Evanescent coupling and virtual photon exchange between two MTs

where $a_{\vec{k},i}$ $(a^{\dagger}_{\vec{k},i})$ is the annihilation (creation) operator for the mode $\vec{k}$ in the i-th MT $(i = 1, 2)$ acting on the vacuum state $|0\rangle$ and γ is a real parameter quantifying the coupling strength. With this notation $|0\rangle_L \rightarrow |\psi_1\rangle = a_1^{\dagger}|0\rangle$, $|1\rangle_L \rightarrow |\psi_2\rangle = a_2^{\dagger}|0\rangle$ and, if γ is approximatively independent of the value of $\vec{k}$, the total Hamiltonian has again the same form of as (37) and we can deduce the (68).

We then easily note that, if we start the systems in the state $|0\rangle_L$, after a time T given by

$$T = \frac{\sqrt{(E_1 - E_2)^2 + 4\gamma^2}}{\pi} \tag{70}$$

it has the maximum probability to be in the state $|1\rangle_L$. For a given value of γ, by suitably choosing the values of E_1, E_2 we can obtain $P(0_L \rightarrow 1_L) = 1$ then realizing the transformation $\hat{X}$ such as

$$\hat{X}|0\rangle_L = |1\rangle_L \tag{71}$$

that switches the quantum state of the system and then acts like a SWAP gate on its logic state, so that we can name $\hat{X} = SWAP$.

Even more interestingly, if $\overline{T} = T/2$, we can define a new operator $\hat{\hat{X}}$ given by

$$\hat{\hat{X}}\left(\hat{\hat{X}}|0\rangle_L\right) = \hat{\hat{X}}^2|0\rangle_L = |1\rangle_L \tag{72}$$

so we can define it as (Fig. 8)

$$\hat{\hat{X}} \equiv \sqrt{SWAP} \tag{73}$$

whose action on the computational basis is defined by

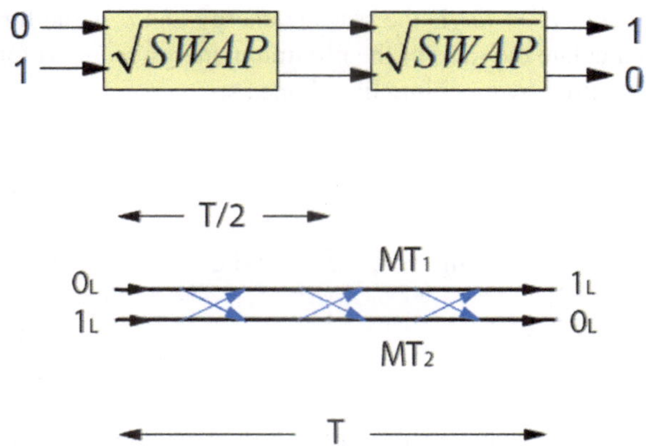

Fig. 8 Realization of quantum SWAP gate by means of virtual photons exchange in MTs

$$\begin{array}{c} |0_A\rangle|0\rangle_B \rightarrow |0_A\rangle|0\rangle_B \\ |0_A\rangle|1\rangle_B \rightarrow \frac{1+i}{2}|0_A\rangle|1\rangle_B + \frac{1-i}{2}|1_A\rangle|0\rangle_B \\ |1_A\rangle|0\rangle_B \rightarrow \frac{1-i}{2}|0_A\rangle|1\rangle_B + \frac{1+i}{2}|1_A\rangle|0\rangle_B \\ |1_A\rangle|1\rangle_B \rightarrow |1_A\rangle|1\rangle_B \end{array} \tag{74}$$

As well-known [41] in the theory of quantum computing, the two-qubits $\sqrt{SWAP}$ gate is universal quantum gate since any generic quantum gate can be constructed by considering a combination of this gate and some one-qubit gates, it then represents the building block of any quantum computer.

How to practically implement and control such quantum gates within the specific biosystem here considered has been analyzed in more detail in other publications [30]. Here we will limit ourselves to the consideration of other two fundamental aspects related to the computational process achievable through brain's MTs: the processing speed, the decoherence time and the data storage and retrieval.

According to Benioff [42], one fundamental limit to the processing speed is given by the Heisenberg time-energy uncertainty principle, namely the requirement for two processing steps to be distinguishable. This limit therefore also defines a minimal duration of the computational step given the available energy E. We have shown [18, 43, 44] that, for a quantum tunneling particle (QTP), as the evanescent photons, this time is given by (from this point on we consider the SI system of unit).

$$\Delta t \approx \frac{N}{\beta(\beta-1)} \frac{\pi\hbar}{2\langle E_*\rangle} \tag{75}$$

where, as usual, $\beta = v/c$, v is the value of the tunneling velocity and $\langle E_*\rangle$ is the averaged energy for QTP gates. An alternative estimation of tunneling velocity for the tunneling photon can be obtained as

$$v \approx c\left(1 + \frac{1}{2\omega d} + \sqrt{\frac{1}{\omega d} + \frac{1}{4\omega^2 d^2}}\right) \tag{76}$$

where $E = \hbar\omega$ is the particle energy and d is the tunneling distance. We can then calculate the ratio of the minimum energy required for standard computation to that required by tunneling based one as

$$R = \frac{\langle E_* \rangle}{\langle E \rangle} \approx \frac{1}{\beta(\beta - 1)} \tag{77}$$

Now, by using (65), we can write

$$\beta_{\max,k} \approx 2 \cdot 10^3 \left(\frac{\omega_k}{\omega_0} \right) \tag{78}$$

where ω_k is the energy of the k-th excited level obtained by (42). As we have seen, the energy accumulated in the water excited vortices can reach even very high values allowing us to obtain R a very small value, so that the energy cost of computation performed through QTP gates could be several orders of magnitude smaller than that of conventional computer systems.

In an ideal system, this energy for computational step is not dissipated during each step but could be available for the next ones. In general, a quantum computer could operate close to this limit if the available energy is equated to the energy range spanned by the Hamiltonian eigenvalues, namely the internal degrees of freedom of the system. In the coherent systems, as water CDs, a characterizing feature is just the possibility to accumulate a lot of environmental energy at high entropy into the low entropy CD inner degrees of freedom, even using its metastable excited levels.

It is commonly believed that quantum computing could be intrinsically much more powerful than classical computation using traditional silicon processors. Nevertheless, interconnect delays inside logical gates cannot be never fully eliminated from any real electronic components even in quantum computers. This effect could be fully compensated, by introducing quantum tunneling processes so meaningfully speed up computation, bringing the overall transmission rate very close to its ultimate speed limit [12] by boosting the speed of signal propagation and, consequently, computers performances.

Apart to this question, one of the most important problem, often called the "enemy" of quantum computing, is the issue of decoherence of quantum states, according to which, in order to quantum computation to take place effectively, it is required that during the calculation time the quantum system remains in a superposition of its quantum states. So, the time required to perform quantum calculation should be much lesser than the decoherence time due to the system—environment interaction. As we have already shown [18, 43, 44], if T_r is the "relaxation" time of a single qubit and Δt the operation time of a single logical gate, the decoherence time of a system of L qubits is $\Delta t_L = 2^L \Delta t$ and we obtain the minimum energy required to perform quantum computation using evanescent tunneling photons is given by

$$E_0 \approx \frac{\hbar \nu_G L}{\beta(\beta - 1)T} 2^L \tag{79}$$

from which follows the use of tunneling evanescent photons will allow an increase of the qubit "size" of the system, with respect to the conventional case, given by

$$\Delta L \approx \frac{\log_2[\beta(\beta - 1)]}{1 + 1/L \log 2} \tag{80}$$

if $\Delta L < L$.

In our system quantum qubits can be codified by the logical states $|0\rangle_L$ and $|1\rangle_L$, corresponding to the photon spatial modes, in turn associated to the excited levels of CDs. As we have shown above, in the coherent system composed by water CDs enclosed inside MTs, the matter-field dynamics stabilized the coherent fraction, even at room temperature, at the value of $F_c(T) \sim 1$ so being able to prevent environmental decoherence. Such excited states (as well as the whole CD structure) cannot in fact decay thermally so they are theoretically unaffected by decoherence. On the other hand, the energy released by excited coherent domains, through non-thermal channels, could pose a noticeable instability question if it would be subsequently converted in heat and not adequately taken off from the system or at least reused to increase the internal coherence in any way. A more detailed analysis of such aspect will be considered in a forthcoming publication, in any case we can estimate, as a meaningful parameter, the ratio of the coherence time τ_D' needed to ensure computation by evanescent photons to that τ_D required by conventional quantum computation as [18, 43, 44]

$$\frac{\tau_D'}{\tau_D} \approx \left(\frac{E_0}{E'}\right) = [\beta(\beta - 1)]^2 \tag{81}$$

from which we could obtain, by choosing a suitable value of $\beta_{\max,k}$, namely a sufficiently high excited energy level of water coherent domains, a value $\tau_D' \gg \tau_D$ namely a very long decoherence time that ensure the possibility to perform, by using evanescent photons, quantum computation even at room temperature and for a long time inside brain's MTs without invoking, for example, the mechanism of quantum gravity proposed by Penrose [45]. The suggested dynamics so completely overcomes the question posed by Tegmark [46] about the MT decoherence time, according to which the warm and wet inner environment of the human brain wouldn't allow any long-time entanglement and superposition effect able to perform quantum computation.

If we consider our network of tunneling-coupled MTs as a quantum processor using evanescent tunneling photons, in which every computational gate could be entangled with any other, according to the non-local feature of the tunneling interaction, we could estimate [18, 43, 44] the coherence time of such quantum processor as

$$\tau_D' \approx \tau_R \left(\frac{1}{\Delta x \sqrt{2m^* k_B T}}\right)^2 [\beta(\beta - 1)]^2 \tag{82}$$

where $\Delta x \sim 1$ cm is a typical hypothetical dimension of the quantum processor, T the absolute temperature of the quantum field, m^* the effective mass of the evanescent photon and $\tau_R \gg 1$ s is the relaxation time of the system. As we have seen, the coherent behavior of our system dynamically corresponds to take $T \to 0$ while the term β could assume very high values $\beta_{\max}$ and the photon-mass term very low ones. We then see that, if the error rate is not too high so that we can use quantum error correction code to eliminate it, the total calculation time of such evanescent quantum processor could be very much shorter than the decoherence time so allowing it to efficiently complete the computational task.

The consideration of evanescent photons to couple brain's MTs opens the door to another very fascinating perspective namely the possibility for the human brain to realize a sort of quantum "hypercomputation". Hypercomputation, that we can define as the possibility to perform infinitely many steps of computation in a finite time, refers to models of computation that go beyond Turing computability. Nevertheless, the realization of an hypercomputer is usually considered not allowed due to the Heisenberg uncertainty principle that limits the energy required by calculation steps. On the other hand, this limitation would affect quantum algorithms as well as the conventional ones. According to Feynman [47], the energy required for computational step is given by

$$\overline{E} = k_B T \frac{f - b}{(f + b)/2} \tag{83}$$

where T is the system temperature, f and b, respectively, the forward and backward computation rate. Lloyd [48] calculated the minimum time Δt required by a quantum system, having average energy ΔE, to evolve to an orthogonal state as

$$\Delta t = \frac{\pi \hbar}{2 \Delta E} \tag{84}$$

from which the total time needed to realize a computation with an infinite number of steps is

$$T = \sum_{n=1}^{\infty} \Delta t_n = \frac{\pi \hbar}{2 E_0} \sum_{n=1}^{\infty} \frac{1}{(1 - k) k^{n-1}} \tag{85}$$

in which Δt_n is the duration of the n-th step of calculation and $k = 2(f - b)\big/(f + b)$. It can be easily verified that, for $0 < k < 1$, the sum diverges so implying that, according to Feynman's model, performing an infinite number of calculation steps requires an infinite amount of time. We have shown [18, 43, 44] that instead, by considering evanescent photons for computation, the time required for the quantum system to complete computation becomes

$$T = \sum_{n=1}^{\infty} \Delta t_i = \frac{\pi \hbar}{2E_0} \sum_{n=1}^{\infty} \frac{1}{\beta_n(\beta_n - 1)(1-k)k^{n-1}} \tag{86}$$

where the coefficients β_n are given by

$$\beta_n = \sqrt{1 + \frac{m_*^2}{E_n^2}} = \sqrt{1 + \frac{m_*^2}{k^{2n} E_0^2}} \tag{87}$$

Numerical calculations have shown that (86) effectively converges to a finite limit when $0 < k < 1$.

These results clearly prove that if we consider evanescent particles, as virtual photons, instead of standard ones, the theoretical questions about the infinite energy required posed by Feynman's model to perform hypercomputation can be overcome so allowing, in principle, the realization of an accelerated Turing machine. Finally, with reference to the important question related to data storage and retrieval to sustain quantum computation, we have already proposed [18, 43, 44] a possible mechanism, based on the possibility to consider MTs as memory units able to store data in the form of holograms, using the tubular substrate having a metamaterial-like framework, composed by sub-wavelength structures. Through this mechanism, evanescent photons, generated inside MTs, can "manipulate" their storage substrate providing for data storage and retrieval.

6 Conclusion and Outlook

It has been shown in brain's microtubules there exist the right conditions to allow, according to the predictions of QED coherence in condensed matter, the water enclosed in their inner hollow volumes to undergo a spontaneous quantum phase transition toward an energetically favored state, in which the electronic clouds of water molecules coherently oscillate in tune with a self-trapped electromagnetic field within well-defined quantum space domains (coherent domains).

Differently from the models of quantum optical coherence in cytoskeletal microtubules proposed so far (that substantially consider the energetic transitions of water molecules associated to rotational energy levels of the order of few meV), the model here discussed assume the coherent system (water + electromagnetic field) to oscillate in phase with the electronic transition of water from the ground state to the level at energy $E = 12.06$ eV, implying the "supperradiant" photons, generated inside the coherence domains, to have a wavelength much smaller than the length of a moderately sized microtubules in brain. We have also discussed as this water coherence domains are characterized by an excited energy spectrum due to the presence, in every domain, of quasi-free electrons organizing themselves into coherent "cold" vortices namely metastable excited states easily excitable, able to accumulate high quantities of energy from environment and characterized by a very long lifetime since they cannot thermally decay.

Furthermore, the coherent electromagnetic field, "condensed" from quantum vacuum, oscillating in tune with water molecules is characterized by a purely imaginary value of wave number. This means this coherent field is "trapped" inside the coherent domains confined in the brain microtubules and is totally reflected at their boundaries. We have also shown brain's microtubules act, with regard to such field, like optical waveguides operating in cutoff regime causing the evanescent tail of the coherent e.m. field to spread across the coherence domains and tunneling through them also considerably amplifying its amplitude.

The tunneling time of such e.m. evanescent waves through optical barriers of various types demands for a superluminal velocity of propagation through these barriers, irrespective of the type of barrier or wave frequency and amplitude.

On the other hand, we can consider tunneling e.m. modes as virtual photons, whose packet velocity is even superluminal, that can interact with the biological structures in the microtubules walls, acting like "evanescent-field-optical couplers" between brain's microtubules.

Furthermore, based on the suggested analogy (both of structural and electromagnetic nature) between microtubules and metamaterials, we have proposed the idea to interpret the inner medium of brain's microtubule as having properties similar to those characterizing metamaterials and so able to generate a nearly perfect tunneling and amplification of the evanescent e.m. field generated inside them by the coherent dynamics of water.

According to the proposed model, brain's microtubules also carry out another function, namely they guarantee the water enclosed in them (interfacial water) to be almost perfectly coherent, that is $F_c(T) \sim 1$, despite the high average temperature inside brain.

We have also proven the very important result according to which the dynamics of excited levels in water coherent domains together with the coupling tunneling interaction between MTs could reproduce a universal two qubits gate as well as different types of one qubit gates, making the brain's MTs able to realize, in principle, any sort of quantum computation.

The discussed model, besides giving a founded physical explanation of the origin of evanescent tunneling photons inside brain microtubules and how these photons could interact with them, also overcomes the issues properly raised by some authors about the actual possibility to consider the generation of such photons in order to explain the occurrence of primary brain functions (as, first of all, the too long wavelength of superradiant photon with respect microtubules size and the too short environmental decoherence time). Related to these aspects we note the proposed model could be able to explain also the formation of brain's memories, as well as some typical non-local features of human brain properties, by considering the mechanism of evanescent wave holography.

We have also seen the consideration of tunneling virtual photons could ensure computational performances in principle much higher than any conventional processor overwhelming, for example, one of the main difficulties related to the practical realization of the so-called hypercomputation (accelerated Turing machine).

In summary, our results suggest the occurrence of a novel and so far unrecognized type of interaction between biological structures based on the tunneling-coherent-coupling between microtubules able to realize a superfast communication network inside the human brain as well as a possible mechanism able to use tunneling photons to perform realistic quantum hypercomputing capabilities.

References

1. Zeilinger et al A (1999) Nature 401:680
2. Marshall IN (1989) New Ideas Psychol 7:73–83
3. Crick F, Koch C (1990) Semin Neurosci 2:263–275
4. Singer W (1993) Annu Rev Physiol 55:349–374
5. Penrose R (1989) The emperor's new mind. Oxford University Press, Oxford
6. Penrose R (1994) Shadows of the mind. Oxford University Press, Oxford
7. Pribram KH (1991) Brain and perception. Lawrence Erlbaum, New Jersey
8. Hameroff SR, Penrose R (1995) J Conscious Stud 2:98–111
9. Hameroff SR, Watt RC (1982) J Theor Biol 98:549–561
10. Del Giudice E, Preparata G, Vitiello G (1988) Phys Rev Lett 61:1085–1088
11. Preparata G (1995) QED coherence in matter. World Scientific, Singapore, London, New York
12. Umezawa H (1993) Advance field theory: micro, macro and thermal physics. American Institute of Physics, New York
13. Jibu M et al (1994) BioSystems 32:195–209
14. Georgiev DD (2002) Quantum computation in the neural microtubules: quantum gates, ordered water and superradiance. https://arxiv.org/abs/quant-ph/0211080
15. Mavromatos NE (2011) J Phys: Conf Ser 329:1–31
16. Caligiuri LM (2015) WSEAS Trans Biol Biomed 12:104–112
17. Caligiuri LM, Musha T (2015) Int J Circuits Syst Signals Process 9:192–204
18. Caligiuri LM, Musha T (2016) The superluminal universe: from quantum vacuum to brain mechanism and beyond. NOVA Science Publishers Inc., New York
19. Caligiuri LM, Musha T (2016) Advanced engineering materials and modeling. Tiwari A, Arul Murugan N, Ahula R (eds). Wiley Scrivener Publishing LLC, New Jersey, pp 291–333
20. Caligiuri LM (2015) Unified field mechanics: natural science beyond the veil of spacetime. In: Amoroso RL, Kauffman LH, Rowlands P (eds) IX symposium honoring noted French mathematical physicist Jean-Pierre Vigier, Cairns, November 2014. World Scientific Publishing, pp 374–396
21. Caligiuri LM (2018) Unified field mechanics II: formulations and empirical tests. In: Amoroso RL, Kauffman LH, Rowlands P, Albertini G (eds) X symposium honoring noted French mathematical physicist Jean-Pierre Vigier, Porto Novo, July 2016. World Scientific Publishing, pp 331–343
22. Arani R, Bono I, Del Giudice E, Preparata G (1995) Int J Mod Phys B 9:1813–1842
23. Bono I, Del Giudice E, Gamberale L, Henry M (2012) Water 4:510–532
24. Clegg JS (1982) Biophysics of water. Franks F, Mathias SF (eds). Wiley, New York, pp 365–385
25. Drost-Hansen W (2006) Water and the cell. Pollack GH, Cameron IL, Wheatley DN (eds). Springer-Verlag, Berlin, pp 175–217
26. Pollack GH, Clegg J (2008) Phase transitions in cell biology. Pollack GH, Chin W (eds). Springer Science & Business Media, Berlin, pp 143–152
27. Buzzacchi M, Del Giudice E, Preparata G (2002) Int J Mod Phys B 16(25):3771–3786
28. Del Giudice E, Tedeschi A, Vitiello G, Voeikov V (2013) J Phys Conf Ser 442:012028

29. Del Giudice E, Tedeschi A (2009) Electromagn Biol Med 28:46–52
30. Caligiuri LM (2020) Submitted for publication
31. Bohm D, Aharonov Y (1959) Phys Rev 115:485–491
32. Del Giudice E, Spinetti PR, Tedeschi A (2010) Water 2:566–586
33. Stahlhofen AA, Nimtz G (2006) Europhys Lett 76:189–196
34. Nimtz G (2009) Found Phys 39:1346
35. Caligiuri LM (2019) J Phys Conf Ser 1251:012009
36. Carniglia CK, Mandel L (1971) Phys Rev D 3:280–296
37. Haibel A, Nimitz G, Stahlhofen AA (2001) Phys Rev E 63:047601
38. Enders A, Nimitz G (1992) J Phys I France 2:1693
39. Baena JD, Jelinek L, Marwues R, Medina F (2005) Phys Rev E 72:075116
40. Satinover J (2001) The quantum brain (The search for freedom and the next generation of man). Wiley, New York
41. Nielsen MA, Chuang IL (2010) Quantum computation and quantum information. Cambridge University Press, Cambridge
42. Benioff P (1982) Phys Rev Lett 48(23):1581–1585
43. Caligiuri LM, Musha T (2019) J Phys Conf Ser 1251:012010
44. Caligiuri LM, Musha T (2019) Proceedings of the 2019 international conference on engineering, science, and industrial applications (ICESI), Tokyo, 22–24 August 2019. https://doi.org/10.1109/ICESI.2019.8862999
45. Penrose R (1999) The large, the small and the human mind. Cambridge University Press, Cambridge
46. Tegmark M (2000) Phys Rev E 61:4194–4206
47. Feynman RP (1999) Feynman lectures on computation. Penguin Books, London
48. Llyod S (2000) Nature 406:1047–1054